ORGANIC SYNTHESES

ORGANIC SYNTHESES

AN ANNUAL PUBLICATION OF SATISFACTORY METHODS FOR THE PREPARATION OF ORGANIC CHEMICALS

VOLUME 76

1999

JOHN WILEY & SONS, INC.

NEW YORK • CHICHESTER • WEINHEIM • BRISBANE • SINGAPORE • TORONTO

The procedures in this text are intended for use only by persons with prior training in the field of organic chemistry. In the checking and editing of these procedures, every effort has been made to identify potentially hazardous steps and to eliminate as much as possible the handling of potentially dangerous materials; safety precautions have been inserted where appropriate. If performed with the materials and equipment specified, in careful accordance with the instructions and methods in this text, the Editors believe the procedures to be very useful tools. However, these procedures must be conducted at one's own risk. Organic Syntheses, Inc., its Editors, who act as checkers, and its Board of Directors do not warrant or guarantee the safety of individuals using these procedures and hereby disclaim any liability for any injuries or damages claimed to have resulted from or related in any way to the procedures herein.

This book is printed on acid-free paper. ♾

Published by John Wiley & Sons, Inc.

"John Wiley & Sons, Inc. is pleased to publish this volume of Organic Syntheses on behalf of Organic Syntheses, Inc. Although Organic Syntheses, Inc. has assured us that each preparation contained in this volume has been checked in an independent laboratory and that any hazards that were uncovered are clearly set forth in the write-up of each preparation. John Wiley & Sons, Inc. does not warrant the preparations against any safety hazards and assumes no liability with respect to the use of the preparations."

For ordering and customer service, call 1-800-CALL-WILEY.

Library of Congress Catalog Card Number: 21-17747
ISBN 0-471-34886-4

Printed in the United States of America

10 9 8 7 6 5 4 3 2 1

ORGANIC SYNTHESES

VOLUME	EDITOR-IN-CHIEF	PAGES
I*	†ROGER ADAMS	84
II*	†JAMES BRYANT CONANT	100
III*	†HANS THACHER CLARKE	105
IV*	†OLIVER KAMM	89
V*	†CARL SHIPP MARVEL	110
VI*	†HENRY GILMAN	120
VII*	†FRANK C. WHITMORE	105
VIII*	†ROGER ADAMS	139
IX*	†JAMES BRYANT CONANT	108
Collective Vol. I	A revised edition of Annual Volumes I–IX †HENRY GILMAN, *Editor-in-Chief* 2nd Edition revised by †A. H. BLATT	580
X*	†HANS THACHER CLARKE	119
XI*	†CARL SHIPP MARVEL	106
XII*	†FRANK C. WHITMORE	96
XIII*	†WALLACE H. CAROTHERS	119
XIV*	†WILLIAM W. HARTMAN	100
XV*	†CARL R. NOLLER	104
XVI*	†JOHN R. JOHNSON	104
XVII*	†L. F. FIESER	112
XVIII*	†REYNOLD C. FUSON	103
XIX*	†JOHN R. JOHNSON	105
Collective Vol. II	A revised edition of Annual Volumes X–XIX †A. H. BLATT, *Editor-in-Chief*	654
20*	†CHARLES F. H. ALLEN	113
21*	†NATHAN L. DRAKE	120
22*	†LEE IRVIN SMITH	114
23*	†LEE IRVIN SMITH	124
24*	†NATHAN L. DRAKE	119
25*	†WERNER E. BACHMANN	120
26*	†HOMER ADKINS	124
27*	†R. L. SHRINER	121
28*	†H. R. SNYDER	121
29*	†CLIFF S. HAMILTON	119
Collective Vol. III	A revised edition of Annual Volumes 20–29 †E. C. HORNING, *Editor-in-Chief*	890
30*	†ARTHUR C. COPE	115
31*	†R. S. SCHREIBER	122

**Out of print.*
† Deceased.

VOLUME	EDITOR-IN-CHIEF	PAGES
32*	Richard T. Arnold	119
33*	Charles C. Price	115
34*	†William S. Johnson	121
35*	†T. L. Cairns	122
36*	N. J. Leonard	120
37*	James Cason	109
38*	†John C. Sheehan	120
39*	†Max Tishler	114
Collective Vol. IV	A revised edition of Annual Volumes 30–39 Norman Rabjohn, *Editor-in-Chief*	1036
40*	†Melvin S. Newman	114
41*	John D. Roberts	118
42*	Virgil Boekelheide	118
43*	B. C. McKusick	124
44*	†William E. Parham	131
45*	†William G. Dauben	118
46*	E. J. Corey	146
47*	William D. Emmons	140
48*	†Peter Yates	164
49*	Kenneth B. Wiberg	124
Collective Vol. V	A revised edition of Annual Volumes 40–49 Henry E. Baumgarten, *Editor-in-Chief*	1234
Cumulative Indices to Collective Volumes, I, II, III, IV, V	†Ralph L. and †Rachel H. Shriner, *Editors*	
50*	Ronald Breslow	136
51*	†Richard E. Benson	209
52*	Herbert O. House	192
53*	Arnold Brossi	193
54*	Robert E. Ireland	155
55*	Satoru Masamune	150
56*	†George H. Büchi	144
57*	Carl R. Johnson	135
58*	†William A. Sheppard	216
59*	Robert M. Coates	267
Collective Vol. VI	A revised edition of Annual Volumes 50–59 Wayland E. Noland, *Editor-in-Chief*	1208
60*	Orville L. Chapman	140
61*	†Robert V. Stevens	165

**Out of print.*
†Deceased.

VOLUME	EDITOR-IN-CHIEF	PAGES
62	MARTIN F. SEMMELHACK	269
63	GABRIEL SAUCY	291
64	ANDREW S. KENDE	308
Collective Vol. VII	A revised edition of Annual Volumes 60–64 JEREMIAH P. FREEMAN, *Editor-in-Chief*	602
65	EDWIN VEDEJS	278
66	CLAYTON H. HEATHCOCK	265
67	BRUCE E. SMART	289
68	JAMES D. WHITE	318
69	LEO A. PAQUETTE	328
Reaction Guide to Collective Volumes I–VII and Annual Volumes 65–68	DENNIS C. LIOTTA AND MARK VOLMER, *Editors*	854
Collective Vol. VIII	A revised edition of Annual Volumes 65–69 JEREMIAH P. FREEMAN, *Editor-in-Chief*	696
Cumulative Indices to Collective Volumes, I, II, III, IV, V, VI, VII, VIII	JEREMIAH P. FREEMAN, *Editor-in Chief*	
70	ALBERT I. MEYERS	305
71	LARRY E. OVERMAN	285
72	DAVID L. COFFEN	333
73	ROBERT K. BOECKMAN, JR.	352
74	ICHIRO SHINKAI	341
Collective Vol. IX	A revised edition of Annual Volumes 70–74 JEREMIAH P. FREEMAN, *Editor-in-Chief*	840
75	AMOS B. SMITH III	257
76	STEPHEN F. MARTIN	340

Collective Volumes, Collective Indices to Collective Volumes I–VIII, Annual Volume 75, and Reaction Guide are available from John Wiley & Sons, Inc.

**Out of print.*
†*Deceased.*

NOTICE

With Volume 62, the Editors of *Organic Synthesis* began a new presentation and distribution policy to shorten the time between submission and appearance of an accepted procedure. The soft cover edition of this volume is produced by a rapid and inexpensive process, and is sent at no charge to members of the Organic Divisions of the American and French Chemical Society. The Perkin Division of the Royal Society of Chemistry, and The Society of Synthetic Organic Chemistry, Japan. The soft cover edition is intended as the personal copy of the owner and is not for library use. A hard cover edition is published by John Wiley & Sons, Inc. in the traditional format, and differs in content primarily in the inclusion of an index. The hard cover edition is intended primarily for library collections and is available for purchase through the publisher. Annual Volumes 70–74 have been incorporated into a new five-year version of the collective volumes of *Organic Syntheses* which has appeared as *Collective Volume Nine* in the traditional hard cover format. It is available for purchase from the publishers. The Editors hope that the new *Collective Volume* series, appearing twice as frequently as the previous decennial volumes, will provide a permanent and timely edition of the procedures for personal and institutional libraries. The Editors welcome comments and suggestions from users concerning the new editions.

NOMENCLATURE

Both common and systematic names of compounds are used throughout this volume, depending on which the Editor-in-Chief felt was more appropriate. The *Chemical Abstracts* indexing name for each title compound, if it differs from the title name, is given as a subtitle. Systematic *Chemical Abstracts* nomenclature, used in both the recent Collective Indexes for the title compound and a selection of other compounds mentioned in the procedure, is provided in an appendix at the end of each preparation. Registry numbers, which are useful in computer searching and identification, are also provided in these appendixes. Whenever two names are concurrently in use and one name is the correct *Chemical Abstracts* name, that name is preferred.

SUBMISSION OF PREPARATIONS

Organic Synthesis welcomes and encourages submission of experimental procedures which lead to compounds of wide interest or which illustrate important new developments in methodology. The Editorial Board will consider proposals in outline format as shown below, and will request full experimental details for those proposals which are of sufficient interest. Submissions which are longer than three steps from commercial sources or from existing *Organic Syntheses* procedures will be accepted only in unusual circumstances.

Organic Synthesis Proposal Format

1) Authors
2) Title
3) Literature reference or enclose preprint if available
4) Proposed sequence
5) Best current alternative(s)
6) a. Proposed scale, final product:
   b. Overall yield:
   c. Method of isolation and purification:
   d. Purity of product (%):
   e. How determined?

7) Any unusual apparatus or experimental technique?
8) Any hazards?
9) Source of starting material?
10) Utility of method or usefulness of product

Submit to: Dr. Jeremiah P. Freeman, Secretary
Department of Chemistry
University of Notre Dame
Notre Dame, IN 46556

Proposals will be evaluated in outline form, again after submission of full experimental details and discussion, and, finally by checking experimental procedures. A form that details the preparation of a complete procedure (Notice to Submitters) may be obtained from the Secretary.

Additions, corrections, and improvements to the preparations previously published are welcomed; these should be directed to the Secretary. However, checking of such improvements will only be undertaken when new methodology is involved. Substantially improved procedures have been included in the Collective Volumes in place of a previously published procedure.

ACKNOWLEDGMENT

Organic Synthesis wishes to acknowledge the contributions of ArQule, Hoffmann-La Roche, Inc. and Merck & Co. to the success of this enterprise through their support, in the form of time and expenses, of members of the Boards of Directors and Editors.

HANDLING HAZARDOUS CHEMICALS
A Brief Introduction

General Reference: *Prudent Practices in the Laboratory*; National Academy Press; Washington, DC, 1995.

Physical Hazards

Fire. Avoid open flames by use of electric heaters. Limit the quantity of flammable liquids stored in the laboratory. Motors should be of the nonsparking induction type.

Explosion. Use shielding when working with explosive classes such as acetylides, azides, ozonides, and peroxides. Peroxidizable substances such as ethers and alkenes, when stored for a long time, should be tested for peroxides before use. Only sparkless "flammable storage" refrigerators should be used in laboratories.

Electric Shock. Use 3-prong grounded electrical equipment if possible.

Chemical Hazards

Because all chemicals are toxic under some conditions, and relatively few have been thoroughly tested, it is good strategy to minimize exposure to all chemicals. In practice this means having a good, properly installed hood; checking its performance periodically; using it properly; carrying out most operations in the hood; protecting the eyes; and, since many chemicals can penetrate the skin, avoiding skin contact by use of gloves and other protective clothing.

a. Acute Effects. These effects occur soon after exposure. The effects include burn, inflammation, allergic responses, damage to the eyes, lungs, or nervous system (e.g., dizziness), and unconsciousness or death (as from overexposure to HCN). The effect and its cause are usually obvious and so are the methods to prevent it. They generally arise from inhalation or skin contact, so should not be a problem if one follows the admonition "work in a hood and keep chemicals off your hands". Ingestion is a rare route, being generally the result of eating in the laboratory or not washing hands before eating.

b. Chronic Effects. These effects occur after a long period of exposure or after a long latency period and may show up in any of numerous organs. Of the chronic effects of chemicals, cancer has received the most attention lately. Several dozen chemicals have been demonstrated to be carcinogenic in man and hundreds to be carcinogenic to animals. Although there is no simple correlation between carcinogenicity in animals and in man, there is little doubt that a significant proportion of the chemicals used in laboratories have some potential for carcinogenicity in man. For this and other reasons, chemists should employ good practices.

The key to safe handling of chemicals is a good, properly installed hood, and the referenced book devotes many pages to hoods and ventilation. It recommends that in a laboratory where people spend much of their time working with chemicals there should be a hood for each two people, and each should have at least 2.5 linear feet (0.75 meter) of working space at it. Hoods are more than just devices to keep undesirable vapors from the laboratory atmosphere. When closed they provide a protective barrier between chemists and chemical operations, and they are a good containment device for spills. Portable shields can be a useful supplement to hoods, or can be an alternative for hazards of limited severity, e.g., for small-scale operations with oxidizing or explosive chemicals.

Specialized equipment can minimize exposure to the hazards of laboratory operations. Impact resistant safety glasses are basic equipment and should be worn at all times. They may be supplemented by face shields or goggles for particular operations, such as pouring corrosive liquids. Because skin contact with chemicals can lead to skin irritation or sensitization or, through absorption, to effects on internal organs, protective gloves are often needed.

Laboratories should have fire extinguishers and safety showers. Respirators should be available for emergencies. Emergency equipment should be kept in a central location and must be inspected periodically.

DISPOSAL OF CHEMICAL WASTE

General Reference: *Prudent Practices in the Laboratory*, National Academy Press, Washington, D.C. 1995

Effluents from synthetic organic chemistry fall into the following categories:

1. **Gases**

   1a. Gaseous materials either used or generated in an organic reaction.
   1b. Solvent vapors generated in reactions swept with an inert gas and during solvent stripping operations.
   1c. Vapors from volatile reagents, intermediates and products.

2. **Liquids**

   2a. Waste solvents and solvent solutions of organic solids (see item 3b).
   2b. Aqueous layers from reaction work-up containing volatile organic solvents.
   2c. Aqueous waste containing non-volatile organic materials.
   2d. Aqueous waste containing inorganic materials.

3. **Solids**

   3a. Metal salts and other inorganic materials.
   3b. Organic residues (tars) and other unwanted organic materials.
   3c. Used silica gel, charcoal, filter aids, spent catalysts and the like.

The operation of industrial scale synthetic organic chemistry in an environmentally acceptable manner* requires that all these effluent categories be dealt with properly. In small scale operations in a research or academic set-

*An environmentally acceptable manner may be defined as being both in compliance with all relevant state and federal environmental regulations *and* in accord with the common sense and good judgement of an environmentally aware professional.

ting, provision should be made for dealing with the more environmentally offensive categories.

1a. Gaseous materials that are toxic or noxious, e.g., halogens, hydrogen halides, hydrogen sulfide, ammonia, hydrogen cyanide, phosphine, nitrogen oxides, metal carbonyls, and the like.

1c. Vapors from noxious volatile organic compounds, e.g., mercaptans, sulfides, volatile amines, acrolein, acrylates, and the like.

2a. All waste solvents and solvent solutions of organic waste.

2c. Aqueous waste containing dissolved organic material known to be toxic.

2d. Aqueous waste containing dissolved inorganic material known to be toxic, particularly compounds of metals such as arsenic, beryllium, chromium, lead, manganese, mercury, nickel, and selenium.

3. All types of solid chemical waste.

Statutory procedures for waste and effluent management take precedence over any other methods. However, for operations in which compliance with statutory regulations is exempt or inapplicable because of scale or other circumstances, the following suggestions may be helpful.

Gases

Noxious gases and vapors from volatile compounds are best dealt with at the point of generation by "scrubbing" the effluent gas. The gas being swept from a reaction set-up is led through tubing to a (large!) trap to prevent suck-back and on into a sintered glass gas dispersion tube immersed in the scrubbing fluid. A bleach container can be conveniently used as a vessel for the scrubbing fluid. The nature of the effluent determines which of four common fluids should be used: dilute sulfuric acid, dilute alkali or sodium carbonate solution, laundry bleach when an oxidizing scrubber is needed, and sodium thiosulfate solution or diluted alkaline sodium borohydride when a reducing scrubber is needed. Ice should be added if an exotherm is anticipated.

Larger scale operations may require the use of a pH meter or starch/iodide test paper to ensure that the scrubbing capacity is not being exceeded.

When the operation is complete, the contents of the scrubber can be poured down the laboratory sink with a large excess (10–100 volumes) of water. If the solution is a large volume of dilute acid or base, it should be neutralized before being poured down the sink.

Liquids

Every laboratory should be equipped with a waste solvent container in which *all* waste organic solvents and solutions are collected. The contents of these containers should be periodically transferred to properly labeled waste solvent drums and arrangements made for contracted disposal in a regulated and licensed incineration facility.**

Aqueous waste containing dissolved toxic organic material should be decomposed *in situ*, when feasible, by adding acid, base, oxidant, or reductant. Otherwise, the material should be concentrated to a minimum volume and added to the contents of a waste solvent drum.

Aqueous waste containing dissolved toxic inorganic material should be evaporated to dryness and the residue handled as a solid chemical waste.

Solids

Soluble organic solid waste can usually be transferred into a waste solvent drum, provided near-term incineration of the contents is assured.

Inorganic solid wastes, particularly those containing toxic metals and toxic metal compounds, used Raney nickel, manganese dioxide, etc. should be placed in glass bottles or lined fiber drums, sealed, properly labeled, and arrangements made for disposal in a secure landfill.** Used mercury is particularly pernicious and small amounts should first be amalgamated with zinc or combined with excess sulfur to solidify the material.

Other types of solid laboratory waste including used silica gel and charcoal should also be packed, labeled, and sent for disposal in a secure landfill.

Special Note

Since local ordinances may vary widely from one locale to another, one should always check with appropriate authorities. Also, professional disposal services differ in their requirements for segregating and packaging waste.

**If arrangements for incineration of waste solvent and disposal of solid chemical waste by licensed contract disposal services are not in place, a list of providers of such services should be available from a state or local office of environmental protection.

PREFACE

The series of annual volumes published by *Organic Syntheses* has provided organic chemists with carefully checked and edited experimental procedures that describe useful synthetic methods, transformations, reagents, and building blocks or intermediates. This, the 76th volume in the series, continues this rich tradition and provides 30 such procedures. Given the diversity of synthetic organic chemistry, there is no underlying theme, but the procedures fall generally into four broad areas: (1) reagents and methods for asymmetric synthesis; (2) useful synthetic transformations; (3) organometallic chemistry and transformations of organometallic reagents; (4) synthetically useful reagents and compounds.

This collection begins with four procedures for the preparation of chiral ligands that have found broad use in organic synthesis. The **RESOLUTION OF 1,1′-BI-2-NAPHTHOL** provides facile access to both enantiomers of this important chiral reagent and ligand. The following procedure for the synthesis of **(R)-(+)- AND (S)-(−)-2,2′-BIS(DIPHENYLPHOSPHINO)-1,1′-BINAPHTHYL (BINAP)** then provides details for the preparation of a chiral bisphosphine ligand that has been widely used in catalytic asymmetric transformations. The TADDOLS constitute an important class of chiral auxiliaries, and the preparation of **(4R,5R)-2,2-DIMETHYL-α,α,α′,α′-TETRA(NAPHTH-2-YL)-1,3-DIOXOLANE-4,5-DIMETHANOL** from dimethyl tartrate and 2-naphthylmagnesium bromide is representative of the general method for their preparation. TADDOL derivatives have been used in asymmetric synthesis in a variety of highly enantioselective processes including lithium aluminum hydride reductions, Michael additions, and hydrosilylations. They have also been used to prepare chiral Lewis acids that serve as catalysts in nucleophilic additions to carbonyl compounds and Diels-Alder reactions. Chiral diamines such as **(R,R)- AND (S,S)-N,N′-DIMETHYL-1,2-DIPHENYLETHYLENE-1,2-DIAMINE** are not only efficacious chiral auxiliaries for asymmetric synthesis, but they may be also used in analytical applications. These compounds are prepared by reductive dimerization of imines followed by resolution.

The series then continues with two procedures for preparing chiral reagents that are used in enantioselective synthesis. Chiral sulfoxides such as

1S-(–)-1,3-DITHIANE 1-OXIDE, which is prepared via an asymmetric oxidation, are useful sources of chirality in asymmetric carbon-carbon bond constructions. A (salen) Mn-catalyzed epoxidation reaction is featured in the enantioselective synthesis of **(1S,2R)-1-AMINOINDAN-2-OL,** a versatile chiral ligand and auxiliary that may be used in a range of asymmetric transformations including Diels-Alder reactions, carbonyl reductions, diethyl zinc additions to aldehydes, and enolate additions.

The series on asymmetric synthesis then concludes with procedures for the preparation of enantiomerically pure products. The asymmetric syntheses of unnatural α-amino acids by the alkylation of pseudoephedrine glycinamide is nicely exemplified by the preparation of **L-ALLYLGLYCINE** and **N-BOC-L-ALLYLGLYCINE.** One of the advantages of this method is the ready availability of the chiral auxiliary and the mildness of the conditions required for the hydrolysis of the pseudophedrine amide to provide the α-amino acid. Biocatalytic transformations are also gaining importance in asymmetric synthesis as illustrated by the preparation of **1-CHLORO-(2S,3S)-DIHYDROXYCYCLOHEXA-4,6-DIENE** by the microbial oxidation of chlorobenzene. Such cyclohexadiene diols are becoming widely used as starting materials in asymmetric synthesis. The synthesis of **(2S,3S)-(+)-(3-PHENYLCYCLOPROPYL)METHANOL** illustrates a powerful method for the enantioselective cyclopropanation of allylic alcohols using an easily-prepared, chiral dioxaborolane ligand in a modification of the Simmons-Smith reaction. A general method for the preparation of chiral non-racemic diols by the opening of **(S,S)-1,2,3,4-DIEPOXYBUTANE** is illustrated by the preparation of **(2S,3S)-DIHYDROXY-1,4-DIPHENYLBUTANE.** Enantiomerically pure α-N,N-dibenzylamino aldehydes undergo a variety of stereoselective nucleophilic additions, and the preparation of **S-2-(N,N-DIBENZYLAMINO)-3-PHENYLPROPANAL** outlines a convenient method for the synthesis of these important intermediates. The use of amino acids as starting materials for the synthesis of other chiral building blocks is exemplified by the synthesis of **METHYL (S)-2-PHTHALIMIDO-4-OXOBUTANOATE.**

The next eight procedures highlight important synthetic transformations. A method for the synthesis of tertiary amines from nitriles is illustrated by the preparation of **N,N-DIMETHYLHOMOVERATRYLAMINE**; this technique may also be applied to the synthesis of secondary amines. The synthesis of **ETHYL 5-CHLORO-3-PHENYLINDOLE-2-CARBOXYLATE** is representative of a general procedure for the reductive cyclization of amino carbonyl derivatives using low-valent titanium reducing agents. Because of their biological significance, the synthesis of fluorine-containing compounds has become increasingly important. The procedure for the synthesis of **4-HYDROXY-1,1,1,3,3-PENTAFLUORO-2-HEXANONE HYDRATE** features a convenient procedure for generating lithium pentafluoropropen-2-olate and its subsequent use in an aldol reaction. The electrophilic bromofluorination of alkenes, which is illustrated by the synthesis of **1-BROMO-2-FLUORO-2-PHENYLPROPANE,** represents another route to a number of monofluorinated compounds. The mono-C-methylation of arylacetonitriles and methyl arylacetates by dimethyl carbonate as a route to 2-arylpropionic acids is exemplified by the synthesis of **2-PHENYLPROPIONIC ACID,** the simplest member of an important class of anti-inflammatory agents.

The next three procedures feature a rearrangement as a key transformation. The first involves the preparation of **(tert-BUTYLDIMETHYLSILYL)ALLENE.** This method, which is performed in a single operation, involves conversion of a propargylic alcohol into a propargylic diazene that then undergoes a sigmatropic elimination of dinitrogen. Because of the mildness of the reaction conditions, the procedure may be applied to preparing allenes bearing sensitive functional groups. Vinylcyclobutenediones are pivotal intermediates for the synthesis of cyclobutenediones and quinones. This chemistry is nicely illustrated by a procedure for preparing **2-BUTYL-6-ETHENYL-5-METHOXY-1,4-BENZOQUINONE** by a sequence that involves the ring expansion of a 1,2-adduct of **3-ETHENYL-4-METHOXYCYCLOBUTENE-1,2-DIONE.** This procedure also provides a convenient method for the preparation of **DIMETHYL SQUARATE**, an important intermediate. The synthesis of **(1R*,6S*,7S*)-4-(tert-BUTYLDIMETHYLSILOXY)-6-(TRIMETHYLSILYL)BICYCLO-[5.4.0]UNDEC-4-EN-2-ONE** is representative of a general protocol for the construction of cycloheptenones by a [3 + 4] annulation. The method features the addition of a lithium enolate to an acryloyl silane to give a 1,2-adduct that undergoes a novel sequence of a concerted Brook rearrangement/cyclopropanation and an anionic oxy-Cope rearrangement.

The next six procedures involve various aspects of organometallic chemistry. The synthesis of **6-PHENYLHEX-2-YN-5-EN-4-OL** features an inexpensive and convenient method for the generation of 1-propynyllithium from (Z/E)-1-bromo-1-propene. The reaction of α,ω-bromochloroalkenes with allylmagnesium bromide provides a convenient route to **6-CHLORO-1-HEXENE** and **8-CHLORO-1-OCTENE.** These halo alkenes are useful intermediates for the synthesis of long-chain alkenols and alkenolic acids. The reactions of carbonyl compounds with organolithium or Grignard reagents to give alcohols is sometimes accompanied by undesired side reactions such as enolization, reduction, condensation, conjugate addition, and pinacol coupling. The suppression of these aberrant transformations using cerium(III) chloride is a generally useful tactic and is illustrated in the synthesis of **1-BUTYL-1,2,3,4-TETRAHYDRO-1-NAPHTHOL.** Two problems that are commonly associated with the classical alkylation of lithium enolates are a loss of regioselectivity and the formation of polyalkylated products. By contrast, the regioselective monoalkylation of the manganese enolates of ketones is normally observed as is illustrated by the efficient synthesis of **2-BENZYL-6-METHYLCYCLOHEXANONE** from 2-methylcyclohexanone. The copper-catalyzed conjugate addition of functionalized organozinc reagents to α,β-unsaturated ketones is exemplified by the preparation of **ETHYL 5-(3-OXOCYCLOHEXYL)PENTANOATE.** This general procedure may be readily applied to a variety of enones and alkyl zinc reagents containing diverse functionality such as chloro, cyano, keto, and ester groups. β-Alkynyl allylic alcohols, which are prepared by the palladium-catalyzed coupling of allylic bromides with acetylenes, may be isomerized using catalytic amounts of silver nitrate on silica gel to give substituted furans as illustrated by an efficient synthesis of **2-PENTYL-3-METHYL-5-HEPTYLFURAN.**

The volume concludes with the preparation of four useful reagents and compounds. Oligonucleoside phosphorothioates may be prepared using **2-CHLOROPHENYL PHOSPHORODICHLORIDOTHIOATE** as a coupling agent. A convenient synthesis of multigram quantities of **VITAMIN D_2 FROM ERGOSTEROL** is detailed, and the syntheses of **5,15-DIPHENYLPORPHYRIN**, and **9,10-DIPHENYLPHENANTHRENE** constitute the final two procedures in this volume.

The continued process of the series *Organic Syntheses* derives from the concerted commitment and dedication of numerous individuals. I am particularly grateful to my colleagues on the Editorial Board for their combined assistance and insight in selecting interesting procedures for checking. I am also grateful to them and the members of their respective research groups for carefully checking and sometimes modifying and even improving the procedures presented herein. Of course, were it not for the synthetic community at large and the submitters in particular who are continuously developing innovative synthetic organic chemistry, there would be no procedures to check. I am especially grateful to Professor Jeremiah P. Freeman, Secretary to the Board of Editors. Through his tireless efforts, all the multifarious aspects of selecting, checking, and publishing procedures proceed smoothly and in an organized manner. I am also grateful to Dr. Theodora W. Greene, the Assistant Editor, who ensured that all of the procedures were complete and properly formatted and who also assembled the index, a truly tedious task.

Finally, I thank the many members of the Martin Research Group at the University of Texas who have checked procedures contained in this and other volumes of *Organic Syntheses* and who also carefully read the procedures of this volume making useful suggestions for revisions.

STEPHEN F. MARTIN

Austin, Texas

CONTENTS

RESOLUTION OF 1,1'-BI-2-NAPHTHOL

(1,1'-Binaphthalene]-2,2'-diol)

OH OH N-Benzylcinchonidinium chloride, CH_3CN → OH OH + OH OH

(R)-(+) (S)-(-)-

Submitted by Dongwei Cai, David L. Hughes, Thomas R. Verhoeven, and Paul J. Reider.[1]

Checked by Rachel van Rijn and Amos B. Smith, III.

1. Procedure

A 500-mL flask, equipped with a magnetic stirring bar and a reflux condenser, is charged with 1,1'-bi-2-naphthol (23.0 g, 80 mmol) and N-benzylcinchonidinium chloride (18.6 g, 44 mmol) (Note 1). Acetonitrile (300 mL) is added, and the resulting suspension is refluxed for 4 hr, cooled and stirred at room temperature overnight. The mixture is then cooled to 0-5°C, kept at that temperature for 2 hr, and filtered (Note 2). The filtrate is concentrated to dryness, redissolved in ethyl acetate (300 mL), and washed with 1 N hydrochloric acid (HCl, 2 x 100 mL) (Note 3) and brine (100 mL). The organic layer is dried over sodium sulfate (Na_2SO_4), filtered, and concentrated to a light brown solid [10.28-10.65 g, mp 205-206°C, 89-93% recovery, 99.0% ee S-enantiomer $[\alpha]_D^{21}$ -27.6~-29.4° (THF, *c* 1)] (Notes 4, 5, 6).

The solid complex is washed with acetonitrile (50 mL). This acetonitrile solution is discarded because of the low ee (~80% ee of the S-enantiomer is contained). The resulting solid complex (96% ee, R-enantiomer) is transferred to a 250-mL flask. Methanol (100 mL) is added, and the resulting suspension is refluxed for 24 hr to upgrade the enantiomeric excess to >99% ee. After the mixture is cooled to room temperature, it is filtered and the solid washed with methanol (20 mL). The solid complex is suspended in a mixture of ethyl acetate (300 mL) and 1 N HCl (150 mL) and stirred until complete dissolution occurs (0.5 hr). The solution is transferred to a separatory funnel, and the organic layer is separated and then washed with 1 N HCl (150 mL) and brine (150 mL). The organic layer is dried over Na_2SO_4, filtered, and concentrated to an off-white crystalline solid [9.83-10.16 g, 85-88% recovery, mp 206-207°C, >99.8% ee of the R-enantiomer $[\alpha]_D^{21}$ 26.2~30.9° (THF, *c* 1)] (Notes 4, 5, 6).

2. Notes

1. Racemic 1,1'-bi-2-naphthol and N-benzylcinchonidinium chloride were purchased from Aldrich Chemical Company, Inc., acetonitrile (LC grade) was obtained from Fisher Scientific.

2. The enantiomeric excess of 1,1'-bi-2-naphthol in the filtrate at room temperature is 98.6% and at 0-5°C 99.0%.

3. These acid washes are to remove residual N-benzylcinchonidinium chloride in the filtrate.

4. Numerous chiral HPLC columns have been used for determination of chiral purity of 1,1'-bi-2-naphthol.[8b,9b] The submitters used Diacel Chiralpak OP(+) column (4.6 mm x 250 mm) at room temperature for their chiral assay. Typical retention times of 1,1'-bi-2-naphthol are 14 min (R-enantiomer) and 20 min (S-enantiomer) using methanol as an eluting solvent at 0.5 mL/min. The submitters' detection limit of minor

enantiomers is about 0.1%. The checkers used a Pirkle covalent D-phenylglycine column using isopropyl alcohol: hexane (5:95) as the eluting solvent at 1.0 mL/min with UV at 312 nm.

5. Enrichment to >99.8% ee is possible by recrystallization from a tert-butyl methyl ether (MTBE)/hexane mixture: 1.0 g of (S)-1,1'-bi-2-naphthol is dissolved in MTBE (10 mL), then hexane is added (20 mL). The resulting solid is stirred at room temperature for 2 hr, then filtered to provide a white crystalline solid (0.65 g, >99.8% ee, >99 wt% purity).

6. Other physical properties of the products are as follows: IR cm^{-1}: 3550 (s), 3050 (m), 1610 (m), 1590 (m), 1390 (m), 1180 (s), 1140 (s); ^{1}H NMR (250 MHz, $CDCl_3$) δ: 5.0 (s, 2 H, O<u>H</u>), 7.16 (d, 2 H, J = 8.3), 7.30 (m, 2 H), 7.38 (m, 4 H), 7.90 (d, 2 H, J = 8.1), 7.99 (d, 2 H, J = 8.9); ^{13}C NMR (62.9 MHz, $CDCl_3$) δ: 110.8, 117.8, 124.0, 124.2, 127.5, 128.4, 129.5, 131.4, 133.4, 152.8; HRMS (FAB, m-nitrobenzyl alcohol): R enantiomer, m/z 304.1335 [($M+NH_4^+$); calcd for $C_{20}H_{14}O_2+NH_4^+$: 304.1337]; S-enantiomer, m/z 304.1331 [($M+NH_4^+$); calcd for $C_{20}H_{14}O_2+NH_4^+$: 304.1337].

Waste Disposal Information

All toxic materials were disposed of in accordance with "Prudent Practices in the Laboratory"; National Academy Press; Washington, DC, 1995.

3. Discussion

Both enantiomers of 1,1'-bi-2-naphthol are widely used for various applications: 1) chiral inducing agents for catalytic, asymmetric reactions such as the Diels-Alder reaction,[2] ene reaction,[3] or as Lewis acids;[4] 2) enantioselective reduction of ketones;[5] 3) synthesis of chiral macrocycles[6] and other interesting compounds.[7] Previously reported resolutions include: 1) making a cyclic phosphate of binaphthol, then resolution and subsequent reduction to release the pure binaphthol;[8] 2) using enzymatic hydrolysis of the diester of binaphthol;[9] and 3) forming inclusion complexes with suitable compounds.[10] The use of N-benzylcinchonidinium chloride to make inclusion complexes was reported by Tanaka and co-workers for obtaining one enantiomer of binaphthol.[11] Using acetonitrile as solvent, in which the inclusion complex has very low solubility, allows for the isolation of both enantiomers with high enantiomeric excess. This simple and efficient procedure represents a much better resolution for 1,1'-bi-2-naphthol.[12]

1. Merck Research Labs, P.O. Box 2000, Rahway, NJ 07065.

2. Bao, J.; Wulff, W. D.; Rheingold, A. L. *J. Am. Chem. Soc.* **1993**, *115*, 3814.
3. (a) Terada, M.; Motoyama, Y.; Mikami, K. *Tetrahedron Lett.* **1994**, *35*, 6693; (b) Mikami, K.; Matsukawa, S. *Tetrahedron Lett.* **1994**, *35*, 3133.
4. (a) Sakane, S.; Maruoka, K.; Yamamoto, H. *Tetrahedron Lett.* **1985**, *26*, 5535; (b) Maruoka, K.; Itoh, T.; Shirasaka, T.; Yamamoto, H. *J. Am. Chem. Soc.* **1988**, *110*, 310.
5. (a) Noyori, R.; Tomino, I.; Tanimoto, Y.; Nishizawa, M. *J. Am. Chem. Soc.* **1984**, *106*, 6709; (b) Noyori, R.; Tomino, I.; Yamada, M.; Nishizawa, M. *J. Am. Chem. Soc.* **1984**, *106*, 6717.
6. (a) Sogah, G. D. Y.; Cram. D. J. *J. Am. Chem. Soc.* **1979**, *101*, 3035; (b) Lehn, J.-M.; Simon, J.; Moradpour, A. *Helv. Chim. Acta* **1978**, *61*, 2407.
7. Miyano, S.; Tobita, M.; Nawa, M.; Sato, S.; Hashimoto, H. *J. Chem. Soc., Chem. Commun.* **1980**, 1233.
8. (a) Fabbri, D.; Delogu, G.; De Lucchi, O. *J. Org. Chem.* **1993**, *58*, 1748; (b) Jacques, J.; Fouquey, C. *Org. Synth., Coll. Vol. VIII* **1993**, 50; (c) Truesdale, L. K. *Org. Synth., Coll. Vol. VIII* **1993**, 46; (d) Kyba, E. P.; Gokel, G. W.; de Jong, F.;

Koga, K.; Sousa, L. R.; Siegel, M. G.; Kaplan, L.; Sogah, G. D. Y.; Cram, D. J. *J. Org. Chem.* **1977**, *42*, 4173; (e) Gong, B.-q.; Chen, W.-y.; Hu, B.-f. *J. Org, Chem.* **1991**, *56*, 423; (f) Brunel, J.-M.; Buono, G. *J. Org. Chem.* **1993**, *58*, 7313.

9. (a) Kazlauskas, R. J. *J. Am. Chem. Soc.* **1989**, *111*, 4953; (b) Kazlauskas, R. J. *Org. Synth., Coll. Vol. IX* **1998**, 77.
10. (a) Toda, F.; Tanaka, K. *J. Org. Chem.* **1988**, *53*, 3607; (b) Kawashima, M.; Hirata, R. *Bull. Chem. Soc. Jpn.* **1993**, *66*, 2002; (c) Periasamy, M., Bhanu Prasad, A. S.; Bhaskar Kanth, J. V.; Reddy, C. K. *Tetrahedron: Asymmetry* **1995**, *6*, 341.
11. (a) Tanaka, K.; Okada, T.; Toda, F. *Angew. Chem., Int. Ed. Engl.* **1993**, *32*, 1147; (b) Toda, F.; Tanaka, K.; Stein, Z.; Goldberg, I. *J. Org. Chem.* **1994**, *59*, 5748.
12. Cai, D.; Hughes, D. L.; Verhoeven, T. R.; Reider, P. J. *Tetrahedron Lett.* **1995**, *36*, 7991.

Appendix

Chemical Abstracts Nomenclature (Collective Index Number); (Registry Number)

1,1'-Bi-2-naphthol: [1,1'-Binaphthalene]-2,2'-diol (8,9); (602-09-5)

N-Benzylcinchonidinium chloride: Cinchonanium, 9-hydroxy-1-(phenylmethyl)-, chloride, (9S)- (10); (69221-14-3)

Acetonitrile: TOXIC (8,9); (75-05-8)

(S)-(-)-1,1'-Bi-2-naphthol: [1,1'-Binaphthalene]-2,2'-diol, (S)-(-)- (8); [1,1'-Binaphthalene]-2,2'-diol, (S)- (9); (18531-99-2)

(R)-(+)-1,1'-Bi-2-naphthol: [1,1'-Binaphthalene]-2,2'-diol, (R)-(+)- (8); [1,1'-Binaphthalene]-2,2'-diol, (R)- (9); (18531-94-7)

(R)-(+)- AND (S)-(-)-2,2'-BIS(DIPHENYLPHOSPHINO)-1,1'-BINAPHTHYL (BINAP)

(Phosphine, [1,1'-binaphthalene]-2,2'-diylbis[diphenyl-, (R)- and (S))

A. (R)-(+) 1,1'-bi-2-naphthol (OH, OH) → Tf_2O, pyridine → ditriflate (OTf, OTf)

B. ditriflate (OTf, OTf) → Ph_2PH (2.2 eq), 10% $NiCl_2$dppe; DMF, DABCO, 100°C → (R)-(+)-BINAP (PPh_2, PPh_2)

Submitted by Dongwei Cai, Joseph F. Payack, Dean R. Bender, David L. Hughes, Thomas R. Verhoeven, and Paul J. Reider.[1]

Checked by Rachel van Rijn and Amos B. Smith, III.

1. Procedure

A. Preparation of ditriflate of 1,1'-bi-2-naphthol.[2] An oven-dried, 100-mL, single-necked flask, equipped with a magnetic stirring bar, is charged with (R)-(+)-1,1'-bi-2-naphthol (8.5 g, >99% ee, 30 mmol) (Note 1). Dry methylene chloride (60 mL) is added followed by dry pyridine (7.2 mL, 90 mmol) and triflic anhydride (20.0 g, 70 mmol) at 5-10°C (bath temperature) under a nitrogen (N_2) atmosphere (Notes 2 and 3). After the addition, the reaction solution is stirred at room temperature overnight (17 hr) (Note 4). Hexane (60 mL) is then added, and the resulting mixture is filtered under reduced pressure through a pad of silica gel [50 g of silica gel (230-400 mesh) in a 150-mL sintered glass funnel]. The silica gel is washed with a 1:1 mixture of hexane and CH_2Cl_2 (200 mL). The resulting filtrate is concentrated under vacuum to provide the ditriflate as a white solid (15.4 g, 94% yield, 99.6 area% by LC at 220 nm, mp 72-75°C) (Notes 5, 6).

B. Preparation of (R)-(+)-BINAP.[3] An oven-dried, 250-mL, single-necked flask, equipped with a magnetic stirring bar, is charged with [1,2-bis(diphenylphosphino)ethane]nickel(II) chloride] ($NiCl_2$dppe, 1.1 g, 2 mmol). The flask is purged with N_2 (using a vacuum and nitrogen manifold) and anhydrous dimethylformamide (DMF, 40 mL) is added via a syringe, followed by diphenylphosphine (2.0 mL, 12 mmol) at room temperature (Note 7). The resulting dark red solution is heated at 100°C for 30 min. A solution of the chiral ditriflate of binaphthol (11.0 g, 20 mmol) and 1,4-diazabicyclo[2.2.2]octane (DABCO, 9.0 g, 80 mmol) in anhydrous, degassed DMF (60 mL) (Note 8) is transferred in one portion to the reaction flask via cannula, and the resulting dark green solution is kept at 100°C. Three additional portions of diphenylphosphine (3 x 2 mL) are added by syringe after 1 hr, 3 hr, and 7 hr. The reaction is heated at 100°C until the ditriflate of binaphthol is completely consumed (2-3 days) (Note 6). The dark brown solution is cooled to -15 ~ -20°C with an ice/acetone bath and stirred for 2 hr. The product is filtered, and the solid is washed with methanol (2 x 20 mL) and dried under vacuum. The isolated product (9.6 g, 77%) is a white to off-white crystalline compound with a chemical purity of ~97 area% (HPLC, 220 nm) containing ~1% of the monooxide of BINAP (Notes 6, 9, 10, 11).[4]

2. Notes

1. Both enantiomers of 1,1'-bi-2-naphthol are available from Aldrich Chemical Company, Inc.

2. Methylene chloride and pyridine were purchased from Aldrich Chemical Company, Inc. and dried over activated 4 Å molecular sieves.

3. Triflic anhydride was purchased from Aldrich Chemical Company, Inc. and used without purification.

4. The reaction is complete in 2 hr, but may be run overnight.

5. The spectral properties of the ditriflate of binaphthol are as follows: ^1H NMR (500 MHz, $CDCl_3$) δ: 7.27 (d, 2 H, J = 8.5), 7.42 (ddd, 2 H, J = 1.1, 6.8, 8.2), 7.59 (ddd, 2 H, J = 1.0, 7.0, 8.1), 7.63 (d, 2 H, J = 9.1), 8.02 (d, 2 H, J = 8.2), 8.15 (d, 2 H, J = 9.1); ^{13}C NMR (125 MHz, $CDCl_3$) δ: 118.3, 119.4, 123.6, 126.8, 127.4, 128.1, 128.5, 132.1, 132.5, 133.2, 145.5.

6. Conditions for the LC assay were as follows: Zorbax Rx-C_8 column, 4.6 mm x 25 cm, room temperature, 1.50 mL/min, linear gradient, 60% CH_3CN/water to 90% CH_3CN/water in 20 min, then hold at 90% CH_3CN/water for 5 min; water contained 0.1% H_3PO_4; UV detection at 220 nm. Typical retention times are 1.75 min (DMF), 2.32 min (Ph_2POH), 5.53 min (dppe), 6.42 min (Ph_2PH), 7.73 min (BINAPO, dioxide of BINAP), 8.55 min [Ar(OTf)-ArP(O)Ph_2], 11.69 min (ditriflate of binaphthol), 14.54 min (monooxide of BINAP), 16.00 min Ar(H)-ArPPh_2, 20.99 min (BINAP); typical LC (area %) at the end of the reaction are DMF (49%), dppe (1.4%) BINAPO (0.6%), ditriflate of bi-2-naphthol (0.5%), monooxide of BINAP (4%), Ar(H)-ArPPh_2 (0.9%) and BINAP (36%).

7. $NiCl_2$dppe, DMF (anhydrous grade) and DABCO were obtained from Aldrich Chemical Company, Inc., and used without further purification. $NiCl_2$dppe can also be obtained from Strem Chemicals Inc. Diphenylphoshine (DPP) was obtained in an

ampoule from Strem Chemicals Inc. When not handled properly, DPP [^{31}P NMR; (200 MHz, $CDCl_3$) δ: -40 ppm] is rapidly oxidized to diphenylphoshine oxide [^{31}P NMR; (200 MHz, $CDCl_3$) δ: 22 ppm]. DPP was transferred directly from the ampoule to a Schlenk flask under an inert atmosphere. Its purity was checked by ^{31}P NMR to ensure that it was free of oxidation products.

8. This solution was degassed via vacuum and nitrogen 3-6 times. Exclusion of air from the reaction is critical to minimize formation of phosphine oxide by-products.

9. Recrystallization of the mother liquor after the first crop was obtained yielded a product with a lower purity.

10. Two impurities were identified as

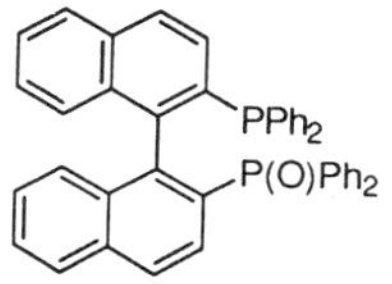

Monooxide of BINAP

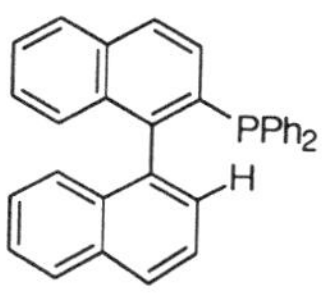

$Ar(H)-ArPPh_2$

11. The submitters' isolated BINAP had $[\alpha]_D^{20}$ +219°, 99% ee, mp 237-238°C (lit ref.[6b] $[\alpha]_D^{20}$ +217°, 98.4% ee). Other physical properties of BINAP are as follows: IR cm^{-1}: 3050 (s), 3010 (s), 1480 (m), 1450 (s), 1310 (m), 1180 (m), 1110 (m), 1090 (m); ^{1}H NMR (500 MHz, $CDCl_3$) δ: 6.83 (d, 2 H, J = 8.4), 6.91 (ddd, 2 H, J = 1.2, 8.2, 6.9), 7.04-7.18 (m, 20 H), 7.34 (ddd, 2 H, J = 1.1, 6.9, 8.0), 7.46 (ddd, 2 H, J = 2.6, 8.5), 7.83 (d, 2 H, J = 8.2), 7.89 (d, 2 H, J = 8.5); ^{13}C NMR (125 MHz, $CDCl_3$) δ: 125.7, 126.5, 127.5, 127.7, 128.0, 128.1, 128.4, 130.5, 132.8, 132.9, 133.0, 133.2, 133.4, 133.5, 134.1, 134.2, 134.3, 135.5, 135.6, 137.4, 137.5, 138.0, 145.1, 145.4; ^{31}P NMR (101 MHz, CDCl) δ: -14.9 ppm; HRMS (FAB, m-nitrobenzyl alcohol): m/z 623.2074 [$(M+H)^+$; calcd for $C_{44}H_{32}P_2$: 623.2058].

Waste Disposal Information

All toxic materials were disposed of in accordance with "Prudent Practices in the Laboratory"; National Academy Press; Washington, DC, 1995.

3. Discussion

Both enantiomers of BINAP are very useful ligands for various catalytic asymmetric reactions.[5] However, the scarce supply and high cost of BINAP somewhat limit their wide application. A previously reported synthesis of BINAP was not easy to scale up because of potentially hazardous conditions (320°C with HBr evolution), and low overall yield.[6] This procedure presents a short and efficient process to chiral BINAP from readily available chiral 1,1'-bi-2-naphthol.

1. Merck Research Labs, P.O. Box 2000, Rahway, NJ 07065.
2. Vondenhof, M.; Mattay, J. *Tetrahedron Lett.* **1990**, *31*, 985.
3. Cai, D.; Payack, J. F.; Bender, D. R.; Hughes, D. L.; Verhoeven, T. R.; Reider, P. J. *J. Org. Chem.* **1994**, *59*, 7180.
4. Ozawa, F.; Kubo, A.; Hayashi, T. *Chem. Lett.* **1992**, 2177.
5. (a) Miyashita, A.; Yasuda, A.; Takaya, H.; Toriumi, K.; Ito, T.; Souchi, T.; Noyori, R. *J. Am. Chem. Soc.* **1980**, *102*, 7932; (b) Ojima, I. In "Catalytic Asymmetric Synthesis"; Ojima, I., Ed.; VCH: New York, NY, 1993; (c) Noyori, R. In "Asymmetric Catalysis in Organic Syntheses"; John Wiley & Sons, Inc.: New York, NY, 1994.
6. (a) Takaya, H.; Mashima, K.; Koyano, K.; Yagi, M.; Kumobayashi, H.; Taketomi, T.; Akutagawa, S.; Noyori, R. *J. Org. Chem.* **1986**, *51*, 629; (b) Takaya, H.; Akutagawa, S.; Noyori, R. *Org. Synth., Coll. Vol. VIII* **1993**, 57.

Appendix

Chemical Abstracts Nomenclature (Collective Index Number);

(Registry Number)

(R)-(+)-2,2'-Bis(diphenylphosphino)-1,1'-binaphthyl [(R)-BINAP]: Phosphine, [1,1'-binaphthalene]-2,2'-diylbis[diphenyl-, (R)- (10); (76189-55-4)

(S)-(-)-2,2'-Bis(diphenylphosphino)-1,1'-binaphthyl [(S)-BINAP]: Phosphine, [1,1'-binaphthalene]-2,2'-diylbis[diphenyl-, (3)- (10); (76189-56-5)

1,1'-Bi-2-naphthol ditriflate: Methanesulfonic acid, trifluoro-, [1,1'-binaphthalene]-2,2'-diyl ester, (±)- (12); (128575-34-8)

(R)-(+)-1,1'-Bi-2-naphthol: [1,1'-Binaphthalene]-2,2'-diol, (R)-(+)- (8); [1,1'-Binaphthalene]-2,2'-diol, (R)- (9); (18531-94-7)

Triflic anhydride: Methanesulfonic acid, trifluoro-, anhydride (8,9); (358-23-6)

[1,2-Bis(diphenylphosphino)ethane]nickel(II) chloride: CANCER SUSPECT AGENT: Nickel, dichloro[ethylenebis[diphenylphosphine]]- (8); Nickel, dichloro[1,2-ethanediylbis[diphenylphosphine]-P,P']- (SP-4-2)- (9); (14647-23-5)

Diphenylphosphine: PYROPHORIC: Phosphine, diphenyl- (8,9); (829-85-6)

1,4-Diazabicyclo[2.2.2]octane (DABCO) (8,9); (280-57-9)

(4R,5R)-2,2-DIMETHYL-α,α,α',α'-TETRA(NAPHTH-2-YL)-1,3-DIOXOLANE-4,5-DIMETHANOL FROM DIMETHYL TARTRATE AND 2-NAPHTHYL-MAGNESIUM BROMIDE

(1,3-Dioxolane-4,5-dimethanol, 2,2-dimethyl-α,α,α',α'-tetra-2-naphthalenyl-, (4R-trans)-)

A. MeO_2C–CH(OH)–CH(OH)–CO_2Me + MeCOMe $\xrightarrow[\text{acetone}]{BF_3 \cdot OEt_2}$ 2,2-dimethyl-1,3-dioxolane-4,5-di(CO_2Me)

B. 2-bromonaphthalene (Br) + Mg $\xrightarrow{\text{THF}}$ 2-naphthyl-MgBr

C. 2-naphthyl-MgBr + 2,2-dimethyl-1,3-dioxolane-4,5-di(CO_2Me) $\xrightarrow{\text{THF}}$ (product: H, H, HO, OH)

Submitted by Albert K. Beck,[1] Peter Gysi,[2] Luigi La Vecchia,[2] and Dieter Seebach.[1]

Checked by Chad E. Bennett and William R. Roush.

1. Procedure

A. (R,R)-Dimethyl O,O-isopropylidenetartrate. Under an inert atmosphere (Note 1) a 2-L, two-necked, round-bottomed flask, equipped with a magnetic stirring bar and a pressure-equalized addition funnel, is charged with (R,R)-dimethyl tartrate (89.1 g, 0.5 mol) (Note 2) dissolved in acetone (900 mL). To the clear solution is added, at room temperature, boron trifluoride diethyl etherate (82.5 mL 48% solution, 0.31 mol) (Note 3) over 30-40 min (Note 4). The resulting yellow solution is stirred for an additional 3 hr during which time the color of the solution becomes red-brown.

For workup the reaction mixture is poured into an aqueous saturated sodium bicarbonate solution (4 L) (Note 5). The turbid mixture is divided into two parts, and each part is extracted three times with ethyl acetate (3 x 500 mL). The combined organic layers are washed twice with water (2 x 1 L) and dried over anhydrous magnesium sulfate. After filtration the solvent is removed by rotary evaporation at ca. 45°C/20 mm. The yellow oil that is obtained (103 g) is purified by fractional distillation using a 15-cm Vigreux column (bp 92-95°C/1.5 mm) to afford 84.3 g (77%) of product as a yellowish oil with a specific rotation of $[\alpha]_D^{RT}$ -62.0° (neat), -44.0° ($CHCl_3$, *c* 1) (Notes 6 and 7).

B. 2-Naphthylmagnesium bromide. Under an inert atmosphere (Note 1) a 4-L, four-necked, round-bottomed flask, fitted with a reflux condenser, pressure-equalized addition funnel, mechanical stirring bar and a thermometer, is charged with magnesium turnings (24.7 g, 1.02 at. equiv) and some iodine crystals. Then 30 mL of a solution of 2-bromonaphthalene (200.7 g, 0.97 mol) (Note 8) in tetrahydrofuran (THF) (675 mL) (Note 9) is added. As soon as the reaction has started (Note 10), the remainder of the tetrahydrofuran solution is added at such a rate that a gentle reflux is maintained (Note 11). After complete addition, reflux is continued for 1 hr by heating with an oil bath (Note 12). Finally, the reaction mixture is allowed to cool to room temperature.

C. (4R,5R)-2,2-Dimethyl-α,α,α',α'-tetra(naphth-2-yl)-1,3-dioxolane-4,5-dimethanol. To the Grignard solution obtained in Part B is added, with stirring and cooling by an ice bath, a solution of (R,R)-dimethyl O,O-isopropylidenetartrate (48.1 g, 0.22 mol) (see Part A) in tetrahydrofuran (480 mL). During the addition the internal temperature should not exceed 20°C (Notes 13 and 14). After completion of the addition, the reaction mixture is heated at reflux for 1.5 hr using an oil bath, then cooled to room temperature.

For workup, an aqueous saturated ammonium chloride solution (1.6 L) is carefully added, cooling the mixture with an ice bath (Notes 15 and 16). The mixture is extracted once with ethyl acetate (500 mL) (Note 17). After separation of the layers, the aqueous phase is extracted twice with ethyl acetate (2 x 250 mL). The combined organic layers are washed twice with brine (2 x 250 mL) and dried over anhydrous magnesium sulfate, and the solvent is evaporated on a rotary evaporator at 45°C/100 mm. The resulting yellowish foam (200 g) (Notes 18 and 19) is dissolved in diethyl ether (100 mL) followed by the addition of ethanol (400 mL). After a few minutes a white solid precipitates that is the clathrate of the product with ethanol (Note 20). After standing for several hours (or overnight), the crystals are filtered, washed with ethanol/diethyl ether (300 mL, 4:1), and ethanol (100 mL), and then dried overnight at 50°C/8 mm to give 125-130 g of colorless crystals (Note 21).

In order to remove the ethanol, the crystals are dissolved in toluene (3 mL per g of clathrate) at 70°C, and the solution is evaporated to dryness on a rotary evaporator at 45°C/100 mm. This procedure is repeated once more. A portion of the solid so obtained (68.5 g) is mixed with toluene (800 mL) at 80°C in a 2-L, two-necked, round-bottomed flask equipped with an overhead, mechanical stirrer, until the TADDOL [(R,R)-2,2-dimethyl-α,α,α',α'-tetra(naphth-2-yl)-1,3-dioxolane-4',5'-dimethanol] is completely dissolved. Hexane (800 mL) is added slowly with the solution being maintained between 65-70°C. On cooling, a white precipiate starts to form at approximately 58°C. The mixture is allowed to cool completely to room temperature, and the resulting thick slurry is stirred overnight. Further solvent (500 mL of a 1 : 1 mixture of toluene-hexane) is added to the now unstirrable suspension. The mixture is shaken vigorously to give a stirrable slurry. The solid is removed by vacuum filtration and washed first with the mother liquor, then with a 1 : 1 toluene/hexane mixture (300 mL), and finally with hexane (300 mL). The resulting solid is dried under high vacuum (0.3 mbar) at 90°C for 10 hr in a vacuum oven to give the title TADDOL ligand as a white solid (42.0 g, 61%). The mother liquor is concentrated under vacuum to give a yellow solid (24.5 g) that is again purified as described above by recrystallization from 300 mL of toluene and 300 mL of hexane to give a second crop of the TADDOL ligand (14.5 g, 21%), for a combined yield of 56.5 g (82%), mp 204-208°C (sintering at 155°C), $[\alpha]_D^{RT}$ -115.4° (ethyl acetate, *c* 1) (Notes 22, 23).

2. Notes

1. The reaction can be carried out either under an argon atmosphere, using a balloon, or under nitrogen, passing a continuous flow of nitrogen over the solution.

2. (R,R)-Dimethyl tartrate is commercially available. The submitters used product donated by the Chemische Fabrik Uetikon, Uetikon, Switzerland, whereas the checkers used material obtained from Lancaster.

3. Boron trifluoride-diethyl etherate solution, 48%, is commercially available and was used without prior purification. Both the submitters and checkers used reagent grade material from Fluka Chemie AG.

4. Boron trifluoride-diethyl etherate should be added at such a rate that the internal temperature does not exceed 30°C.

5. *Caution!* Vigorous evolution of carbon dioxide takes place, which causes foaming. Use an appropriately sized beaker (10 L).

6. The product has the following spectral properties: ^{1}H NMR (200 MHz, $CDCl_3$) δ: 1.50 (s, 2 OCH_3), 3.85 (s, O-$C(CH_3)_2$-O), 4.80 (s, 2 H-C-O); ^{13}C NMR (100 MHz, $CDCl_3$) δ: 26.2, 52.7, 76.9, 113.8, 170.0; IR (neat) cm^{-1}: 2992, 2957, 1761. For comparison with literature data see Refs. 3. and 4.

7. For an alternative synthesis using tartaric acid and 2,2-dimethoxypropane in methanol see Ref. 4.

8. 2-Bromonaphthalene is commercially available or can be prepared according to an Organic Syntheses Procedure, see Ref. 5.

9. The submitters used tetrahydrofuran p.a. quality.

10. Usually warming with a heat gun is necessary to start the reaction.

11. The addition takes about 45 min.

12. After completion of the reaction some magnesium particles remain (an ca. 5% excess was employed).

13. The addition usually takes about 20 min.

14. It is important to keep the internal temperature below 20°C; otherwise the yield decreases. Efficient cooling is therefore necessary.

15. At the beginning, the addition of the aqueous saturated ammonium chloride solution is highly exothermic and efficient cooling is required. During the addition, precipitation of magnesium salts takes place but they dissolve as addition progresses.

16. At the end of the addition the reaction mixture should have a pH of 7-8; if not, hydrochloric acid (HCl) 10% aqueous solution (the checkers needed 600 mL) should be added to reach this pH range.

17. The submitters removed most of the tetrahydrofuran using a rotary evaporator at 45°C/100 mm before extracting the residue with either isopropyl or ethyl acetate. However, the checkers experienced significant bumping of the solution

during the rotary evaporation step, and found it advantageous simply to subject the entire aqueous THF mixture to the extraction procedure.

18. *Caution!* Towards the end, the evaporation apparatus must be watched carefully because of foaming (adjust the pressure).

19. Small samples of the crude product can also be purified by flash chromatography using toluene as eluent, product $R_f = 0.1$ (toluene).

20. It is well known that TADDOLs easily form clathrates, (see Ref. 6).

21. The clathrate obtained contained 1 equiv of ethanol, as shown by ^1H NMR spectroscopy. However, the submitters obtained 158 g of clathrate that contained 2 equiv of ethanol per mol of TADDOL. For the crystal structure of the clathrate of this TADDOL with piperidine see Ref. 7.

22. The checkers obtained a 73% yield when following this crystallization procedure using crude, unpurified clathrate, and an 89% yield when once crystallized TADDOL was recrystallized using this procedure.

23. The product has the following spectral properties: ^1H NMR (200 MHz, $CDCl_3$) δ: 1.18 (s, 2 CH_3), 4.22 (s, 2 OH), 4.98 (s, 2 H-C-O), 7.21-8.19 (m, arom. H); ^{13}C NMR (100 MHz, $CDCl_3$) δ: 27.4, 78.6, 81.4, 109.9, 125.8, 125.89, 125.92, 126.0, 126.09, 126.13, 126.8, 127.1, 127.2, 127.3, 127.4, 127.9, 128.5, 128.6, 132.5, 132.62, 132.63, 132.7, 140.5, 142.5. For comparison with literature data see Ref. 3.

Waste Disposal Information

All toxic materials were disposed of in accordance with "Prudent Practices in the Laboratory"; National Academy Press; Washington, DC, 1995.

3. Discussion

The procedure described here is a typical one for the preparation of α,α,α',α'-tetraaryl 2,2-disubstituted 1,3-dioxolane-4,5-dimethanols (TADDOLs, **1**), a class of diols of which ca. 50 representatives have been synthesized.[7] They have become useful chiral auxiliaries for the preparation of enantiomerically enriched or pure compounds and for analytical purposes. The diols themselves have been employed

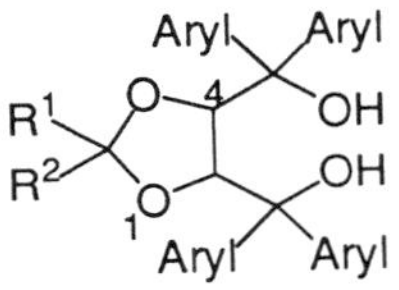

1 (4R,5R- or 4S,5S-) **2**

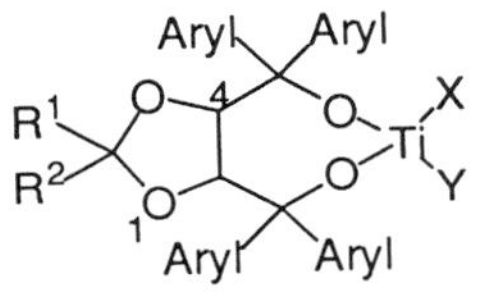

as clathrate-forming compounds (resolutions and solid-state reactions[6]) and NMR shift reagents.[6d,8] The main applications involve metal complexes such as the Ti-TADDOLates **2** and their equally readily available enantiomers ent-**2**. The substituents X and Y on Ti may be OR, Cl, Br, or cyclopentadienyl. The reactions for which these chiral Lewis acids have been used in equimolar or catalytic amounts include: nucleophilic additions (of alkyl, aryl, or allyl groups) to aldehydes, ketones and nitroolefins, aldol additions, hydrophosphonylations, cyanohydrin reactions, intra-, intermolecular and hetero Diels-Alder additions, [2+2]- and [3+2]-cycloadditions, intra- and intermolecular ene reactions, iodolactonizations, transesterifications, and anhydride alcoholyses. TADDOL derivatives have also been applied in enantioselective lithium aluminium hydride reductions, Michael additions of Li enolates, hydrosilylations, Pd-catalyzed allylations and metatheses. Polymer-bound

TADDOLs have also been used, and review articles with leading references have been published recently.[9]

In many instances, the tetranaphthyl-substituted diol, the preparation of which is described here, gave the most effective chiral titanium catalysts. Three examples are the nucleophilic addition of diethylzinc to aldehydes (eq 1),[10] the Diels-Alder reaction of 3-crotonoyl-1,3-oxazolidin-2-one to cyclopentadiene (eq 2),[7] and the ring opening of a meso anhydride to a half ester (eq 3).[11]

(1)

H

+ Et_2Zn

– 25°C
0.2 eq **2**
($R^1 = R^2 =$ Me
Aryl = 2-naph
X = Y = $OCHMe_2$
1.8 eq $Ti(OCHMe_2)_4$
in toluene

OH

83%
(S/R > 99.5:0.5)

(2)

O O

N O

+

–75°C to –15°C
0.15 eq **2**
($R^1 = R^2 =$ Me
Aryl = 2-naph
X = Y = Cl
in toluene

1 COOX
2

Me

96% (all isomers)
(endo/exo 9:1,
2S/2R 94:6)

(3)

O
O
O

– 35°C, 5 days
1 eq **2**
($R^1 = R^2 =$ Me
Aryl = 2-naph
X = Y = $OCHMe_2$
in THF

CO_2H
CO_2CHMe_2

65%
er > 98:2

Note that the Ti-complexes of (R,R)-TADDOLs, (P)-BINOL, and the (R,R)-N,N'-1,2-cyclohexanediylbistrifluoromethanesulfonamide (CYDIS) usually give the same product enantiomer in a large variety of reactions, suggesting related transition states of the corresponding various reactions![7]

(R,R)-TADDOL (P)-BINOL (R,R)-CYDIS

1. Laboratorium für Organische Chemie, ETH-Zentrum, Universitätstrasse 16, CH-8092 Zürich, Switzerland.
2. Preclinical Research, Novartis Pharma Ltd., Basle, Switzerland.
3. Beck, A. K.; Bastani, B.; Plattner, D. A.; Petter, W.; Seebach, D.; Braunschweiger, H.; Gysi, P.; La Vecchia, L. *Chimia* **1991**, *45*, 238.
4. (a) Carmack, M.; Kelley, C. J. *J. Org. Chem.* **1968**, *33*, 2171; (b) Seebach, D.; Kalinowski, H.-O.; Bastani, B.; Crass, G.; Daum, H.; Dörr, H.; DuPreez, N. P.; Ehrig, V.; Langer, W.; Nüssler, C.; Oei, H.-A.; Schmidt M. *Helv. Chim. Acta* **1977**, *60*, 301; (c) Seebach, D.; Beck, A. K.; Imwinkelried, R.; Roggo, S.; Wonnacott, A. *Helv. Chim. Acta* **1987**, *70*, 954; (d) Mash, E. A.; Nelson, K. A.; Van Deusen, S.; Hemperly S. B. *Org. Synth., Coll. Vol. VIII* **1993**, 155.
5. Schaefer, J. P.; Higgins, J.; Shenoy, P. K. *Org. Synth., Coll. Vol. V* **1973**, 142.
6. (a) Toda, F. *Yuki Gosei Kagaku Kyokaishi* **1994**, *52*, 923; (b) Kaupp, G. *Angew. Chem.* **1994**, *106*, 768; *Angew. Chem., Int. Ed. Engl.* **1994**, *33*, 728; (c) Weber, E.; Dörpinghaus, N.; Wimmer C.; Stein, Z.; Krupitsky, H.; Goldberg I. *J. Org. Chem.* **1992**, *57*, 6825; (d) von dem Bussche-Hünnefeld, C.; Beck, A. K.; Lengweiler, U.; Seebach, D. *Helv. Chim. Acta* **1992**, *75*, 438.

7. Seebach, D.; Dahinden, R.; Marti, R. E.; Beck, A. K.; Plattner, D. A.; Kühnle, F. N. M. *J. Org. Chem.* **1995**, *60*, 1788.

8. Tanaka, K.; Ootani, M.; Toda, F. *Tetrahedron: Asymmetry* **1992**, *3*, 709.

9. (a) Dahinden, R.; Beck, A. K.; Seebach, D. *Encyclopedia of Reagents for Organic Synthesis*, Paquette, L. A., Ed.; John Wiley & Sons; Chichester, **1995**, *3*, 2167; (b) Seebach, D.; Beck, A. K. *Chimia* **1997**, *51*, 293.

10. Seebach, D.; Beck, A. K.; Schmidt, B.; Wang, Y. M. *Tetrahedron* **1994**, *50*, 4363.

11. (a) Seebach, D., Jaeschke, G.; Wang, Y. M. *Angew. Chem.* **1995**, *107*, 2605; *Angew. Chem., Int. Ed. Engl.* **1995**, *34*, 2395; (b) Jaeschke, G.; Seebach, D. *J. Org. Chem.* **1998**, *63*, 1190.

Appendix
Chemical Abstracts Nomenclature (Collective Index Number); (Registry Number)

(4R,5R)-2,2-Dimethyl-α,α,α',α'-tetra(naphth-2-yl)-1,3-dioxolane-4,5-dimethanol: 1,3-Dioxolane-4,5-dimethanol, 2,2-dimethyl-α,α,α',α'-tetra-2-naphthalenyl-, (4R-trans)- (12); (137365-09-4)

(R,R)-Dimethyl tartrate: Tartaric acid, dimethyl ester, (+)- (8); Butanedioic acid, 2,3-dihydroxy-, [R-(R*,R*)]-, dimethyl ester (9); (608-68-4)

2-Naphthylmagnesium bromide: Magnesium, bromo-2-naphthyl- (8); Magnesium, bromo-2-naphthalenyl- (9); (21473-01-8)

(R,R)-Dimethyl O,O-isopropylidenetartrate: 1,3-Dioxolane-4,5-dicarboxylic acid, 2,2-dimethyl-, dimethyl ester, (4R-trans)- (9); (37031-29-1)

Boron trifluoride etherate: Ethyl ether, compd. with boron fluoride (1:1) (8); Ethane, 1,1'-oxybis-, compd. with trifluoroborane (1:1) (9); (109-63-7)

Magnesium (8,9); (7439-95-4)

Iodine (8,9); (7553-56-2)

2-Bromonaphthalene: Naphthalene, 2-bromo- (8,9); (580-13-2)

(R,R)- AND (S,S)-N,N'-DIMETHYL-1,2-DIPHENYLETHYLENE-1,2-DIAMINE

(1,2-Ethanediamine, N,N'-dimethyl-1,2-diphenyl-, [R-(R*,R*)]- and [S-(R*,R*)-])

A. PhCHO + $MeNH_2$ ⟶ PhCH=NMe

B. PhCH=NMe ⟶ (Zn, Me_3SiCl, CH_3CN) PhCH(N(SiMe₃)Me)CH(N(SiMe₃)Me)Ph ⟶ (H_2O) PhCH(NHMe)CH(NHMe)Ph

C. PhCH(NHMe)CH(NHMe)Ph ⟶ (Li, isoprene) PhCH(NHMe)CH(NHMe)Ph (racemic)

D. PhCH(NHMe)CH(NHMe)Ph ⟶ (Resolution, Tartaric acid) (R,R) + (S,S)

Submitted by Alex Alexakis, Isabelle Aujard, Tonis Kanger, and Pierre Mangeney.[1]
Checked by Nabi Magomedov and David J. Hart.

1. Procedure

A. N-Methylbenzimine. Under a well-ventilated hood, a 250-mL Erlenmeyer flask is equipped with a magnetic stirrer and charged with an aqueous solution (40% w/w, 100 mL) of methylamine. Freshly distilled benzaldehyde (26.5 g, 0.25 mol) is added to the stirred solution of methylamine at room temperature. A mildly exothermic reaction occurs resulting in a milky white emulsion. The Erlenmeyer flask is stoppered and the mixture is stirred overnight (15 hr). The milky emulsion is transferred to a separatory funnel, diethyl ether (200 mL) is added, and the organic phase is separated and dried over potassium carbonate (K_2CO_3). The solids are removed by filtration, washed with ether (50 mL), and the combined filtrate and washings are concentrated on a rotary evaporator. The residue is distilled through a 10-cm Vigreux column using a water aspirator to give 28.3 g (95%) of imine as a water white liquid (bp 99-100°C/25 mm) (Note 1).

B: meso- and dl-N,N'-Dimethyl-1,2-diphenylethylenediamine. A flame-dried, three-necked, round-bottomed flask (500 mL) is equipped with a mechanical or magnetic stirrer, a thermometer, water condenser connected to a nitrogen inlet, and a septum, and then charged with zinc (Zn) powder (13.1 g, 0.2 mol) and anhydrous acetonitrile (50 mL) under an atmosphere of nitrogen (Note 2). The zinc powder is activated by the addition of 1,2-dibromoethane (1.5 mL, 3.5 g, 0.02 mol) via syringe followed by warming the mixture at reflux for 1 min and then allowing the mixture to cool to room temperature. A small amount of chlorotrimethylsilane (Me_3SiCl) is added via syringe, whereupon evolution of ethylene gas is observed (Note 3). The mixture is stirred for 45 min, and then N-methylbenzimine (23.8 g, 0.2 mol) is added in one portion via syringe, followed by anhydrous acetonitrile (100 mL). The septum is replaced with an addition funnel (100 mL), and chlorotrimethylsilane (Note 4) (32.5 g, 38 mL, 0.3 mol) is slowly added (30 min) at a rate that maintains the interna temperature below 30°C. The reaction mixture is stirred for 2 hr and then cooled to 0°C with an ice bath. The vigorously stirred mixture is hydrolyzed by *cautiously* adding, via the addition funnel, a solution prepared by mixing concentrated aqueous ammonium hydroxide (60 mL) with saturated aqueous ammonium chloride (140 mL). The excess zinc is removed by filtration at 1 atm, washed with ether (200 mL), and the residual zinc is covered with water (Note 5). The organic layer is separated, and the aqueous phase is extracted with diethyl ether (1 x 200 mL) and dichloromethane (2 x 200 mL) (Note 6). The combined organic phases are dried over anhydrous K_2CO_3. The salts are removed by filtration, washed with diethyl ether (50 mL), and the solvent is removed on a rotary evaporator to afford 24.3 g of a semi-solid residue that is used directly in the next reaction (Notes 7 and 8).

C. Isomerization of the diamine: dl-N,N'-Dimethyl-1,2-diphenylethylenediamine. A solution of the crude diamine (24.3 g) in anhydrous tetrahydrofuran (THF) (200 mL) is placed in a three-necked, round-bottomed flask equipped with a mechanical or magnetic stirrer, a thermometer, an addition funnel, and a water condenser fitted with a nitrogen inlet. The addition funnel is removed, and small pieces of freshly cut lithium wire (1.8 g, 0.26 mol) are added in one portion (Note 9). Isoprene (10.2 g, 15 mL, 0.15 mol) is slowly added from the addition funnel while the solution is stirred vigorously. An exothermic reaction occurs, and the internal temperature is maintained below 40°C with the aid of a water bath. The lithium pieces become brilliant and the solution turns dark red. The course of the reaction is carefully monitored by TLC (Note 10) until the meso-diamine is isomerized into the dl-isomer. This takes approximately 1-2 hr. The reaction mixture is then cooled to 0°C, and the excess lithium metal is removed by gravity filtration through a funnel containing a plug of glass wool. The excess lithium is stored under oil. The filtrate is cooled in an ice bath and hydrolyzed by slow addition of aqueous 2.5 N hydrochloric acid (HCl, 200 mL). The layers are separated, and the aqueous phase is extracted with ether (Et_2O,

2 x 100 mL). The organic layer is discarded, and the aqueous layer is made basic with aqueous 35% (w/w) sodium hydroxide (50 mL). The aqueous solution is extracted with Et_2O (2 x 200 mL). The ether extracts are dried over K_2CO_3, the solids are removed by filtration, and the solvent is removed using a rotatory evaporator to give 23.4 g (95%) of crude dl-diamine as a brown oil (Note 11). Pure dl-diamine is isolated as follows. A 1-L, round-bottomed flask, equipped with a magnetic stirrer and a water condenser, is charged with the above crude diamine, racemic dl-tartaric acid (14.4 g, 0.096 mol) (Note 12) and absolute ethanol (700 mL). The heterogeneous mixture is brought to reflux at which point the precipitate is mainly dissolved. After 10 min at reflux, the mixture is left at room temperature for 2-4 hr. The resulting precipitate is collected by suction filtration and rinsed with ethanol (50 mL). The collected salt is added to a mixture of aqueous 35% (w/w) sodium hydroxide (60 mL), demineralized water (200 mL) and Et_2O (200 mL). The mixture is stirred for 30 min, the phases are separated and the aqueous layer is extracted with Et_2O (2 x 200 mL). The combined organic phases are dried over K_2CO_3, the salts are removed by filtration and the solvent is removed on a rotatory evaporator to give 12.2 g (51%) of pure dl-diamine as a colorless oil (Note 13).

D. Resolution of the dl-diamine: (R,R)-(+)-N,N'-Dimethyl-1,2-diphenylethylenediamine and (S,S)-(-)-N,N'-Dimethyl-1,2-diphenylethylenediamine. A 1-L, round-bottomed flask, equipped with a magnetic stirring bar and a water condenser, is charged with pure dl-diamine (12 g, 0.05 mol), natural L-(+)-tartaric acid (7.1 g, 0.05 mol) (Note 14) and absolute ethanol (350 mL). The heterogeneous mixture is brought to reflux and the precipitate dissolves completely (Note 15). After 30 min the solution is allowed to cool to room temperature and stand overnight (17 hr). The precipitate is collected by filtration and washed with ethanol (2 x 50 mL). The salt is added to a mixture of aqueous 35% (w/w) NaOH (30 mL), demineralized water (100 mL) and Et_2O (100 mL). The solution is stirred for 30 min, and the phases are separated. The aqueous layer is extracted with Et_2O (2 x 100 mL), and the combined extracts are dried over K_2CO_3. The solids are removed by filtration, and the solvent is removed on a rotary evaporator to give 5.5 g (90%) of crude (R,R)-diamine as a white powder, mp 49-50°C (Note 16). The mother liquor is concentrated under reduced pressure, and the residue is stirred with a mixture of aqueous 35% (w/w) NaOH (30 mL), demineralized water (100 mL) and Et_2O (100 mL). The solution is stirred for 30 min, and the phases are separated. The aqueous layer is extracted with Et_2O (2 x 100 mL), and the combined extracts are dried over K_2CO_3. The solids are removed by filtration, and the solvent is removed on a rotary evaporator to give 6.15 g (101%) of crude (S,S)-diamine as an oil that solidifies upon drying under reduced pressure, mp 45-48°C (Note 17).

2. Notes

1. Spectral data for the imine follow: ^1H NMR ($CDCl_3$) δ: 3.5 (d, 3 H, J = 1.7), 7.4 (m, 3 H), 7.7 (m, 2 H), 8.2 (q, 1 H, J = 1.7); ^{13}C NMR ($CDCl_3$) δ: 48.1, 127.9, 128.5, 130.5, 136.4, 162.3.

2. Zinc powder (325 mesh) was purchased from Aldrich Chemical Company, Inc., or Prolabo. Anhydrous acetonitrile was purchased directly from Aldrich Chemical Company, Inc., and used without purification.

3. Activation of zinc is not always needed. For zinc purchased from Prolabo, the submitters report that the coupling reaction proceeds directly. The checkers note that gas evolution begins upon addition of the 1,2-dibromoethane and increases upon addition of the chlorotrimethylsilane.

4. Commercial chlorotrimethylsilane (Janssen Chimica or Aldrich Chemical Company, Inc.) is used without any purification.

5. The excess zinc is highly reactive and when dry could be exothermically oxidized by air.

6. The meso-diamine is less soluble in Et_2O than the dl-diamine. Therefore, dichloromethane is used to extract both isomers completely.

7. This crude diamine is a 50/50 mixture of meso- and dl-isomers. The mixture also contains 2-5% of N-methylbenzylamine and small amounts of other unidentified contaminants.

8. ^1H NMR analysis of the mixture ($CDCl_3$) shows the following distinct signals due to the meso-isomer δ: 2.09 (s, 6 H), 3.62 (s, 2 H); the dl-isomer δ: 2.25 (s, 6 H), 3.53 (s, 2 H); and N-methylbenzylamine δ: 2.45 (s, 3 H), 3.75 (s, 2 H).

9. The submitters indicate that the isomerization rate is faster when lithium containing 0.5-1.0% sodium is used.

10. An aliquot of approximately 0.5 mL is withdrawn from the reaction using a Pasteur pipette and mixed with an equal volume of aqueous saturated ammonium chloride (NH_4Cl). Approximately 0.5 mL of ethyl acetate is added followed by brief mixing and withdrawal of the upper organic layer via pipette. TLC analysis of this sample is performed over silica gel using diethyl ether-triethylamine (95:5) as the eluant with a UV lamp for visualization. The R_f values of the dl-diamine and meso-diamine are 0.8 and 0.6, respectively. Close monitoring of the reaction is important as reduction of the diamine to afford N-methylbenzylamine ($R_f = 0.35$) can occur. The reaction can also be monitored by ^1H NMR analysis of the aforementioned aliquot after drying and concentration under reduced pressure.

11. This material is a mixture of dl-diamine:meso-diamine:N-methylbenzylamine (approximately 82.6:4.4:13.0) as shown by ^1H NMR (see Note 8).

12. Any commercial source of dl-tartaric acid works well. The ratio of tartaric acid to diamine is 1:1. The checkers found that if isomerization of the meso-diamine is incomplete, this purification procedure provides a mixture of meso- and dl-diamines.

13. Spectral data for the dl-diamine follow: ^{1}H NMR ($CDCl_3$) δ: 1.95 (b s, 2 H), 2.25 (s, 6 H), 3.53 (s, 2 H), 7.05 (m, 4 H), 7.15 (m, 6 H); ^{13}C NMR ($CDCl_3$) δ: 34.7, 71.4, 127.0, 128.0, 128.1, 141.2.

14. Any commercial source of (+)-tartaric acid works well.

15. The shape of the precipitate becomes lighter and finer until total dissolution. Total dissolution is achieved only with absolute ethanol. With 95% grade ethanol the fine precipitate does not dissolve completely, but the resolution proceeds equally well.

16. The crude (R,R)-diamine has $[\alpha]_D$ +19.2 ($CHCl_3$, *c* 0.01) and an enantiomeric excess of >99%. Recrystallization of a 1.0-g sample from pentane at 0°C gave 0.66-0.72 g of (R,R)-diamine with mp 50-51°C; $[\alpha]_D$ +20.0 ($CHCl_3$, *c* 0.01) and an enantiomeric excess of >99%. The enantiomeric excess was measured as follows:[2] Into a dry NMR tube under nitrogen is placed a solution of the (R,R)-diamine (32 mg, 0.13 mol) in 0.8 mL of dry deuterochloroform ($CDCl_3$). To this solution are added dry N,N-diethylaniline (80 μL) and a solution of freshly distilled phosphorus trichloride (PCl_3, 50 μL of a 3.2 M solution in dry $CDCl_3$, 0.16 mmol). The solution is shaken vigorously. ^{31}P NMR analysis indicates the formation of a "P–Cl" species at δ 176.2. Enantiomerically pure l-menthol (25 mg, 0.16 mmol) is dissolved in 0.8 mL of dry $CDCl_3$, and the resulting solution is added to the NMR tube under nitrogen. The tube is shaken, and ^{31}P NMR analysis indicates the formation of only one "P–OR" species at δ 143.5. The diastereomeric "P–OR" species derived from the (S,S)-diamine appears at δ 141.2. If the sample is contaminated with meso-diamine, the signal for the corresponding "P–OR" species appears at δ 138.9. Depending on the quality of reagents, some additional peaks may appear in the spectrum, but they do not interfere with the measurement of % ee.

17. The crude (S,S)-diamine has $[\alpha]_D$ -17.1 ($CHCl_3$, *c* 0.01) and an enantiomeric excess of approximately 96% using the method described above. Recrystallization of a 1.0-g sample from pentane at 0°C gave 0.71 g of (S,S)-diamine with mp 49.5-50.5°C,

$[\alpha]_D$ -19.7 ($CHCl_3$, *c* 0.01) and an enantiomeric excess of >99% using the method described above. A second recrystallization of this material gave 0.5 g of material with little change in physical properties: mp 50-50.5°C, $[\alpha]_D$ -20.0 ($CHCl_3$, *c* 0.01).

Waste Disposal Information

All toxic materials were disposed of in accordance with "Prudent Practices in the Laboratory"; National Academy Press; Washington, DC, 1995.

3. Discussion

There are a few reports on the synthesis of N,N'-dimethyl-1,2-diphenylethylenediamine by pinacol couplings[3-8] or by N-methylation of the primary diamine.[9,10] Pinacol type couplings of imines have been reported for the synthesis of other C_2 symmetrical N-substituted or unsubstituted 1,2-diarylethylenediamines using various reducing metals in different solvents.[11-30] However, the present synthesis of N,N'-dimethyl-1,2-diphenylethylenediamine is the most suitable for large scale preparation. There are no expensive or hazardous starting materials, and the procedure does not require tedious filtrations or chromatographic separation of diastereomers.

N,N'-Dimethyl-1,2-diphenylethylenediamine and related chiral diamines are not only useful chiral auxiliaries in asymmetric synthesis, but they have also found applications as analytical reagents. They allow the resolution and the determination of enantiomeric composition of aldehydes by formation of diastereomeric aminals.[31-34] Combined with PCl_3, they allow determination of the enantiomeric composition of alcohols, biphenols, thiols and amines.[35,36] The diamine described here is used most extensively.

Formation of aminals (the nitrogen equivalent of acetals) is a very easy process and occurs without any catalyst and at room temperature. The chiral imidazolidine ring thus formed is a powerful stereodifferentiating group, acting either by steric effects or through chelation control. Aminals also serve as efficient protective groups of aldehyde functionality, allowing the easy recovery of this functionality without epimerization, even in the α-position. Several applications of this concept are summarized in Scheme 1.[37-51]

The present method, with additional examples, was recently published.[52,53]

1. Laboratoire de Chimie des Organoelements, CNRS UA 473, Universite Pierre et Marie Curie, 4, Place Jussieu F-75252, Paris Cedex 05, France.

2. Alexakis, A.; Frutos, J. C.; Mutti, S.; Mangeney, P. *J. Org. Chem.* **1994**, *59*, 3326-3334.
3. Neumann, W. P.; Werner, F. *Chem. Ber.* **1978**, *111*, 3904-3911.
4. Thies, H.; Schönenberger, H.; Bauer, K. H. *Arch. Pharm.* **1958**, *291/63*, 373-375.
5. Baruah, B.; Prajapati, D.; Sandhu, J. S. *Tetrahedron Lett.* **1995**, *36*, 6747-6750.
6. Nosek, J. *Collect. Czech. Chem. Comm.* **1967**, *32*, 2025-2028.
7. Betschart, C.; Seebach, D. *Helv. Chim. Acta* **1987**, *70*, 2215-2231.
8. Mangeney, P.; Tejero, T.; Alexakis, A.; Grosjean, F.; Normant, J. *Synthesis* **1988**, 255-257.
9. Fiorini, M.; Giongo, G. M. *J. Mol. Catal.* **1979**, *5*, 303-310.
10. Mikiciuk-Olasik, E.; Glinka, R.; Kotelko, B. *Rocz. Chem.* **1977**, *51*, 907-913.
11. Anselmino, O. *Ber. Deutsch. Chem. Ges.* **1908**, *41*, 623.
12. Stühmer, W.; Messwarb, G. *Arch. Pharm.* **1953**, *286*, 221.
13. Tanaka, H.; Dhimane, H.; Fujita, H.; Ikemoto, Y.; Torii, S. *Tetrahedron Lett.* **1988**, *29*, 3811-3814.

14. Khan, N. H.; Zuberi, R. H.; Siddiqui, A. A. *Synth. Commun.* **1980**, *10*, 363-371.

15. Smith, J. G.; Boettger, T. J. *Synth. Commun.* **1981**, *11*, 61-64.

16. Shimizu, M.; Iida, T.; Fujisawa, T. *Chem. Lett.* **1995**, 609-610.

17. Smith, J. G.; Veach, C. D. *Can. J. Chem.* **1966**, *44*, 2497-2502.

18. Smith, J. G.; Ho, I. *J. Org. Chem.* **1972**, *37*, 653-656.

19. Eisch, J. J.; Kaska, D. D.; Peterson, C. J. *J. Org. Chem.* **1966**, *31*, 453-456.

20. Schlenk, W.; Appenrodt, J.; Michael, A.; Thal, A. *Ber. Deutsch. Chem. Ges.* **1914**, *47*, 473.

21. Schlenk, W.; Bergmann, E. *Liebigs Ann. Chem.* **1928**, *463*, 281.

22. Jaunin, R. *Helv. Chim. Acta* **1956**, *39*, 111-116.

23. Bachmann, W. E. *J. Am. Chem. Soc.* **1931**, *53*, 2672-2676.

24. Kalyanam, N.; Venkateswara Rao, G. *Tetrahedron Lett.* **1993**, *34*, 1647-1648.

25. Roskamp, E. J.; Pedersen, S. F. *J. Am. Chem. Soc.* **1987**, *109*, 3152-3154.

26. Takaki, K.; Tsubaki, Y.; Tanaka, S.; Beppu, F.; Fujiwara, Y. *Chem. Lett.* **1990**, 203-204.

27. Imamoto, T.; Nishimura, S. *Chem. Lett.* **1990**, 1141-1142.

28. Enholm, E. J.; Forbes, D. C.; Holub, D. P. *Synth. Comm.* **1990**, 20, 981-987.

29. Shono, T.; Kise, N.; Oike, H.; Yoshimoto, M.; Okazaki, E. *Tetrahedron Lett.* **1992**, *33*, 5559-5562.

30. Buchwald, S. L.; Watson, B. T.; Wannamaker, M. W.; Dewan, J. C. *J. Am. Chem. Soc.* **1989**, *111*, 4486-4494.

31. Mangeney, P.; Alexakis, A.; Normant, J. F. *Tetrahedron Lett.* **1988**, *29*, 2677-2680.

32. Cuvinot, D.; Mangeney, P.; Alexakis, A.; Normant, J.-F. *J. Org. Chem.* **1989**, *54*, 2420-2425.

33. Pinsard, P.; Lellouche, J.-P.; Beaucourt, J.-P.; Grée, R. *Tetrahedron Lett.* **1990**, *31*, 1137-1140.

34. Barrett, A. G. M.; Doubleday, W. W.; Tustin, G. J.; White, A. J. P.; Williams, D. J. *J. Chem. Soc., Chem. Commun.* **1994**, 1783-1784.

35. Alexakis, A.; Frutos, J. C.; Mutti, S.; Mangeney, P. *J. Org. Chem.* **1994**, *59*, 3326-3334.

36. Alexakis, A.; Frutos, J. C.; Mangeney, P.; Meyers, A. I.; Moorlag, H. *Tetrahedron Lett.* **1994**, *35*, 5125-5128.

37. Alexakis, A.; Sedrani, R.; Mangeney, P. Normant, J. F. *Tetrahedron Lett.* **1988**, *29*, 4411-4414; Alexakis, A.; Sedrani, R.; Mangeney, P. *Tetrahedron Lett.* **1990**, *31*, 345-348.

38. Alexakis, A.; Sedrani, R.; Normant, J. F.; Mangeney, P. *Tetrahedron: Asymmetry* **1990**, *1*, 283-286.

39. Commercon, M.; Mangeney, P.; Tejero, T.; Alexakis, A. *Tetrahedron: Asymmetry* **1990**, *1*, 287-290.

40. Gosmini, R.; Mangeney, P.; Alexakis, A.; Commercon, M.; Normant, J. F. *Synlett* **1991**, 111-113.

41. Mangeney, P.; Gosmini, R.; Alexakis, A. *Tetrahedron Lett.* **1991**, *32,* 3981-3984.

42. Mangeney, P.; Gosmini, R.; Raussou, S.; Commercon, M.; Alexakis, A. *J. Org. Chem.* **1994**, *59*, 1877-1888.

43. Raussou, S.; Urbain, N.; Mangeney, P.; Alexakis, A. *Tetrahedron Lett.* **1996**, *37*, 1599-1602.

44. Alexakis, A.; Mangeney, P.; Marek, I.; Rose-Munch, F.; Rose, E.; Semra, A.; Robert, F. *J. Am. Chem. Soc.* **1992**, *114*, 8288-8290.

45. Alexakis, A.; Kanger, T.; Mangeney, P.; Rose-Munch, F.; Perrotey, A.; Rose, E. *Tetrahedron: Asymmetry* **1995**, *6*, 47-50.

46. Alexakis, A.; Kanger, T.; Mangeney, P.; Rose-Munch, F.; Perrotey, A.; Rose, E. *Tetrahedron: Asymmetry* **1995**, *6*, 2135-2138.

47. Kanemasa, S.; Hayashi, T.; Tanaka, J.; Yamamoto, H.; Sakurai, T. *J. Org. Chem.* **1991**, *56*, 4473-4481.

48. Alexakis, A.; Sedrani, R.; Lensen, N.; Mangeney, P. In "Organic Synthesis via Organometallics (OSM4)"; Enders, D.; Gais, H.-J.; Keim, W.; Eds.; Vieweg: Wiesbaden, **1993**; pp. 1-9.

49. Alexakis, A.; Tranchier, J.-P.; Lensen, N.; Mangeney, P. *J. Am. Chem. Soc.* **1995**, *117*, 10767-10768.

50. Alexakis, A.; Lensen, N.; Tranchier, J.-P.; Mangeney, P.; Feneau-Dupont, J.; Declersq, J. P.; *Synthesis* **1995**, 1038-1050.

51. Alexakis, A.; Sedrani, R.; Mangeney, P. *unpublished results.*

52. Alexakis, A.; Aujard, I.; Mangeney, P. *Synlett* **1998**, 873-874

53. Alexakis, A.; Aujard, I.; Mangeney, P. *Synlett* **1998**, 875-876.

Appendix

Chemical Abstracts Nomenclature (Collective Index Number); (Registry Number)

(R,R)-N,N'-Dimethyl-1,2-diphenylethylene-1,2-diamine: 1,2-Ethanediamine, N,N'-dimethyl-1,2-diphenyl-, [R-(R*,R*)]- (12); (118628-68-5)

(S,S)-N,N'-Dimethyl-1,2-diphenylethylene-1,2-diamine: 1,2-Ethanediamine, N,N'-dimethyl-1,2-diphenyl-, [S-(R*,R*)]- (10); (70749-06-3)

N-Methylbenzimine: Methylamine, N-benzylidene- (8); Methanamine, N-(phenylmethylene)- (9); (622-29-7)

Methylamine (8); Methanamine (9): (74-89-5)

Benzaldehyde (8,9); (100-52-7)

meso-N,N'-Dimethyl-1,2-diphenylethylenediamine: 1,2-Ethanediamine, N',N'-dimethyl-1,2-diphenyl-, (R*,S*)- (9); (60509-62-8) dl-N,N'-Dimethyl-1,2-diphenylethylenediamine: 1,2-Ethanediamine, N,N'-dimethyl-1,2-diphenyl-, (R*,S*)- (9); (60508-97-6)

Zinc (8,9); (7440-66-6)

Acetonitrile: TOXIC: (8,9); (75-05-8)

1,2-Dibromoethane: Ethane, 1,2-dibromo- (8,9); (106-93-4)

Chlorotrimethylsilane: Silane, chlorotrimethyl- (8,9); (75-77-4)

Ethylene (8); Ethene (9); (74-85-1)

Lithium (8,9); (7439-93-2)

Isoprene (8); 1,3-Butadiene, 2-methyl- (9); (78-79-5)

DL-Tartaric acid: Tartaric acid, DL- (8); Butanedioic acid, 2,3-dihydroxy-, (R*,R*)-(±)- (9); (133-37-9)

L-Tartaric acid: Tartaric acid, L- (8); Butanedioic acid, 2,3-dihydroxy-, [R-(R*,R*)]- (9); (87-69-4)

N-Methylbenzylamine: Aldrich: See: Benzylmethylamine: Benzylamine, N-methyl- (8); Benzenemethanamine, N-methyl- (9); (103-67-3)

N,N-Diethylaniline: Aniline, N,N-diethyl- (8); Benzeneamine, N,N-diethyl- (9); (91-66-7)

Phosphorus trichloride (8,9); (7719-12-2)

Scheme 1

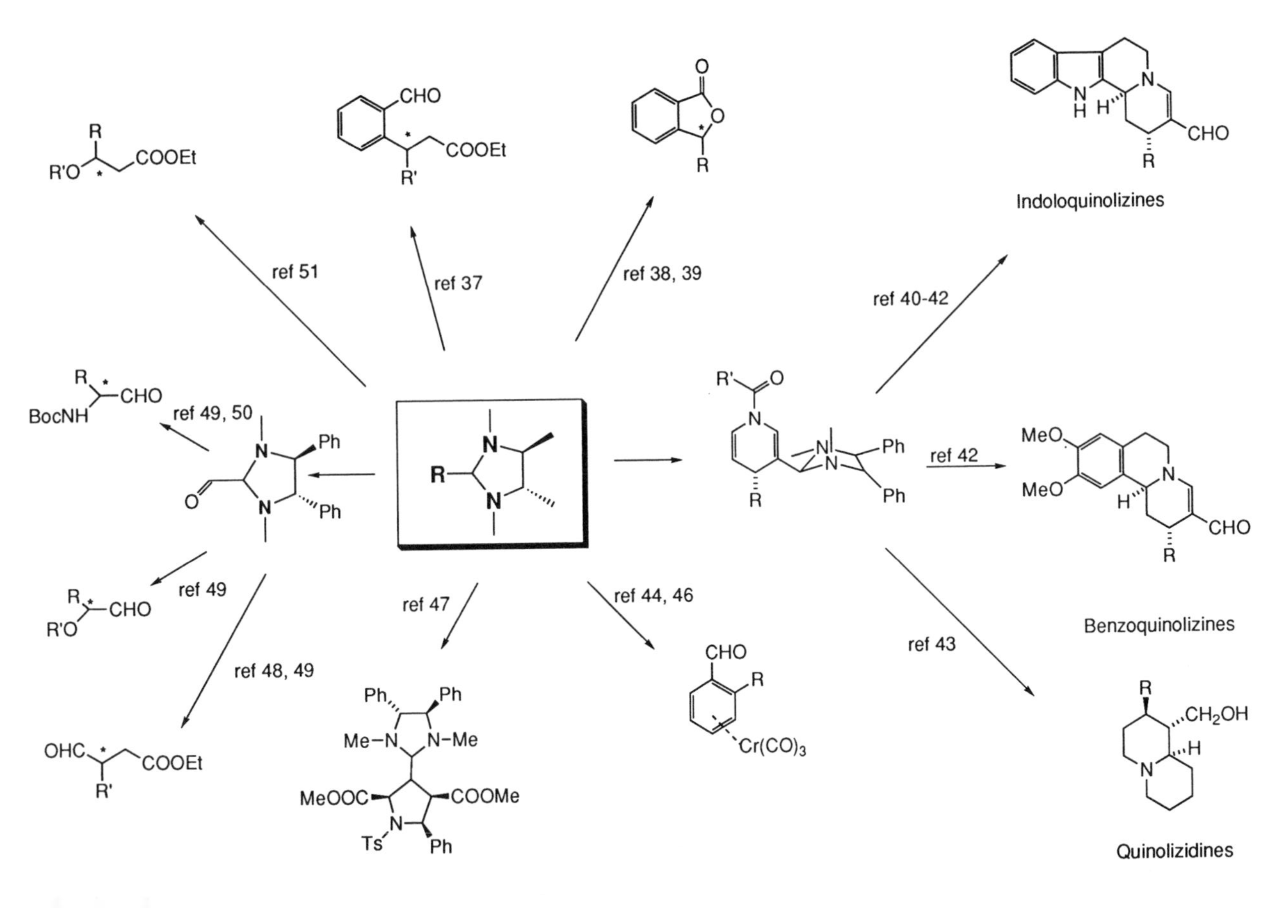

ref 51
ref 37
ref 38, 39
ref 40-42
Indoloquinolizines
ref 49, 50
ref 42
Benzoquinolizines
ref 49
ref 47
ref 44, 46
ref 43
ref 48, 49
Quinolizidines

1S-(-)-1,3-DITHIANE 1-OXIDE

(1,3-Dithiane, 1-oxide, (S)-)

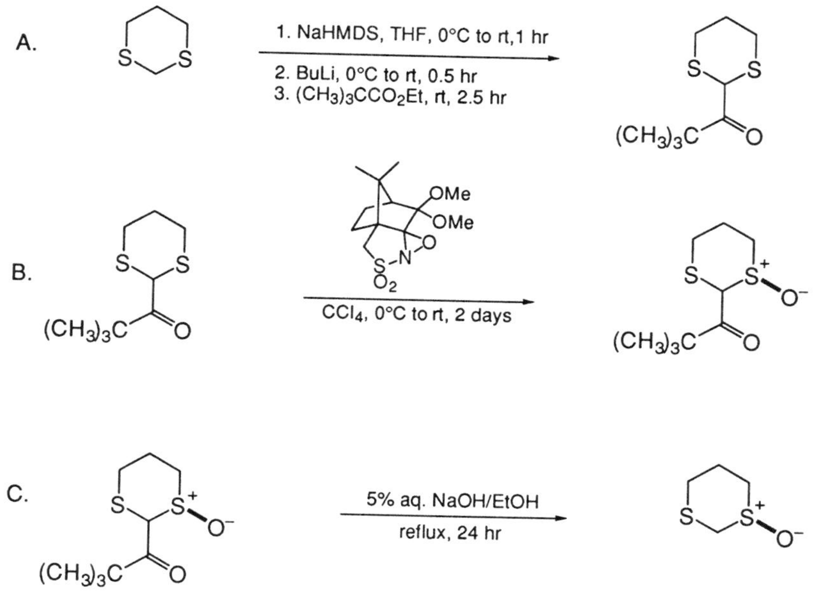

A. 1. NaHMDS, THF, 0°C to rt,1 hr; 2. BuLi, 0°C to rt, 0.5 hr; 3. $(CH_3)_3CCO_2Et$, rt, 2.5 hr

B. CCl_4, 0°C to rt, 2 days

C. 5% aq. NaOH/EtOH, reflux, 24 hr

Submitted by Philip C. Bulman Page,[1] Jag P. Heer,[1] Donald Bethell,[1] Eric W. Collington,[2] and David M. Andrews.[2]

Checked by William Moser and Amos B. Smith, III.

1. Procedure

A. 2-(2,2-Dimethylpropanoyl)-1,3-dithiane. To 43.0 g (0.358 mol) of 1,3-dithiane (Note 1) at 0°C under a nitrogen atmosphere is added 396 mL of a 1 M solution of sodium hexamethyldisilazide in tetrahydrofuran (THF) (Notes 2 and 3). The resulting yellow solution is allowed to reach room temperature and then stirred at room temperature for 1 hr. The solution is cooled to 0°C, and 172 mL of a 2.5 M solution of butyllithium in hexanes (Note 4) is added. The reaction mixture is allowed to warm to room temperature and then stirred at room temperature for 30 min. Ethyl 2,2-dimethylpropanoate (65.0 mL, 0.427 mol) (Note 5) is added, and the mixture is stirred at room temperature for 2.5 hr. An aqueous saturated solution of ammonium chloride (200 mL) is added, and the aqueous phase is extracted three times with 200 mL of dichloromethane (Note 6). The combined organic extracts are washed with 100 mL of water and dried over anhydrous magnesium sulfate. The solvents are removed under reduced pressure to give a yellow solid. Repeated trituration with petroleum ether 40-60°C (Note 7) followed by filtration gives 51.0-55.1 g (70-75%) of 2-(2,2-dimethylpropanoyl)-1,3-dithiane as colorless needles, mp 97-99°C (Note 8).

B. anti- and syn-1S-(2,2-Dimethylpropanoyl)-1,3-dithiane 1-oxide. (+)-[(8,8-Dimethoxycamphoryl)sulfonyl]oxaziridine (51.0 g, 0.176 mol) (Note 9) is added to a cooled, stirred solution of 36.0 g (0.176 mol) of 2-(2,2-dimethylpropanoyl)-1,3-dithiane in 1000 mL of carbon tetrachloride (Note 10) at 0°C. The reaction mixture is allowed to reach room temperature, and stirring is continued at room temperature for a further 48 hr. The reaction mixture is filtered to remove the bulk of the (+)-[(8,8-dimethoxycamphoryl)sulfonyl]imine, and the filtrate is evaporated to dryness under reduced pressure. The residue is purified by passage through a short column of silica gel using dichloromethane as initial eluant to remove residual (+)-[(8,8-dimethoxycamphoryl)sulfonyl]imine. The column is then flushed with ethyl acetate to give 29.8-33.5 g (77-86%) of an ca. 3:1 mixture of anti- and syn-1S-(2,2-dimethylpropanoyl)-1,3-dithiane 1-oxide as a colorless crystalline solid, mp 103-105°C (Note 11).

C. (1S)-(-)-1,3-Dithiane 1-oxide. A mixture of anti- and syn-1S-2-(2,2-dimethylpropanoyl)-1,3-dithiane 1-oxide (33 g, 0.150 mol) is dissolved in 500 mL of ethanol (Note 12), and 200 mL of aqueous 5% sodium hydroxide is added. The mixture is heated under reflux for 24 hr. The mixture is allowed to cool, and 500 mL of dichloromethane is added. The organic layer is separated, and the aqueous phase is extracted four times, with 100 mL of dichloromethane. The combined organic extracts are dried over anhydrous magnesium sulfate and evaporated to dryness under reduced pressure to give a beige solid. The solid is triturated with diethyl ether to give 13 g (64%) of 1S-(-)-1,3-dithiane 1-oxide as a colorless solid, mp 90-92°C (Notes 13 and 14).

2. Notes

1. 1,3-Dithiane was stored in a desiccator over self-indicating silica gel.

2. Tetrahydrofuran was distilled under nitrogen from the benzophenone ketyl radical.

3. Sodium hexamethyldisilazide [sodium bis(trimethylsilyl)amide] was purchased from the Aldrich Chemical Company, Inc., in 100- or 800-mL bottles as a 1 M solution in tetrahydrofuran. Glassware used for moisture sensitive reactions was dried at 180°C and allowed to cool in a desiccator over self-indicating silica gel. Reactions were carried out under a slight positive static pressure of argon.

4. Butyllithium was purchased from the Aldrich Chemical Company, Inc., in 800-mL bottles as a 2.5 M solution in hexanes; the molarity was determined by titration against a solution of diphenylacetic acid.

5. Commercially available reagents were used as supplied unless otherwise stated.

6. Dichloromethane was dried by distillation from calcium hydride.

7. Petroleum ether (40-60°C) was distilled prior to use.

8. The analytical data for 2-(2,2-dimethylpropanoyl)-1,3-dithiane are as follows: Found: C, 52.73; H, 7.87. $C_9H_{16}OS_2$ requires C, 52.90; H, 7.89%; IR (Nujol) cm^{-1}:

2900, 1673; ^1H NMR (400 MHz, $CDCl_3$) δ: 1.24 (s, 9 H), 1.95-2.09 (m, 1 H), 2.13-2.23 (m, 1 H), 2.56 (ddd, 2 H, J = 2.4, 7.0, 12.5), 3.43 (dt, 2 H, J = 2.4, 12.5), 4.51 (s, 1 H); m/z (EI) 204.06445 (M^+); $C_9H_{16}OS_2$ requires 204.06425.

9. For the preparation of (+)-[(8,8-dimethoxycamphoryl)sulfonyl]oxaziridine see: Chen, B.-C.; Murphy, C. K.; Kumar, A.; Reddy, R. T.; Clark, C.; Zhou, P.; Lewis, B. M.; Gala, D.; Mergelsberg, I.; Scherer, D.; Buckley, J.; DiBenedetto, D.; Davis, F. A. *Org. Synth., Coll. Vol. IX* **1998**, 212. A somewhat modified procedure[11] is as follows: *(+)-[(8,8-Dimethoxycamphoryl)sulfonyl]oxaziridine.* Aliquat 336® (tri-n-octyl-methyl-ammonium chloride) (5.0 mL, 10.9 mmol) is added to a stirred solution of 50.0 g (183 mmol) of (+)-[(8,8-dimethoxycamphoryl)sulfonyl]imine in 250 mL of dichloromethane at 0°C. A solution of 50.0 g (362 mmol) of potassium carbonate in 100 mL water is added and the biphasic reaction mixture is stirred for 5 min. A commercial solution (30% w/v) of hydrogen peroxide (83.0 mL, 732 mmol) is added dropwise over 30 min. The reaction is then allowed to warm to room temperature and stirred for about 6-7 hr (Note 15). The organic layer is separated and the aqueous phase extracted three times, each with 100 mL of dichloromethane. Residual hydrogen peroxide in the aqueous phase is carefully destroyed by the addition of saturated aqueous sodium sulfite. The combined organic extracts are rapidly washed with an aqueous solution of 5.0 g of sodium sulfite in 100 mL water and 100 mL of saturated brine and dried over anhydrous magnesium sulfate. Removal of the solvent under reduced pressure, at a bath temperature not exceeding 40°C, gives a white solid consisting of (+)-[(8,8-dimethoxycamphoryl)sulfonyl]oxaziridine contaminated with (+)-[(8,8-dimethoxy-camphoryl)sulfonyl]imine. Recrystallization from absolute ethanol furnishes 51.3 g (97%) of (+)-[(8,8-dimethoxycamphoryl)sulfonyl]oxaziridine, mp 188-190°C (Note 16).

10. Carbon tetrachloride was used as supplied without further purification.

11. The analytical data for anti- and syn-1S-(2,2-dimethylpropanoyl)-1,3-dithiane 1-oxides are as follows: Found: C, 48.91; H, 7.35. $C_9H_{16}O_2S_2$ requires C,

49.06; H, 7.32; IR (Nujol) cm^{-1}: 2900, 1706, 1030; ^{1}H NMR (400 MHz, $CDCl_3$) δ for anti-: 1.26 (s, 9 H), 2.04-2.15 (m, 1 H), 2.45-2.70 (m, 2 H), 2.75-2.90 (m, 2 H), 3.44-3.56 (m, 1 H), 4.72 (s, 1 H); for syn-: 1.24 (s, 9 H), 2.22-2.35 (m, 1 H), 2.45-2.55 (m, 2 H), 3.00-3.15 (m, 2 H), 3.97 (dt, 1 H, J = 3.5, 13.8), 4.98 (s, 1 H); m/z (EI) 220.05931 (M^+); $C_9H_{16}OS_2$ requires 220.059187; ee (anti) = 87%, ee (syn) = 88% from ^{1}H NMR studies (Note 17).

12. Ethanol was used as supplied without further purification.

13. The analytical data for 1S-(-)-1,3-dithiane 1-oxide are as follows: Found: C, 35.18; H, 5.93. $C_4H_8OS_2$ requires C, 35.27; H, 5.89; IR (Nujol) cm^{-1}: 2927, 1047 ; ^{1}H NMR (400 MHz, $CDCl_3$) δ: 2.10-2.35 (m, 1 H), 2.45-2.77 (m, 4 H), 3.35 (ddd, 1 H, J = 3.0, 6.0, 9.5), 3.66 (d, 1 H, J = 12.7), 4.03 (d, 1 H, J = 12.7); m/z (EI) 136.00151 (M^+); $C_4H_8OS_2$ requires 136.00166; ee = 87% from ^{1}H NMR studies (Note 17).

14. The checkers obtained the product in about 54% yield and found flash chromatography to be more effective in its purification. This was accomplished using a 16-cm x 5-cm column of silica gel and $CHCl_3$/MeOH (96:4) as the eluant. With collection of ca. 50-mL fractions, the product was observed in fractions 12-21. Visualization of the product was accomplished by TLC (product R_f = 0.4 in $CHCl_3$/MeOH 96:4, anisaldehyde stain).

15. The checkers noted that complete oxidation typically required ca. 6-7 hr and recommend checking the progress of the reaction in the following way: a 1-mL aliquot is removed from the organic layer, diluted with 2 mL of methylene chloride, and analyzed by TLC eluting with methylene chloride (I_2 visualization); imine R_f = 0.34, oxaziridine R_f = 0.51

16. The analytical data for (+)-[(8,8-dimethoxycamphoryl)sulfonyl]oxaziridine are as follows: Found: C, 49.77; H, 6.62; N, 4.88. $C_{12}H_{19}NO_5S$ requires C, 49.83; H, 6.57; N, 4.84; IR (CH_2Cl_2 film) cm^{-1}: 1367, 1345, 1165; ^{1}H NMR (400 MHz, $CDCl_3$) δ: 1.06 (s, 3 H), 1.32 (s, 3 H), 1.75-2.30 (m, 5 H), 3.08 (d, 1 H, J = 12.0), 3.29 (d, 1 H, J =

12.0), 3.27 (s, 3 H), 3.34 (s, 3 H, CH_3); ^{13}C NMR (100 MHz, $CDCl_3$) δ: 20.5, 21.6, 28.1, 29.3, 45.1, 47.4, 52.9, 50.5, 50.8, 54.6, 97.6, 102.8; m/z (CI) 290.10619 (MH^+); $C_{12}H_{20}NO_5S$ requires 290.10622; $[\alpha]_D^{20}$ +91° ($CHCl_3$, *c* 3.00) (Note 17).

17. Optical rotations were measured on Optical Activity AA-1000 or polAAr 2001 polarimeters operating at 589 nm, corresponding to the sodium D line. Enantiomeric excesses were determined by ^1H NMR chiral shift reagent studies using 10 equiv of (R)-(-)- or (S)-(+)-2,2,2-trifluoro-1-(9-anthryl)ethanol (Pirkle reagent).

Waste Disposal Information

All toxic materials were disposed of in accordance with "Prudent Practices in the Laboratory"; National Academy Press; Washington, DC, 1995.

3. Discussion

Non-racemic chiral sulfoxides have become important as sources of chirality for asymmetric carbon-carbon bond formation.[3] For example, we have developed 1,3-dithiane 1-oxide (DiTOX) units as effective moieties for stereocontrol of a range of carbonyl group reactions, including enolate alkylation and amination, Mannich reaction, reduction, and heterocycloaddition.[4] While we have been able to prepare several 2-monosubstituted[5] and 2,2-disubstituted-1,3-dithiane 1-oxides[6] in high enantiomeric excesses (ee) on scales of a few grams, we had difficulty until recently in preparing the parent compound, 1,3-dithiane 1-oxide, with very high ee in quantities of more than ca. 5 g.[7] Enantiomerically pure 1,3-dithiane 1-oxide has previously been prepared via adducts with (+)-camphor,[8] and, by ourselves, using modified Sharpless oxidation techniques.[7,9,10]

We have recently reported that [(8,8-dimethoxycamphoryl)sulfonyl]oxaziridine is a particularly effective reagent for asymmetric sulfide oxidation, especially in non-aryl sulfide substrates.[11] Here we report a three-step chemical synthesis of 1,3-dithiane 1-oxide with very high ee that is based upon such an oxidation as the key step. The procedure is effective for production of multigram quantities of material of either absolute configuration. The sequence is illustrated for the preparation of 1S-(-)-1,3-dithiane 1-oxide.

The route is based upon an acylation-oxidation-deacylation sequence, with commercially available, inexpensive 1,3-dithiane employed as the starting material. 2-Acyl-1,3-dithianes have proved to be particularly effective substrates for asymmetric oxidation in our hands,[7,9,11,12] and as 2-(2,2-dimethylpropanoyl)-1,3-dithiane undergoes this asymmetric oxidation most efficiently (ca. 90% ee), it was chosen as the intermediate.

1. Robert Robinson Laboratories, Department of Chemistry, University of Liverpool, Oxford Street, Liverpool L69 3BX, England. This investigation has enjoyed the support of the EPSRC and Glaxo Research & Development (CASE award to JPH).
2. Glaxo Research & Development, Gunnels Wood Road, Stevenage, Hertfordshire SG13 9NJ, England.
3. Solladié, G. *Synthesis*, **1981**, 185; Posner, G. H. In "Asymmetric Synthesis"; Morrison, J. D., Ed.; Academic Press: New York, 1983; Vol. 2; Barbachyn, M. R.; Johnson, C. R. In "Asymmetric Synthesis"; Morrison, J. D.; Scott, J. W., Eds.; Academic Press: New York, 1983; Vol. 4; Nudelman, A. In "The Chemistry of Optically Active Sulfur Compounds", Gordon and Breach: New York, 1984; Posner, G. H. In "The Chemistry of Sulphones and Sulphoxides"; Patai, S.; Rappoport, Z.; Stirling, C., Eds.; Wiley, 1988; p. 823; Andersen, K. K. In "The

Chemistry of Sulphones and Sulphoxides", Patai, S.; Rappoport, Z.; Stirling, C., Eds.; Wiley, 1988; p. 55; Kagan, H. B.; Rebiere, F. *Synlett* **1990**, 643; Aggarwal, V. K.; Thomas, A.; Franklin, R. J. *J. Chem. Soc., Chem. Commun.* **1994**, 1653; Solladié, G.; Carreño, M. E. In "Organosulfur Chemistry: Synthetic Aspects"; Page, P., Ed.; Adademic Press: London, 1995; Vol. 1.

4. Page, P. C. B.; McKenzie, M. J.; Buckle, D. R. *J. Chem. Soc., Perkin Trans. I* **1995**, 2673, and references therein.
5. Page, P. C.; Namwindwa, E. S.; Klair, S. S.; Westwood, D. *Synlett* **1990**, 457; Page, P. C. B.; Wilkes, R. D.; Barkley, J. V.; Witty, M. J. *Synlett* **1994**, 547.
6. Page, P. C. B.; Wilkes, R. D.; Namwindwa, E. S.; Witty, M. J. *Tetrahedron* **1995**, *51*, 2125.
7. Page, P. C. B.; Wilkes, R. D.; Witty, M. J. *Org. Prep. Proced. Int.* **1994**, *26*, 702.
8. Bryan, R. F.; Carey, F. A.; Dailey, Jr., O. D.; Maher, R. J.; Miller, R. W. *J. Org. Chem.* **1978**, *43*, 90.
9. Page, P. C. B.; Gareh, M. T.; Porter, R. A. *Tetrahedron: Asymmetry* **1993**, *4*, 2139.
10. Kagan, H. B.; Rebiere, F. *Synlett* **1990**, 643, and references therein.
11. Page, P. C. B.; Heer, J. P.; Bethell, D.; Collington, E. W.; Andrews, D. M. *Tetrahedron: Asymmetry* **1995**, *6*, 2911. For a convenient procedure for the preparation of camphorsulfonyloxaziridines see: Page, P. C. B.; Heer, J. P.; Bethell, D.; Lund, A.; Collington, E. W.; Andrews, D. M. *J. Org. Chem.* **1997**, *62*, 6093-6094.
12. Page, P. C. B.; Heer, J. P.; Bethell, D.; Collington, E. W.; Andrews, D. M. *Synlett* **1995**, 773.

Appendix

Chemical Abstracts Nomenclature (Collective Index Number); (Registry Number)

1S-(-)-1,3-Dithiane 1-oxide: 1,3-Dithiane, 1-oxide, (S)- (10); (63865-78-1)

2-(2,2-Dimethylpropanoyl)-1,3-dithiane: 1-Propanone, 1-(1,3-dithian-2-yl)-2,2-dimethyl- (10); (73119-31-0)

1,3-Dithiane: m-Dithiane (8); 1,3-Dithiane (9); (505-23-7)

Sodium hexamethyldisilazide (NHMDS): Aldrich: Sodium bis(trimethylsilyl)amide: Disilazane, 1,1,1,3,3,3-hexamethyl-, sodium salt (8); Silanamine, 1,1,1-trimethyl-N-(trimethylsilyl)-, sodium salt (9); (1070-89-9)

Butyllithium: Lithium, butyl- (8,9); (109-72-8)

Ethyl 2,2-dimethylpropanoate: Aldrich: See: Ethyl trimethylacetate: Propanoic acid, 2,2-dimethyl-, ethyl ester (9); (3938-95-2)

anti-1S-(2,2-Dimethylpropanoyl)-1,3-dithiane 1-oxide: 1-Propanone, 2,2-dimethyl-1-(1-oxido-1,3-dithian-2-yl)-, (1S-trans)- (13); (160496-17-3)

(+)-[(8,8-Dimethoxycamphoryl)sulfonyl]oxaziridine: 4H-4a,7-Methanooxazirino[3,2-i][2,1]benzisothiazole, tetrahydro-8,8-dimethoxy-9,9-dimethyl-, 3,3-dioxide, [2R-(2α,4aα,7α, 8aR*)]- (12); (131863-82-6)

Aliquat 336: Methyltri-n-octylammonium chloride: Ammonium, methyltrioctyl-, chloride (8); 1-Octanaminium, N-methyl-N,N-dioctyl-, chloride (9); (5137-55-3)

Hydrogen peroxide (8,9); (7722-84-1)

(1S,2R)-1-AMINOINDAN-2-OL

(1H-Inden-2-ol, 1-amino-2,3-dihydro-(1S-cis)-)

A. indene + NaOCl → (Mn-salen catalyst; Ph-pyridine N^+-O^-, CH_2Cl_2, 0°C) → indene oxide, 84-86% ee

B. indene oxide + CH_3CN → (H_2SO_4 (fuming), hexanes, 0-5°C) → [oxazoline, CH_3] → (H_2O) → aminoindanol NH_3^+ HSO_4^-

C. aminoindanol NH_3^+ HSO_4^- → (1. NaOH; 2. L-tartaric acid) → aminoindanol NH_3^+ tartrate → (NaOH (aq)) → aminoindanol, >99% ee

Submitted by Jay F. Larrow,[1] Ed Roberts,[2] Thomas R. Verhoeven,[2] Ken M. Ryan,[2,3a] Chris H. Senanayake,[2,3b] Paul J. Reider,[2] and Eric N. Jacobsen.[1]

Checked by Stephanie A. Lodise and Amos B. Smith, III.

1. Procedure

A. (1S,2R)-Indene oxide. A 500-mL, three-necked, round-bottomed flask equipped with an overhead mechanical stirrer, a 125-mL addition funnel, and a thermocouple is charged with indene (Note 1, 29.0 g, 0.25 mol, 1 equiv), dichloromethane (CH_2Cl_2) (30 mL), (S,S)-(N,N')-bis(3,5-di-tert-butylsalicylidene)-1,2-cyclohexanediaminomanganese(III) chloride (0.953 g, 1.5 mmol, 0.6 mol%, Note 2), and 4-phenylpyridine N-oxide (Note 3, 1.28 g, 7.5 mmol, 3.0 mol%) under a nitrogen (N_2) atmosphere. The resulting brown mixture is cooled to -5°C, and then a cold sodium hypochlorite (NaOCl) solution (191 mL, 1.7 M, 1.3 equiv, Note 4) is added slowly with vigorous stirring while maintaining the reaction temperature between 0°C and 2°C (Note 5). Upon complete addition of the bleach, the reaction is stirred for an additional 1 hr at 0°C. At this point, hexanes (200 mL) are added in one portion with stirring, and the reaction mixture is filtered through a pad of Celite on a large Büchner funnel. The filter cake is washed with dichloromethane (2 x 50 mL), and the filtrate is transferred to a 500-mL separatory funnel. The lower aqueous layer is removed, and the brown organic layer is washed with aqueous saturated sodium chloride (NaCl) solution (100 mL). The organic layer is dried over sodium sulfate (Na_2SO_4), filtered, and concentrated by rotary evaporation. A small amount of calcium hydride (CaH_2) (100 mg) is added to the brown residue, and the epoxide is isolated by short path vacuum distillation, bp 58-60°C (0.025 mm), to yield 24.0 g of epoxyindane (84-86% ee) as a colorless to slightly yellow liquid (0.197 mol, 71% yield, Notes 6, 7, and 8).

B. (1S,2R)-1-Aminoindan-2-ol (crude). A dry, 1000-mL, three-necked, round-bottomed flask equipped with a large magnetic stir bar, a 250-mL addition funnel, a 50-mL addition funnel, and a thermocouple is charged with dry acetonitrile (100 mL, Note 9) and cooled to -5°C under a N_2 atmosphere. The mixture is stirred vigorously, and slow addition of fuming sulfuric acid (20 mL, 0.4 mol, 2 equiv, 27-33% SO_3, Note

10) is begun, followed by dropwise addition of a solution of the epoxide (26.0 g, 0.197 mol) in dry hexanes (200 mL, Note 9). The reagents are added simultaneously at a rate such that the reaction temperature is maintained between 0 and 5°C. After the additions are complete, the reaction mixture is warmed to room temperature and stirred for 1 hr. Water (100 mL) is added via the addition funnel over 10-15 min, and the resulting biphasic mixture is stirred for an additional 30 min. The lower aqueous phase is separated, diluted with 100 mL of water and concentrated by distillation at atmospheric pressure to a head temperature of 100°C. The mixture is heated at reflux for 3 hr, after which time the mixture is cooled to room temperature. The crude aqueous solution of aminoindanol is used without further purification in the next step (Note 11).

C. (1S,2R)-1-Aminoindan-2-ol (100% ee). A 500-mL, three-necked, round-bottomed flask equipped with a large magnetic stir bar, a 125-mL addition funnel, a pH probe, and a thermocouple is charged with the hydrolysis solution from Part B and 1-butanol (100 mL). Sodium hydroxide (80 mL of an aqueous 50% solution, Note 12) is added slowly with external ice bath cooling to maintain the temperature below 30°C until the reaction mixture reaches a pH of 12-13. The upper 1-butanol layer is separated, and the aqueous layer is extracted with another 100 mL of 1-butanol. The combined butanol extracts are diluted with methanol (200 mL), and vacuum-filtered through a Büchner funnel into a 2000-mL, three-necked, round-bottomed flask equipped with a mechanical overhead stirrer, 250-mL addition funnel, and reflux condenser. The reaction mixture is stirred vigorously and heated to reflux, and a solution of L-tartaric acid (35.5 g, 0.24 mol, 1.2 equiv) in methanol (200 mL) is added over 15-30 min while reflux is maintained. Heating is discontinued until a thick but stirrable slurry is formed (Note 13). The suspension is reheated to reflux for 2 hr (Note 14). At this point, the mixture is cooled to room temperature and allowed to stand for 1 hr. The resulting solids are collected on a Büchner funnel and washed with methanol

(2 x 100 mL). The white solid thus obtained is then dried under reduced pressure to yield 47.4 g of the 1:1 tartrate salt (Note 15).

A 300-mL, three-necked, round-bottomed flask equipped with a large magnetic stir bar, a 125-mL addition funnel, a pH probe, and a thermocouple is charged with the aminoindanol-tartrate salt. Water (95 mL, 2:1 v/w) is added, and the mixture is stirred under a N_2 atmosphere. Aqueous sodium hydroxide (50 wt-%, 23 mL, 2 equiv, Notes 12 and 16) is added with external ice bath cooling until the reaction mixture reaches pH 12-13, resulting in precipitation of the aminoindanol free base. The mixture is cooled to 0°C and allowed to stand at that temperature for an additional 30 min. The white to tan solid is collected by vacuum filtration, washed with ice-cold water (20 mL), and air-dried on the filter. The solid is dissolved in hot toluene (1:10 w/v, Note 17), and the resulting solution is allowed to cool to room temperature, then further cooled to 4°C for 1 hr. The resulting white solid is collected by vacuum filtration, washed with cold toluene (20 mL), and dried under reduced pressure. The total yield of aminoindanol (100% ee, Note 18) is 17.2 g, mp 122-124°C (Note 19).

2. Notes

1. Technical grade indene (92%) was obtained from Aldrich Chemical Company, Inc., and was passed through a 10 x 10-cm column of basic alumina to remove highly colored impurities. The compound was then distilled under a N_2 atmosphere from a small amount of CaH_2. The distillate was stored under N_2 at 4°C.

2. The epoxidation catalyst was prepared according to the published procedures.[4] Alternatively, research quantities can be purchased from Aldrich Chemical Company, Inc., or bulk quantities can be purchased from ChiRex Ltd, Dudley, UK.

3. 4-Phenylpyridine N-oxide was purchased from Aldrich Chemical Company, Inc., and used as received. Other pyridine N-oxide derivatives have been used with success in the epoxidation reaction. The choice of the pyridine N-oxide derivative has been demonstrated to have a small yet measurable impact on both the rate of reaction and the enantiomeric excess of the product epoxide.[5]

4. Commercial 10-13% NaOCl was purchased from Aldrich Chemical Company, Inc., and stored at 4°C. The concentration of bleach was determined according to the method of Kolthoff and Belcher.[6]

To a 250-mL Erlenmeyer flask containing a magnetic stir bar was added 2 mL of the commercial NaOCl solution, 100 mL of water, 1.5 mL of concd HCl, and 7 g of potassium iodide. The resulting dark brown solution was titrated with a 1 M solution of sodium thiosulfate ($Na_2S_2O_3$). The endpoint is reached when the solution becomes colorless. Using the following equations, the concentration of the NaOCl solution can be calculated.

$$OCl^- + 2I^- + 2H^+ \rightarrow H_2O + Cl^- + I_2$$

$$2S_2O_3 + I_2 \rightarrow S_4O_6 + 2I^-$$

The solution (191 mL) used by the submitters was found to be 1.7 M in NaOCl and was then made 0.2 M in sodium hydroxide (NaOH) by the addition of 1.52 g of solid NaOH.

5. The epoxidation reaction is exothermic. The bleach is added over a period of 2-2.5 hr in order to maintain the desired temperature range. If the rate of addition appears to be faster, then the rate of stirring should be increased to ensure proper mixing of the biphasic mixture.

6. The physical properties are as follows: ^1H NMR (400 MHz, $CDCl_3$) δ: 2.95 (dd, 1 H, J = 2.9, 18.2), 3.19 (d, 1 H, J = 18.2), 4.10 (t, 1 H, J = 2.9), 4.25 (m, 1 H), 7.15-7.25 (m, 3 H), 7.48 (d, 1 H, J = 7.4); ^{13}C NMR (100 MHz, $CDCl_3$) δ: 34.5, 57.5, 59.0, 125.0, 126.0, 126.1, 128.4, 140.8, 143.4; IR (NaCl): υ (cm^{-1}) 3045, 3029, 2913, 1474,

1466, 1419, 1372, 1230, 1175, 1000, 983; $[\alpha]_D^{23}$ +23.3° (hexanes, *c* 1.31); HRMS (CI, cool probe): calcd for $C_9H_{12}NO$ [M+NH_4] 150.0919, observed 150.0913.

7. The enantiomeric excess of the epoxide is determined by HPLC analysis using a Chiracel OB column (25 cm x 4.6 mm, Daicel) eluted with EtOH/hexanes (5:95) at 1 mL/min, while monitoring at 254 nm. The retention times of the epoxide enantiomers are 11.1 (1S,2R) and 15.3 (1R,2S) min.

8. The submitters distilled the epoxyindane at lower pressure, bp 47-48°C (0.005 mm).

9. Solvents were freshly distilled from CaH_2 prior to reaction.

10. Fuming sulfuric acid was purchased from Aldrich Chemical Company, Inc., and used as received. Approximately 10% of the acid is added before addition of the epoxide solution is begun.

11. The weight of the hydrolysis mixture is 225-250 g.

12. Slightly more or less NaOH solution may be required to reach the desired pH range.

13. The precipitation of the diastereomeric salt is rapid, and the solid may need to be broken up with a spatula in order to maintain proper mixing. Additional methanol may be added if needed.

14. The salt that precipitates initially is not diastereomerically pure, and the additional reflux period is necessary to allow equilibration to the diastereomerically pure material.

15. The salt is isolated as a methanol solvated complex, FW = 331.

16. The pH of the initial mixture is approximately 3.5. The free base begins to precipitate around pH 8.5.

17. The product is recrystallized from toluene in order to remove any of the trans isomer, as well as to dry the product. The hot mixture should be heated long enough to azeotropically remove water from the product.

18. The stereochemical purity of the product is determined by reacting a sample of the product (15 mg) with 2,4-dinitrofluorobenzene (13 μL) in CH_2Cl_2 (5 mL). The yellow solution is diluted with ethanol (1:10), then analyzed by HPLC on an N-naphthylleucine column (4.6 x 25 mm, Regis) eluted with IPA/hexanes (8:92) at 1 mL/min while monitoring at 350 nm. The retention times of the trans enantiomers are 17.4 and 19.1 min, while those of the cis enantiomers are 24.4 (1R,2S) and 27.1 (1S,2R) min. The product is enantio- and diastereomerically pure after recrystallization from toluene.

19. The physical properties are as follows: ^{1}H NMR (400 MHz, CD_3OD) δ: 2.88 (dd, 1 H, J = 2.9, 16.1), 3.05 (dd, 1 H, J = 5.4, 16.1), 4.13 (d, 1 H, J = 5.0), 4.39 (m, 1 H), 7.17-7.22 (m, 3 H), 7.38 (m, 1 H); ^{13}C NMR (100 MHz, CD_3OD) δ: 40.0, 60.4, 75.2, 125.3, 126.1, 127.8, 128.6, 141.8, 145.1; IR (NaCl): υ (cm^{-1}) 3343, 3290, 3170-3022, 2918, 1618, 1605, 1474, 1344, 1180, 1058; $[\alpha]_D^{23}$ -41.2° (MeOH, *c* 1.00, MeOH); HRMS (CI, cool probe): calcd for $C_9H_{12}NO$ [M+H] 150.0919, observed 150.0913.

Waste Disposal Information

All toxic materials were disposed of in accordance with "Prudent Practices in the Laboratory"; National Academy Press; Washington, DC, 1995.

3. Discussion

The development of practical routes to the title compound has been the focus of intensive research effort since cis-aminoindanol was identified as a critical component of the highly effective HIV protease inhibitor indinavir (Crixivan®).[7,8] Reported routes include racemate synthesis followed by resolution via diastereomeric salts,[8] enzymatic resolution,[9] and asymmetric hydroxylation.[10] However, the use of a modified Ritter reaction to convert indene oxide to the corresponding cis-amino alcohol as described in this procedure constitutes the most direct and economical route devised thus far.[11] The application of the (salen)Mn-catalyzed epoxidation reaction[12] in the first step allows access to the requisite epoxide in good yield and good enantiomeric excess, thus rendering the overall process highly efficient. A final purification of the amino alcohol product involving formation of the L-tartrate salt serves to enhance both the chemical and stereochemical purity of the final product.

In addition to serving as a key stereochemical controlling element for synthesis of indinavir,[5] the title compound has proven to be a remarkably versatile chiral ligand and auxiliary for a range of asymmetric transformations including Diels-Alder reactions,[13] carbonyl reductions,[14] diethylzinc additions to aldehydes,[15] and enolate additions.[16]

1. Department of Chemistry and Chemical Biology, Harvard University, Cambridge, MA 02138.
2. Merck Research Laboratories, Department of Process Research, Rahway, NJ 07065.
3. (a) Present address: Merck & Co., Stonewall Plant, Route 340 South, Elkton, VA 22827; (b) Present address: Sepracor, Inc., 33 Locke Drive, Marlboro, MA 01752.
4. (a) Larrow, J. F.; Jacobsen, E. N.; Gao, Y.; Hong, Y.; Nie, X.; Zepp, C. M. *J. Org. Chem.* **1994**, *59*, 1939-1942; (b) Larrow, J. F.; Jacobsen, E. N. *Org. Synth.* **1998**, *75*, 1.
5. (a) Senanayake, C. H.; Smith, G. B.; Ryan, K. M.; Fredenburgh, L. E.; Liu, J.; Roberts, F. E.; Hughes, D. L.; Larsen, R. D.; Verhoeven, T. R.; Reider, P. J. *Tetrahedron Lett.* **1996**, *37*, 3271-3274; (b) Hughes, D. L.; Smith, G. B.; Liu, J.; Dezeny, G. C.; Senanayake, C. H.; Larsen, R. D.; Verhoeven, T. R.; Reider, P. J.

J. Org. Chem. **1997**, *62*, 2222-2229; (c) Bell, D.; Davies, M. R.; Finney, F. J. L.; Geen, G. R.; Kincey, P. M.; Mann, I. S. *Tetrahedron Lett.* **1996**, *37*, 3895-3898.

6. Kolthoff, I. M.; Belcher, R. "Volumetric Analysis, Volume III"; Interscience Publishers: New York, 1957, pp 262-263.

7. Vacca, J. P.; Dorsey, B. D.; Schleif, W. A.; Levin, R. B.; McDaniel, S. L.; Darke, P. L.; Zugay, J.; Quintero, J. C.; Blahy, O. M.; Roth, E.; Sardana, V. V.; Schlabach, A. J.; Graham, P. I.; Condra, J. H.; Gotlib, L.; Holloway, M. K.; Lin, J.; Chen, I.-W.; Vastag, K.; Ostovic, D.; Anderson, P. S.; Emini, E. A.; Huff, J. R. *Proc. Natl. Acad. Sci. U.S.A.* **1994**, *91*, 4096-4100.

8. Thompson, W. J.; Fitzgerald, P. M. D.; Holloway, M. K.; Emini, E. A.; Darke, P. L.; McKeever, B. M.; Schlief, W. A.; Quintero, J. C.; Zugay, J. A.; Tucker, T. J.; Schwering, J. E.; Homnick, C. F.; Nunberg, J.; Springer, J. P.; Huff, J. R. *J. Med. Chem.* **1992**, *35*, 1685-1701.

9. (a) Didier, E.; Loubinoux, B.; Ramos Tombo, G. M.; Rihs, G. *Tetrahedron* **1991**, *47*, 4941-4958; (b) Takahashi, M.; Ogasawara, K. *Synthesis* **1996**, 954-958.

10. Boyd, D. R.; Sharma, N. D.; Bowers, N. I.; Goodrich, P. A.; Groocock, M. R.; Blacker, A. J.; Clarke, D. A.; Howard, T.; Dalton, H. *Tetrahedron: Asymmetry* **1996**, *7*, 1559-1562.

11. (a) Senanayake, C. H.; Roberts, F. E.; DiMichele, L. M.; Ryan, K. M.; Liu, J.; Fredenburgh, L. E.; Foster, B. S.; Douglas, A. W.; Larsen, R. D.; Verhoeven, T. R.; Reider, P. J. *Tetrahedron Lett.* **1995**, *36*, 3993-3996; (b) Senanayake, C. H.; DiMichele, L. M.; Liu, J.; Fredenburgh, L. E.; Ryan, K. M.; Roberts, F. E.; Larsen, R. D.; Verhoeven, T. R.; Reider, P. J. *Tetrahedron Lett.* **1995**, *36*, 7615-7618.

12. (a) Jacobsen, E. N.; Zhang, W.; Muci, A. R.; Ecker, J. R.; Deng, L. *J. Am. Chem. Soc.* **1991**, *113*, 7063-7064; (b) Jacobsen, E. N.; Deng, L.; Furukawa, Y.; Martínez, L. E. *Tetrahedron* **1994**, *50*, 4323-4334. For reviews, see: (c) Jacobsen, E.N. In "Catalytic Asymmetric Synthesis"; Ojima, I., Ed.; VCH: New

York, 1993; Chapter 4.2; (d) Jacobsen, E. N. In "Comprehensive Organometallic Chemistry II"; Wilkinson, G.; Stone, F. G. A.; Abel, E. W.; Hegedus, L. S., Eds.; Pergamon: New York, 1995; Vol. 12, Chapter 11.1; (e) Katsuki, T. *Coord. Chem. Rev.* **1995**, *140*, 189-214.

13. (a) Davies, I. W.; Senanayake, C. H.; Castonguay, L.; Larsen, R. D.; Verhoeven, T. R., Reider, P. J. *Tetrahedron Lett.* **1995**, *36*, 7619-7622; (b) Davies, I. W.; Gerena, L.; Castonguay, L.; Senanayake, C. H.; Larsen, R. D.; Verhoeven, T. R., Reider, P. J. *J. Chem. Soc.,Chem. Commun.* **1996**, 1753-1754; (c) Davies, I. W.; Senanayake, C. H.; Larsen, R. D.; Verhoeven, T. R., Reider, P. J. *Tetrahedron Lett.* **1996**, *37*, 1725-1726; (d) Ghosh, A. K.; Mathivanan, P.; Cappiello, J. *Tetrahedron Lett.* **1996**, *37*, 3815-3818; (e) Ghosh, A. K.; Mathivanan, P. *Tetrahedron: Asymmetry* **1996**, *7*, 375-378.

14. (a) Hong, Y.; Gao, Y.; Nie, X.; Zepp, C. M. *Tetrahedron Lett.* **1994**, *35*, 6631-6634; (b) DiSimone, B.; Savoia, D.; Tagliavini, E.; Umani-Ronchi, A. *Tetrahedron: Asymmetry* **1995**, *6*, 301-306; (c) Ghosh, A. K.; Chen, Y. *Tetrahedron Lett.* **1995**, *36*, 6811-6814.

15. Solà, L.; Vidal-Ferran, A.; Moyano, A.; Pericàs, M. A.; Riera, A. *Tetrahedron: Asymmetry* **1997**, *8*, 1559-1568. See also ref. 7b.

16. (a) Askin, D.; Eng, K. K.; Rossen, K.; Purick, R. M.; Wells, K. M.; Volante, R. P.; Reider, P. J. *Tetrahedron Lett.* **1994**, *35*, 673-676; (b) Askin, D.; Wallace, M. A.; Vacca, J. P.; Reamer, R. A.; Volante, R. P.; Shinkai, I. *J. Org. Chem.* **1992**, *57*, 2771-2773; (c) Maligres, P. E.; Upadhyay, V.; Rossen, K.; Cianciosi, S. J.; Purick, R. M.; Eng, K. K.; Reamer, R. A.; Askin, D.; Volante, R. P.; Reider, P. J. *Tetrahedron Lett.* **1995**, *36*, 2195-2198; (d) Ghosh, A. K.; Onishi, M. *J. Am. Chem. Soc.* **1996**, *118*, 2527-2528; (e) Ghosh, A. K.; Duong, T. T.; McKee, S. P. *J. Chem. Soc., Chem. Commun.* **1992**, 1673-1674.

Appendix

Chemical Abstracts Nomenclature (Collective Index Number); (Registry Number)

(1S,2R)-1-Aminoindan-2-ol: 1H-Inden-2-ol, 1-amino-2,3-dihydro-, (1S-cis)- (12); (126456-43-7)

(1S,2R)-Indene oxide: Indan, 1,2-epoxy- (8); 6H-Indeno[1,2-b]oxirene, 1a,6a-dihydro- (9); (768-22-9)

Indene (8); 1H-Indene (9); (95-13-6)

(R,R)-N,N'-Bis(3,5-di-tert-butylsalicylidene)-1,2-cyclohexanediaminomanganese(III) chloride: Manganese, chloro[[2,2'-[1,2-cyclohexanediylbis(nitrilomethylidyne)]bis[4,6-bis(1,1-dimethylethyl)phenolato]](2-)-N,N',O,O']-, [SP-5-13-(1S-trans)]- (12); (135620-04-1)

4-Phenylpyridine N-oxide: Pyridine, 4-phenyl-, 1-oxide (8,9); (1131-61-9)

Sodium hypochlorite solution: Hypochlorous acid, sodium salt (8,9); (7681-52-9)

Acetonitrile TOXIC (8,9); (75-05-8)

Sufuric acid, fuming: Sulfuric acid, mixt. with sulfur trioxide (9); (8014-95-7)

L-Tartaric acid: Tartaric acid, L- (8); Butanedioic acid, 2,3-dihydroxy-, [R-(R*,R*)]- (9); (87-69-4)

ASYMMETRIC SYNTHESIS OF α-AMINO ACIDS BY THE ALKYLATION OF PSEUDOEPHEDRINE GLYCINAMIDE: L-ALLYLGLYCINE AND N-BOC-L-ALLYLGLYCINE

(Acetamide, 2-amino-N-(2-hydroxy-1-methyl-2-phenylethyl)-N-methyl-, [R-(R*,R*)]-, 4-Pentenoic acid, 2-amino-, (R)- and 4-Pentenoic acid, 2-[[(1,1-dimethylethoxy)carbonyl]amino]-, (R)-)

A. CH_3 NHCH$_3$ OH → Gly-OMe / LiOMe, LiCl → CH_3 O N NH_2 OH CH_3

B. → LDA, LiCl; Br →

C. → H_2O, Δ → HO NH_2

D. → NaOH; Boc_2O → HO NHBoc

Submitted by Andrew G. Myers and James L. Gleason.[1]

Checked by Evan G. Antoulinakis and Robert K. Boeckman, Jr.

1. Procedure

A. (R,R)-(–)-Pseudoephedrine glycinamide. An oven-dried, 3-L, three-necked, round-bottomed flask is equipped with an argon inlet adapter, a rubber septum, a 150-mL pressure-equalizing addition funnel fitted with a rubber septum, and a Teflon-coated magnetic stirring bar. The flask is flushed with argon and charged with 30.8 g (0.726 mol, 2 equiv) of anhydrous lithium chloride (Note 1), 60.0 g (0.363 mol, 1 equiv) of (R,R)-(–)-pseudoephedrine (Note 2), and 500 mL of dry tetrahydrofuran (THF) (Note 3). The resulting slurry is cooled in an ice bath. After 15 min, 6.89 g (0.182 mol, 0.5 equiv) of solid lithium methoxide (Note 4) is added to the reaction flask in one lot. The resulting mixture is stirred at 0°C for 10 min, after which time the pressure-equalizing addition funnel is charged with a solution of 40.4 g (0.454 mol, 1.25 equiv) of glycine methyl ester (Note 5) in 100 mL of dry THF (Note 3), and dropwise addition of this solution is initiated. The addition is completed within 1 hr, and the reaction flask is maintained at 0°C for an additional 7 hr. The reaction is terminated by the addition of 500 mL of water. The bulk of the THF is removed from the resulting colorless solution by concentration under reduced pressure. An additional 250 mL of water is added to the aqueous concentrate, and the resulting aqueous solution is transferred to a 2-L separatory funnel and extracted sequentially with one 500-mL and four 250-mL portions of dichloromethane. The combined organic extracts are dried over anhydrous potassium carbonate and filtered, and the filtrate is concentrated under reduced pressure. The clear, colorless, oily residue is dissolved in 300 mL of warm (50°C) THF (Note 3), 10 mL of water is added, and the resulting solution is allowed to cool to 23°C, whereupon the product crystallizes as its monohydrate within 1 hr. The crystallization process is completed by cooling the crystallization flask to -20°C. After standing for 2 hr at -20°C, the crystals are collected by filtration and rinsed with 200 mL of ether. The crystals are dried under reduced pressure (0.5 mm) at 23°C for 2 hr to provide 62.8 g (72%) of (R,R)-(–)-pseudoephedrine glycinamide monohydrate (Note 6).

Dehydration of the monohydrate is initiated by suspending the crystalline solid (62.8 g) in 1.2 L of dichloromethane; the resulting suspension is stirred vigorously for 1 hr to break up any large lumps of solid. After 1 hr of vigorous stirring, 60 g of anhydrous potassium carbonate is added to the fine dispersion (Note 7). After the suspension is stirred for 10 min, it becomes translucent. At this point the mixture is filtered through 40 g of Celite in a 10 cm-i.d. Büchner funnel fitted with a Whatman #1 filter paper. The clear, colorless filtrate is concentrated under reduced pressure. The oily residue is dissolved in 200 mL of toluene, and the resulting solution is concentrated to remove any residual dichloromethane. The oily concentrate is then dissolved in 175 mL of hot (60°C) toluene, and the resulting solution is allowed to cool slowly to 23°C. Crystallization of the product may occur spontaneously within 1-3 hr at 23°C; however, if necessary, it can be initiated by scratching the side of the flask until crystals are observed. Once crystallization is initiated, the crystals are broken up periodically with a spatula to obtain a fine powder that is easily manipulated. After 30 min from the onset of crystallization, the flask is cooled to -20°C under an argon atmosphere to complete the crystallization process. After standing at -20°C for 2 hr, the crystals are collected by filtration and rinsed with 200 mL of ether. The product is dried by transferring the solid to a 500-mL, round-bottomed flask fitted with a vacuum adapter and evacuating the flask (0.5 mm). After 1 hr at 23°C, the flask is immersed in an oil bath at 60°C (Note 8). After 12 hr, the flask is cooled to 23°C to afford 53.8 g (67% overall) of anhydrous (R,R)-(–)-pseudoephedrine glycinamide (Note 9).

B. Pseudoephedrine L-allylglycinamide. A 1-L, single-necked, round-bottomed flask is equipped with a Teflon-coated magnetic stirring bar and a rubber septum through which is placed a needle connected to a source of vacuum and argon. The system is evacuated, the flask is flame-dried and then allowed to cool to 23°C under reduced pressure. When the reaction flask has cooled to 23°C, it is flushed with argon and charged with 200 mL of dry THF (Note 3) and 63.0 mL (0.450 mol, 1.025 equiv) of diisopropylamine (Note 10). The resulting solution is cooled to 0°C in an ice bath. With efficient stirring, the solution is deoxygenated at 0°C by alternately evacuating the reaction vessel and flushing with argon three times. After the solution is deoxygenated, 167 mL (0.439 mol, 1 equiv) of a 2.63 M solution of butyllithium in hexanes (Notes 11 and 12) is added via syringe over a 20-min period. After the addition is complete, the solution is stirred at 0°C for 15 min.

Separately, a 2-L, three-necked, round-bottomed flask is equipped with an inlet adapter connected to a source of vacuum and argon, two rubber septa, and a Teflon-coated magnetic stirring bar. The flask is charged with 57.2 g (1.35 mol, 6 equiv) of anhydrous lithium chloride (Note 1) and, with efficient stirring of the solid, the reaction vessel is evacuated and flame-dried. The flask and its contents are allowed to cool to 23°C under reduced pressure. When the flask has cooled to 23°C, it is flushed with argon, 50.0 g (0.225 mol, 1 equiv) of solid (R,R)-(−)-pseudoephedrine glycinamide is added, and one of the septa is replaced with a Teflon thermometer adapter fitted with a thermometer for internal measurement of the reaction temperature. The solids are suspended in 500 mL of dry THF (Note 3) and the resulting milky-white slurry is cooled to an internal temperature of 0°C in an ice bath. With efficient stirring, the slurry is deoxygenated by alternately evacuating the reaction vessel and flushing with argon three times.

The two reaction flasks are connected via a wide-bore (14 gauge) cannula so that one end of the cannula is immersed in the lithium diisopropylamide solution and the other is suspended above the (R,R)-(−)-pseudoephedrine glycinamide-lithium chloride slurry. The flask containing the lithium diisopropylamide solution and its ice bath are raised to a height just above that of the flask containing the glycinamide slurry. The reaction flask containing the glycinamide slurry is very briefly evacuated to initiate siphon transfer of the lithium diisopropylamide solution. Once the siphon flow is established, the flask containing the glycinamide slurry is flushed with argon. By raising or lowering the height of the flask containing the lithium diisopropylamide solution, the rate of addition is modulated so that the temperature of the reaction mixture does not rise above 5°C (approximately 45 min addition time) (Note 13). After the addition is complete, the reaction mixture is stirred at 0°C for 30 min (Note 14). To the resulting pale yellow suspension is added 19.5 mL (0.225 mol) of allyl bromide (Note 15) via syringe over a 20-min period. The rate of addition of allyl bromide is also modulated to prevent the internal reaction temperature from rising above 5°C (Note 16). After the addition of allyl bromide is complete, the reaction mixture is stirred for 45 min at 0°C. The reaction is terminated by the addition of 500 mL of water. The resulting biphasic mixture is slowly acidified by the addition of 300 mL of 3 M aqueous hydrochloric acid solution. The biphasic mixture is transferred to a 2-L separatory funnel and is extracted with 1 L of ethyl acetate. The organic layer is separated and extracted sequentially with 300 mL of 3 M aqueous hydrochloric acid solution and 300 mL of 1 M aqueous hydrochloric acid solution. The aqueous layers are combined and cooled to an internal temperature of 5°C by stirring in an ice bath. The cold aqueous solution is then cautiously made basic (pH 14) by the slow addition of 120 mL of aqueous 50% sodium hydroxide solution. The temperature of the solution should not be allowed to rise above 25°C during the addition of base. The basic solution is extracted sequentially with one 500-mL portion and four 250-mL portions of dichloromethane (Note 17). The combined organic layers are dried over anhydrous potassium carbonate and filtered, and the filtrate is concentrated under reduced pressure. The oily residue is dissolved in 200 mL of toluene, and the resulting solution is concentrated under reduced pressure to remove residual dichloromethane and diisopropylamine. The solid residue is recrystallized by suspending it in 100 mL of toluene and heating the resulting suspension until the solids dissolve (ca. 70°C). The recrystallization mixture is allowed to cool to 23°C. After 3 hr, when crystallization of the product is nearly complete, the recrystallization flask is cooled to 0°C in an ice bath to complete the recrystallization process. After standing at 0°C for 1 hr, the crystals are collected by filtration and rinsed sequentially with two 50-mL portions of cold (0°C) toluene and one 100-mL portion of ether at 23°C. The crystals are dried under reduced pressure (0.5 mm) at 23°C for 2 hr to provide ~31.3 g (~53%) of diastereomerically pure pseudoephedrine L-allylglycinamide (Note 18). The mother liquors are concentrated and the oily residue is dissolved in 50 mL of toluene at 23°C. The resulting solution is cooled to -20°C and seeded with authentic pseudoephedrine L-allylglycinamide. After standing at -20°C for 6 hr, the crystals that have formed are collected by filtration and rinsed with 25 mL of cold (0°C) toluene and 50 mL of ether at 23°C. The product is dried under reduced pressure (0.5 mm) at 23°C for 2 hr to afford a second crop of the alkylation product. The second crop of crystals (~4.8 g) is recrystallized a second time by suspending it in 20 mL of toluene and warming to ca. 70°C to dissolve the solids (Note 19). The resulting solution is allowed to cool slowly to 23°C, whereupon the product crystallizes within 1 hr. The recrystallization flask is cooled to -20°C to complete the crystallization process. After standing at -20°C for 90 min, the crystals are collected by filtration and washed sequentially with two 10-mL portions of cold (0°C) toluene and one 25-mL portion of ether. The crystals are dried under reduced pressure (0.5 mm) at 23°C for 2 hr to afford ~3.6 g (~6%) of diastereomerically pure pseudoephedrine L-allylglycinamide. To obtain additional product, the mother liquors are concentrated under reduced pressure and the oily residue is purified by chromatography on silica gel (100 g, 5-cm i.d. column) eluting with 4% methanol, 4% triethylamine and 92% dichloromethane. The oily residue obtained after concentration of the appropriate fractions is dissolved in 25 mL of warm (50°C) toluene. The resulting solution is cooled to -20°C and held at that temperature for 12 hr. The crystals that form are collected by filtration and rinsed with 20 mL of cold

(0°C) toluene and 30 mL of ether at 23°C. The crystals are dried under reduced pressure (0.5 mm) at 23°C for 2 hr to provide an additional ~5.0 g (~8%, total yield: 39.1-42.0 g, 66-71%) of diastereomerically pure pseudoephedrine L-allylglycinamide.

C. L-Allylglycine. A 1-L, single-neck, round-bottomed flask equipped with an efficient reflux condenser, a Teflon-coated magnetic stirring bar and a heating mantle is charged with 25.0 g (0.0953 mol) of pseudoephedrine L-allylglycinamide and 500 mL of water. The resulting suspension is heated to reflux, causing the solids to dissolve to afford a colorless, homogeneous solution. After 10 hr at reflux, the reaction mixture is allowed to cool to 23°C, whereupon (R,R)-(–)-pseudoephedrine is observed to crystallize (Note 20). Concentrated aqueous ammonium hydroxide solution (10 mL) is added (Note 21), whereupon the resulting aqueous slurry is transferred to a 1-L separatory funnel and extracted with three 200-mL portions of dichloromethane, reserving the aqueous layer. The three organic layers are individually and sequentially extracted with a single aqueous solution prepared by combining 250 mL of water and 5 mL of concentrated aqueous ammonium hydroxide solution. The aqueous extract is combined with the aqueous extract reserved earlier and the resulting solution is concentrated under reduced pressure to provide a white solid residue. The solid is triturated, sequentially, with one 100-mL and one 50-mL portion of absolute ethanol. The triturated solid is collected by filtration and dried under reduced pressure (0.5 mm) at 23°C for 2 hr to afford 10.2 g (93%) of L-allylglycine of ≥99% ee (Note 22). If desired, (R,R)-(–)-pseudoephedrine can be recovered from the organic extracts. The organic extracts are combined and dried over anhydrous potassium carbonate and filtered, and the filtrate is concentrated under reduced pressure to afford a solid. The solid is recrystallized by dissolving it in a minimum volume of hot water (ca. 350 mL). The resulting solution is allowed to cool slowly to 23°C, by which time extensive crystallization of (R,R)-(–)-pseudoephedrine has occurred. The recrystallization flask is cooled to 0°C in an ice bath. After standing at

0°C for 1 hr, the crystals are collected by filtration and dried under reduced pressure (0.5 mm) at 23°C for 2 hr to afford 10.8 g of pure (R,R)-(–)-pseudoephedrine (mp 116-117°C). The mother liquors are concentrated and a second crop of crystals (2.6 g, total yield 13.4 g, 85%) is obtained in a similar manner by recrystallization from ca. 75 mL of water.

D. N-Boc-L-allylglycine. A 1-L, single-neck, round-bottomed flask equipped with an efficient reflux condenser, a Teflon-coated magnetic stirring bar and a heating mantle is charged with 14.7 g (0.056 mol, 1 equiv) of pseudoephedrine L-allylglycinamide and 224 mL (0.112 mol, 2 equiv) of 0.5 M aqueous sodium hydroxide solution. The resulting slurry is heated to reflux whereupon a clear, colorless homogeneous solution is obtained. After 2 hr at reflux, the reaction mixture is cooled to 23°C, inducing the crystallization of (R,R)-(–)-pseudoephedrine (Note 20). The reaction suspension is transferred to a 1-L separatory funnel and is extracted sequentially with one 200-mL and one 100-mL portion of dichloromethane, reserving the aqueous layer. The two organic layers are individually and sequentially extracted with a single 150-mL portion of water. The aqueous layer is combined with the earlier aqueous extract and the resulting solution is reserved. If desired, (R,R)-(–)-pseudoephedrine can be recovered from the organic extracts, as follows. The organic extracts are combined and dried over anhydrous potassium carbonate and filtered, and the filtrate is concentrated under reduced pressure to afford a solid. The solid is recrystallized by dissolving it in a minimum volume of hot water (ca. 250 mL). The resulting solution is allowed to cool slowly to 23°C, by which time extensive crystallization of (R,R)-(–)-pseudoephedrine has occurred. The recrystallization flask is cooled to 0°C in an ice bath. After standing at 0°C for 1 hr, the crystals are collected by filtration and are dried under reduced pressure (0.5 mm) at 23°C for 2 hr to afford 6.2 g (67%) of pure (R,R)-(–)-pseudoephedrine (mp 116-117°C). The mother liquors are concentrated and a second crop of crystals (1.5 g, total yield 7.7 g, 83%) is obtained in a similar manner by recrystallization from ca. 50 mL of water.

The combined aqueous layers are transferred to a 1-L, round-bottomed flask and 9.40 g (0.112 mol, 2 equiv) of solid sodium bicarbonate is added. The resulting solution is reduced to a volume of approximately 150 mL by concentration under reduced pressure. A Teflon-coated magnetic stirring bar is added, and the aqueous mixture is cooled to 0°C in an ice bath. To the cooled solution is added, sequentially, 150 mL of p-dioxane (Note 23) and 13.4 g (0.0615 mol, 1.1 equiv) of di-tert-butyl dicarbonate (Note 24). The reaction mixture is stirred for 90 min at 0°C, at which time the ice bath is removed and the solution is allowed to warm to 23°C. After stirring for 90 min at 23°C, the reaction mixture is diluted with 200 mL of water and the resulting solution is transferred to a 1-L separatory funnel and extracted sequentially with one 400-mL and one 200-mL portion of ethyl acetate, reserving the aqueous layer. The two organic layers are individually and sequentially extracted with a single 100-mL portion of 0.1 M aqueous sodium hydroxide solution. The aqueous layer is combined with the aqueous extract reserved earlier and the resulting solution is stirred while cooling in an ice bath. Before acidification of the aqueous layer, 100 mL of ethyl acetate is added to prevent excessive frothing. The resulting biphasic mixture is carefully acidified by the slow addition of 250 mL of a 1 M aqueous hydrochloric acid solution until the aqueous layer is pH 1. The biphasic mixture is transferred to a 2-L separatory funnel, 400 mL of ethyl acetate is added, and, after thorough mixing, the layers are separated. The organic layer is extracted with 200 mL of water. The two aqueous layers are individually and sequentially extracted with a single 200-mL portion of ethyl acetate. The organic layers are combined, and the resulting solution is dried over anhydrous sodium sulfate and filtered. The filtrate is concentrated under reduced pressure. The residue is dissolved in 100 mL of toluene, and the resulting solution is concentrated. The residue is then sequentially dissolved in and then concentrated from 100 mL of toluene, 100 mL of dichloromethane, and two 100-mL portions of ether, in order to remove residual dioxane and ethyl acetate. The oily residue is dried under reduced pressure (55°C, 0.2 mm) for 12 hr to afford 11.8 g (97%) of analytically pure N-Boc-L-allylglycine as a viscous oil (Note 25).

2. Notes

1. Reagent-grade anhydrous lithium chloride (Mallinckrodt Inc.) is further dried by transferring the solid to a flask equipped with a vacuum adapter. The flask is evacuated (0.5 mm) and immersed in an oil bath at 150°C. After 12 hr at 150°C, the flask is allowed to cool to 23°C and is flushed with argon for storage.

2. (R,R)-(−)-Pseudoephedrine was used as received from Aldrich Chemical Company, Inc.

3. Tetrahydrofuran was obtained from EM Science and was distilled under nitrogen (atmospheric pressure) from sodium benzophenone ketyl.

4. Lithium methoxide was purchased from Aldrich Chemical Company, Inc., and used as received. Butyllithium (BuLi) (10 M in hexanes) may be substituted for lithium methoxide in this reaction and produces a more rapid reaction. For example, the use of 0.25 equiv of 10 M BuLi requires only 1-2 hr for complete reaction and affords 65-69% yield of anhydrous pseudoephedrine glycinamide on a 40-60-g scale.[2] The submitters describe the use of lithium methoxide as a less hazardous alternative to the highly pyrophoric 10 M BuLi.

5. Glycine methyl ester is prepared by the method of Almeida et al.[3] In a mortar and pestle, 80 g of glycine methyl ester hydrochloride (used as received from Aldrich Chemical Company, Inc.) is ground to a fine powder. The powder is suspended in 600 mL of dry ether in a 1-L Erlenmeyer flask equipped with a Teflon-coated magnetic stirring bar. Gaseous ammonia is bubbled rapidly through the

vigorously stirred suspension. After 2 hr, the addition of ammonia is discontinued, the product slurry is filtered through a coarse-fritted glass filter, and the filtrate is concentrated under reduced pressure at 23°C. The liquid residue is distilled under reduced pressure (54-55°C at 18 mm) to provide 51.3 g (90%) of glycine methyl ester as a colorless liquid. Glycine methyl ester will polymerize upon storage at room temperature, but may be stored at -20°C for short periods (up to two weeks) without significant decomposition.

6. The monohydrate and anhydrous product show identical spectroscopic properties (Note 9). The monohydrate exhibits the following physical properties: mp 83-85°C; Anal. Calcd for $C_{12}H_{18}N_2O_2 \cdot H_2O$, C, 59.93; H, 8.32; N, 11.66; Found C, 59.81; H, 8.42; N, 11.51.

7. Alternatively, azeotropic drying with acetonitrile may be employed in lieu of dichloromethane/potassium carbonate.[2] A solution of 50.3 g of (R,R)-(–)-pseudoephedrine glycinamide monohydrate in ca. 200 mL of acetonitrile is concentrated under reduced pressure. The oily residue is dissolved in 250 mL of toluene and the resulting solution is concentrated under reduced pressure. The oily residue obtained may be carried on directly in the alkylation procedure with only a slight decrease in yield from the procedure described above. Alternatively, anhydrous (R,R)-(–)-pseudoephedrine glycinamide may be precipitated and the resulting solid dried and carried forward as outlined above.

8. Proper drying of (R,R)-(–)-pseudoephedrine glycinamide is essential to achieve high yields in the subsequent alkylation step. Complete drying may not be achieved at temperatures below 50°C. To prevent melting of the solid product, it should not be heated above 65°C. A preliminary indication of the hydration state of the product is its melting point. Material that is partially hydrated routinely has a melting point that is depressed relative to that of pure anhydrous product (mp 78-80°C). A more accurate determination of the water content may be obtained either

from C,H,N analysis or by Karl Fischer titration. The product is somewhat hygroscopic. It may be weighed on the open benchtop without significant hydration; however, it should be stored under argon. The glycinamide should be redried at 60°C under reduced pressure (0.5 mm) if it has been stored for an extended period, or if the yield of the subsequent alkylation reaction is lower than expected.

9. The product shows the following physical and spectroscopic properties: mp 78-80°C; $[\alpha]_D^{23}$ -101.2° (CH_3OH, *c* 1.2); TLC R_f = 0.18 (5% CH_3OH, 5% NEt_3, 90% CH_2Cl_2); IR (neat) cm^{-1}: 3361, 2981, 1633, 1486, 1454, 1312, 1126, 1049, 926, 760, 703; ^1H NMR (1:1 ratio of rotamers, $CDCl_3$) δ: 0.99 (d, 1.5 H, J = 6.7), 1.09 (d, 1.5 H, J = 6.7), 2.11 [s(br), 3 H], 2.79 (s, 1.5 H), 2.97 (s, 1.5 H), 3.37 (d, 0.5 H, J = 17.1), 3.46 [d(obs)], 1 H, J = 16.6), 3.72 (d, 0.5 H, J = 15.5), 3.88 (m, 0.5 H), 4.53-4.63 (m, 1.5 H), 7.29-7.40 (m, 5 H); ^{13}C NMR ($CDCl_3$) δ: 14.4, 15.3, 27.1, 30.1, 43.4, 43.7, 57.2, 57.5, 74.9, 75.8, 126.7, 126.9, 127.9, 128.2, 128.5, 128.7, 142.1, 142.3, 173.5, 174.1. Anal. Calcd for $C_{12}H_{18}N_2O_2$: C, 64.84; H, 8.16; N, 12.60. Found: C, 64.65; H, 8.25; N, 12.53.

10. Diisopropylamine was purchased from Aldrich Chemical Company, Inc., and distilled under nitrogen (atmospheric pressure) from calcium hydride prior to use.

11. It is absolutely imperative that the solution of butyllithium be accurately titrated. If an excess of butyllithium (or LDA) is used, reduced yields will result as a consequence of a decomposition reaction that releases pseudoephedrine. This is easily monitored by TLC analysis (5% methanol, 5% triethylamine, and 90% dichloromethane eluent; UV and ninhydrin visualization). It should be noted that even optimal reaction conditions produce small amounts of this cleavage product (2-4%); however, the amount of cleavage is greatly enhanced in the presence of excess base. To titrate the alkyllithium solution we recommend the method of Watson and Eastham.[4] A standard solution of 1.00 M 2-butanol (freshly distilled from calcium hydride) in toluene (freshly distilled from calcium hydride) is prepared in a volumetric

flask. A 50-mL, round-bottomed flask equipped with a Teflon-coated magnetic stirring bar and a rubber septum is charged with 5 mg of 2,2'-dipyridyl and 20 mL of ether. The flask is flushed with argon, and a small amount (ca. 0.5 mL) of the standard 1.00 M solution of 2-butanol in toluene is added to the solution. The butyllithium solution to be titrated is added slowly, dropwise, to a single-drop end point that turns the solution dark red. This initial titration eliminates complications due to moisture or oxygen and should not be used in the calculation of the titer of the butyllithium solution. Several repetitions of the titration cycle are conducted using the same indicator solution by using accurate, air-tight syringes and alternately adding aliquots of 1.00-M 2-butanol solution (1-2.5 mL) followed by titration of the butyllithium to a dark-red end point.

12. The checkers employed 163 mL of a 2.70-M solution of butyllithium in hexanes whose titer was determined by the procedure given in Note 11.

13. The addition required 80 min in the hands of the checkers.

14. The reaction mixture was stirred for 40 min at 0°C by the checkers.

15. Allyl bromide was purchased from Aldrich Chemical Company, Inc., and was distilled under argon (atmospheric pressure) from calcium hydride immediately prior to use.

16. The allyl bromide addition required 30 min in the hands of the checkers.

17. The pH of the aqueous layer is checked after each extraction to ensure that it is >12. If necessary, the pH of the aqueous layer is readjusted to 14 by the addition of aqueous 50% sodium hydroxide solution.

18. The product shows the following physical and spectroscopic properties: mp 71-73°C; $[\alpha]_D^{23}$ -86.4° (CH_3OH, *c* 1.1); TLC R_f = 0.59 (5% CH_3OH, 5% NEt_3, 90% CH_2Cl_2); IR (neat) cm^{-1}: 3354, 3072, 2978, 1632, 1491, 1453, 1109, 1051, 918, 762, 703; ^1H NMR (3:1 rotamer ratio, $CDCl_3$) major rotamer δ: 1.03 (d, 3 H, J = 6.4), 2.13 (m, 1 H), 2.23 (m, 1 H), 2.87 (s, 3 H), 3.65 (dd, 1 H, J = 7.5, 5.3), 4.55-4.59 (m, 2 H), 5.07-5.14 (m, 2 H), 5.64-5.85 (m, 1 H), 7.23-7.38 (m, 5 H); minor rotamer δ: 0.96 (d, 3

H, J = 6.7), 2.61-2.66 (m, 2 H), 2.93 (s, 3 H), 3.69 (m, 1 H), 4.03 (m, 1 H); ^{13}C NMR ($CDCl_3$) major rotamer δ: 14.4, 31.4, 39.6, 51.2, 57.6, 75.5, 118.1, 126.5, 127.6, 128.2, 133.7, 142.1, 176.1; minor rotamer δ: 15.5, 27.0, 39.8, 51.0, 74.9, 117.9, 126.8, 128.1, 128.5, 134.7, 141.8, 175.1. Anal. Calcd for $C_{15}H_{22}N_2O_2$: C, 68.67; H, 8.45; N, 10.68. Found: C, 68.57; H, 8.59; N, 10.70. Determination of the diastereomeric purity of the product by NMR is complicated by the presence of amide rotamers. The diastereomeric purity of the product may be determined accurately and conveniently by preparing the corresponding diacetate and analyzing by capillary gas chromatography. To prepare the diacetate, a 10-mL, round-bottomed flask equipped with a Teflon-coated magnetic stirring bar and a rubber septum is charged with a 16-mg sample of the alkylation product to be analyzed and 1 mL of pyridine. The product is acetylated by adding 1 mL of acetic anhydride and a catalytic amount (5 mg) of 4-(N,N-dimethylamino)pyridine. The reaction mixture is stirred under argon for 1 hr and excess acetic anhydride is quenched by addition of 15 mL of water. The reaction mixture is extracted sequentially with one 30-mL portion and one 20-mL portion of ethyl acetate. The two organic extracts are individually and sequentially extracted with a single 15-mL portion of aqueous saturated sodium bicarbonate solution; the organic extracts are combined, dried over anhydrous sodium sulfate and filtered. The filtrate is concentrated under reduced pressure, and the residue is dissolved in ethyl acetate for capillary gas chromatographic analysis. Analysis was carried out using a Chirasil-Val capillary column (25 m x 0.25 mm x 0.16 μm, available from Alltech Inc.) under the following conditions: oven temp. 220°C, injector temp. 250°C, detector temp. 275°C. The following retention times were observed: (R,R)-(−)-pseudoephedrine glycinamide diacetate, 6.69 min; (R,R)-(−)-pseudoephedrine L-allylglycinamide diacetate, 6.94 min; (R,R)-(−)-pseudoephedrine D-allylglycinamide diacetate, 6.32 min. Note that the retention times can vary greatly with the age and condition of the column. The checkers obtained the following values using an identical new column from Alltech with a flow rate of 4 mL/min, split ratio of 50:1, and an injection volume of 1 μL: retention times (min) 18.33 (D-allyl isomer), 19.24 (glycinamide SM), 20.24 (L-allyl isomer).

19. The second crop of product crystals (mp 69-71°C) was contaminated with 2% of the starting material, (R,R)-(–)-pseudoephedrine glycinamide (as determined by GC analysis, Note 18), and was recrystallized to provide analytically pure product.

20. Although the pseudoephedrine may be recovered by filtration at this stage, the recovery is not quantitative (ca. 50-60%). A more efficient recovery is achieved by the extraction procedure described.

21. Ammonium hydroxide is added to decrease the solubility of pseudoephedrine in the aqueous phase and to minimize the formation of emulsions.

22. The product shows the following spectroscopic and physical properties: mp 275-280°C (dec.); lit.[5] 241-243°C (dec.); lit.[6] 283°C (dec.); $[\alpha]_D^{23}$ -37.2° (H_2O, *c* 4); lit.[7] $[\alpha]_D^{23}$ -37.1 (H_2O, *c* 4) (Note 26); IR (KBr) cm^{-1}: 3130, 2605, 1586, 1511, 1406, 1363, 1345, 1307, 919, 539; ^1H NMR (D_2O) δ: 2.50 (m, 2 H), 3.67 (dd, 1 H, J = 7.0, 5.1), 5.13 (d, 1 H, J = 10.0), 5.14 (d, 1 H, J = 18.6), 5.64 (m, 1 H); ^{13}C NMR (D_2O) δ: 35.6, 54.6, 120.9, 132.0, 175.1. Anal. Calcd for $C_5H_9NO_2$: C, 52.16; H, 7.88; N, 12.17. Found: C, 52.15; H, 7.74; N, 12.03.

The product is determined to be ≥99% ee by HPLC analysis on a Crownpak CR(+) column (pH 1.5 $HClO_4$ mobile phase, 0.4 mL/min, 205 nm detection). The minor enantiomer was identified by comparison with an authentic sample prepared from (S,S)-(+)-pseudoephedrine glycinamide. The following retention times are observed: D-allylglycine, 4.68 min; L-allylglycine, 5.45 min. Using an identical new column and the identical eluent at a flow rate of 0.8 mL/min, the checkers observed retention times of 13.86 min for D-allylglycine and 15.19 min for L-allylglycine.

23. Reagent grade p-dioxane was used as received from Mallinckrodt Inc.

24. Di-tert-butyl dicarbonate was used as received from Aldrich Chemical Company, Inc.

25. If necessary, residual ether may be removed by placing the oily product under reduced pressure (0.5 mm) and warming briefly with a heat gun. The oily residue is typically found to be analytically pure product and requires no purification. The product shows the following physical and spectroscopic characteristics: $[\alpha]_D^{23}$ +11.9° (CH_3OH, c 1.4), $[\alpha]_D^{23}$ -2.5° (CH_2Cl_2, c 1.1); lit.[8] $[\alpha]_D^{23}$ -3.9° (CH_2Cl_2, c 1) (Note 27); IR (neat) cm^{-1}: 3324, 3081, 2980, 2932, 1715, 1513, 1395, 1369, 1251, 1163, 1053, 1025, 922; ^1H NMR (2:1 rotamer ratio, $CDCl_3$) major rotamer δ: 1.44 (s, 9 H), 2.57 (m, 2 H), 4.40 (m, 1 H), 5.14-5.19 (m, 3 H), 5.73 (m, 1 H), 8.86 [s(br), 1 H]; minor rotamer δ: 4.19 (m, 1 H), 6.37 (d, 1 H, J = 5.2); ^{13}C NMR ($CDCl_3$) major rotamer δ: 28.1, 36.3, 52.7, 80.1, 119.1, 132.1, 155.4, 176.0; minor rotamer δ: 54.2, 81.7, 156.7. Anal. Calcd for $C_{10}H_{17}NO_4$: C, 55.80; H, 7.96; N, 6.51. Found: C, 55.71; H, 8.14; N, 6.56.

In order to determine the enantiomeric excess of the product, the Boc protective group must be removed prior to HPLC analysis. The sample is prepared by dissolving 23 mg of N-Boc allylglycine in 1 mL of tetrahydrofuran and adding 2 mL of a 3 M aqueous hydrochloric acid solution. The mixture is allowed to stir at 23°C for 1 hr and then is concentrated under reduced pressure to provide a solid residue. The solid is dissolved in water for HPLC analysis. The product is determined to be ≥99% ee by analysis on a Crownpak CR(+) column (Notes 22 and 27).

26. The checkers obtained material having mp 240-242°C and $[\alpha]_D^{23}$ -37.2° (H_2O, c 4), and ≥99% ee by HPLC analysis on a Crownpak CR(+) column (Note 22) in good agreement with the cited literature values.[5,6]

27. The checkers obtained samples of material having rotations in methanol in the range $[\alpha]_D^{23}$ +8.6° to +11.4° (CH_3OH, c 1.4), and $[\alpha]_D^{23}$ -3.7° to -3.8° (CH_2Cl_2, c

1.1), all of which were determined to be ≥99% ee by HPLC analysis on a Crownpak CR(+) column (Note 22).

Waste Disposal Information

All toxic materials were disposed of in accordance with "Prudent Practices in the Laboratory"; National Academy Press; Washington, DC, 1995.

3. Discussion

This procedure describes a highly practical method for the asymmetric synthesis of α-amino acids by the alkylation of the chiral glycine derivative, pseudoephedrine glycinamide.[9] This methodology has been used for the synthesis of a wide variety of α-amino acids and is distinguished by the fact that the glycine amino group is not protected in the alkylation reaction. The method employs pseudoephedrine as a chiral auxiliary. Pseudoephedrine is readily available and inexpensive in both enantiomeric forms, and many of its N-acyl derivatives are crystalline solids. The procedure that is described here for the enolization of pseudoephedrine glycinamide is modified from our previously reported metalation conditions[8] by reaction temperature (0°C versus -78°C employed earlier) and the order of mixing of reagents (addition of lithium diisopropylamide to pseudoephedrine glycinamide versus addition of pseudoephedrine glycinamide to lithium diisopropylamide). This modified procedure is more convienient for large-scale synthesis and is less sensitive to small errors in the titer of the butyllithium solution. The alkylation reaction proceeds in high yield using a wide variety of electrophiles and with excellent diastereoselectivity. Like the alkylation substrates, the products of the alkylation reaction are frequently crystalline and are readily recrystallized to ≥99% de.

The preparation of the alkylation substrate, pseudoephedrine glycinamide, is achieved in a single step from readily available and inexpensive reagents. This reaction accomplishes amide bond formation between the secondary amino group of pseudoephedrine and the carboxyl group of glycine methyl ester without protection of the glycine amino group. This is possible, it is speculated, by the operation of a base-catalyzed mechanism involving transesterification of the methyl ester with the hydroxyl group of pseudoephedrine, followed by intramolecular O→N acyl transfer. Pseudoephedrine glycinamide of both enantiomeric forms is easily prepared in large quantities by this procedure.

A particularly advantageous feature of this method for the synthesis of α-amino acids is the simplicity and mildness of the hydrolysis of the pseudoephedrine amide bond to reveal the α-amino acid. Two hydrolysis protocols are described, one for the isolation of enantiomerically pure α-amino acids, and the other for the preparation of N-acyl-α-amino acids of ≥99% ee. Simply heating aqueous solutions of the alkylation products results in spontaneous cleavage of the amide bond (presumably by intramolecular N→O acyl transfer, followed by hydrolysis of the resulting α-amino ester) and is ideal for isolation of the free α-amino acid under salt-free conditions, thus obviating the need for ion-exchange chromatography. Heating the alkylation products in the presence of alkali accelerates the cleavage reaction and allows the direct N-acylation of the hydrolysis products by the addition of an acylating agent to the aqueous alkaline α-amino acid solution. N-Protected α-amino acids are thus prepared in a single synthetic operation. Both hydrolysis procedures are highly efficient and proceed without significant racemization (≤1%). In both procedures, the chiral auxiliary is easily recovered in crystalline form in high yield.

1. Division of Chemistry and Chemical Engineering, California Institute of Technology, Pasadena, CA 91125.

2. Myers, A. G.; Yoon, T.; Gleason, J. L. *Tetrahedron Lett.* **1995**, *36*, 4555.
3. Almeida, J. F.; Anaya, J.; Martin, N.; Grande, M.; Moran, J. R.; Caballero, M. C. *Tetrahedron: Asymmetry* **1992**, *3*, 1431.
4. (a) Watson, S. C.; Eastham, J. F. *J. Organomet. Chem.* **1967**, 9, 165; (b) Gall, M.; House, H. O. *Org. Synth., Coll. Vol. VI* **1988**, 121.
5. Broxterman, Q. B.; Kaptein, B.; Kamphius, J.; Schoemaker, H. E. *J. Org. Chem.* **1992**, *57*, 6286.
6. Fluka Chemical Guide, 1995-1996, 70.
7. Black, S.; Wright, N. G. *J. Biol. Chem.* **1955**, *213*, 39.
8. Williams, R. M.; Im, M.-N. *J. Am. Chem. Soc.* **1991**, *113*, 9276.
9. Myers, A. G.; Gleason, J. L.; Yoon, T. *J. Am. Chem. Soc.* **1995**, *117*, 8488.

Appendix

Chemical Abstracts Nomenclature (Collective Index Number); (Registry Number)

(R,R)-(–)-Pseudoephedrine glycinamide: Acetamide, 2-amino-N-(2-hydroxy-1-methyl-2-phenylethyl)-N-methyl-, [R-(R*,R*)]- (13); (170115-98-7)

L-Allylglycine: 4-Pentenoic acid, 2-amino-, (R)- (9); (54594-06-8)

N-Boc-L-allylglycine: 4-Pentenoic acid, 2-[[(1,1-dimethylethoxy)carbonyl]amino]-, (R)- (13); (170899-08-8)

Lithium chloride (8,9); (7447-41-8)

(R,R)-(–)-Pseudoephedrine: Pseudoephedrine, (–)- (8); Benzenemethanol, α-[1-(methylamino)ethyl]-. [R-(R*,R*)]- (9); (321-91-1)

Lithium methoxide: Methanol, lithium salt (8,9); (865-34-9)

Glycine methyl ester (8,9); (616-34-2)

Pseudoephedrine L-allylglycinamide: 4-Pentenamide, 2-amino-N-(2-hydroxy-1-methyl-2-phenylethyl)-N-methyl-, [1S-[1R*(S*),2R*]]- (13); (170642-23-6)

Diisopropylamine (8); 2-Propanamine, N-(1-methylethyl)- (9); (108-18-9)

Butyllithium: Lithium, butyl- (8,9); (109-72-8)

Lithium diisopropylamide: Butylamine, N,N-dimethyl-, lithium salt (8); 2-Propanamine, N-(1-methylethyl)-, lithium salt (9); (4111-54-0)

Allyl bromide: 1-Propene, 3-bromo- (8,9); (106-95-6)

Ethyl acetate: Acetic acid, ethyl ester (8,9); (141-78-6)

Dichloromethane: Methane, dichloro- (8,9); (75-09-2)

Ammonium hydroxide (8,9); (1336-21-6)

p-Dioxane: CANCER SUSPECT AGENT (8); 1,4-Dioxane (9); (123-91-1)

Di-tert-butyl dicarbonate: Formic acid, oxydi-, di-tert-butyl ester (8); Dicarbonic acid, bis(1,1-dimethylethyl) ester (9), (24424-99-5)

Glycine methyl ester hydrochloride: Glycine methyl ester, hydrochloride (8,9); (5680-79-5)

Acetonitrile: TOXIC (8,9); (75-05-8)

2-Butanol: sec-Butyl alcohol (8); 2-Butanol (9); (78-92-2)

2,2'-Dipyridyl: 2,2'-Bipyridine (8,9); (366-18-7)

Acetic anhydride (8); Acetic acid anhydride (9); (108-24-7)

4-(N,N-Dimethylamino)pyridine: Pyridine, 4-(dimethylamino)- (8); 4-Pyridinamine, N,N-dimethyl- (9); (1122-58-3)

(S,S)-(+)-Pseudoephedrine glycinamide: Acetamide, 2-amino-N-(2-hydroxy-1-methyl-2-phenylethyl)-N-methyl-, [S-(R*,R*)]- (13); (170115-96-5)

1-CHLORO-(2S,3S)-DIHYDROXYCYCLOHEXA-4,6-DIENE
(3,5-Cyclohexadiene-1,2-diol, 3-chloro-, (1S-cis)-)

Cl

Pseudomonas putida 39/D

Cl OH OH

Submitted by Tomas Hudlicky,*[1a] Michele R. Stabile,[1a] David T. Gibson,[1b] and Gregory M. Whited.[1c]

Checked by Michel Chartrain, Norihiro Ikemoto, and Ichiro Shinkai.

1. Procedure

Caution! Chlorobenzene is an irritant and possible carcinogen.

A. Preculture preparation. A 250-mL Erlenmeyer flask containing 50 mL of mineral salt broth (MSB) (Note 1) with 0.2% L-arginine hydrochloride (L-arginine·HCl) and fitted with a cotton plug is sterilized (Note 2). In a laminar flow hood, a single colony of *Pseudomonas putida* 39/D (*Pp* 39/D) (Note 3) is selected from a fully grown plate (Notes 4, 5, 6) and transferred to the solution with a sterile loop taking care not to place the hand over the plate or flask. The flask is then placed in a benchtop orbital incubator shaker at 30°C and 200 rpm for 24 hr.

B. Shake flask. A 2800-mL Fernbach flask fitted with a vapor bulb and an air inlet (see Figure 1) containing 500 mL of MSB solution with 0.2% L-arginine·HCl is sterilized (Note 2). After cooling, the 50-mL preculture is added via aseptic transfer (Note 7). The vapor bulb is filled with 10 mL of chlorobenzene, and the flask is carefully placed in an orbital incubator shaker at 150 rpm and 30°C for 48 hr. The excess chlorobenzene is removed from the central chamber with a pipette, and the pH of the solution is adjusted, if necessary, to approximately 8 or 9. The suspension is poured into two large centrifuge tubes, and the cells are removed by centrifugation for 30 min at 10°C and 8000 rpm. The supernatant solution is saturated with salt and extracted with ethyl acetate (4 x 100 mL). The organic layers are combined, dried with magnesium or sodium sulfate, and filtered (Note 8). The solvents are evaporated under reduced pressure to afford a tan-colored solid (190 mg). The diene diol is further purified by recrystallization from methylene chloride/hexanes to yield an off-white solid (160 mg) (Note 9). The material has a reasonable shelf life (Note 10) when kept acid-free to inhibit decomposition to chlorophenols.

2. Notes

1. MSB components[2] are as follows: 4% of solution A, 2% of solution B, 1.5% of solution C. Mix solutions A, B, C and add distilled water to 90% volume. Adjust pH to 7.2 with 10 M KOH, and adjust to final volume with distilled water. Solution A (for 1 L): 1 M KH_2PO_4 (136 g/L) and 1 M $Na_2HPO_4 \cdot 7H_2O$ (268.1 g/L). Adjust the solution to pH 7.2 with 10 M KOH. Solution B (Hunter's Base - Vitamin-free, for 1 L): Nitrilotriacetic acid (NTA), 10 g/L; KOH, 7.5 g/L; $MgSO_4 \cdot 7H_2O$, 29.6 g/L; $CaCl_2 \cdot 2H_2O$, 3.3 g/L; $(NH_4)_6Mo_7O_{24} \cdot 4H_2O$, 9.3 mg/L; $FeSO_4 \cdot 7H_2O$, 99 mg/L; Metals 44 solution 50 mL/L. [NOTE: 10.0 g of nitrilotriacetic acid (NTA) is dissolved with stirring in 150 mL of distilled H_2O containing 7.5 g of KOH. Next, 29.6 g of $MgSO_4 \cdot 7H_2O$ is dissolved separately in 150 mL of distilled H_2O. A solution of 3.3 g of $CaCl_2 \cdot 2H_2O$ in 150 mL of distilled water is prepared. The $MgSO_4$ solution is then added to the NTA solution; add *slowly*, to avoid any clouding of the mixture. Once the solutions are mixed, the $CaCl_2$ solution is *gradually* added with stirring. Again, do NOT allow the mixture to become cloudy as this will result in the formation of insoluble precipitates. Prepare a solution from 9.3 mg of $(NH_4)_6Mo_7O_{24} \cdot 4H_2O$ and 99 mg of $FeSO_4 \cdot 7H_2O$ in 150 mL of

distilled water. Add this solution with stirring to the mixture already prepared. A pale yellow color should appear when the solutions have been mixed. Add the Metals 44 solution and bring the total volume to 1.0 L with distilled H_2O. The pH of the solution should be adjusted to 6.8 CAREFULLY with 10 N NaOH - preferably in 0.2-mL aliquots - otherwise, insoluble precipitates will form. Store refrigerated at 4-10°C]. Metals 44 solution (for 100 mL): ethylenediaminetetraacetic acid (EDTA), 250 mg; $ZnSO_4 \cdot 7H_2O$, 1.095 g; $FeSO_4 \cdot 7H_2O$, 500 mg; $MnSO_4 \cdot H_2O$, 154 mg; $CuSO_4 \cdot 5H_2O$, 39.2 mg; $Co(NO_3)_2 \cdot 6H_2O$, 24.8 mg; $Na_2B_4O_7 \cdot 10H_2O$, 17.7 mg. [The solids are dissolved one at a time in 100 mL of distilled H_2O and ~1 drop of sulfuric acid (1 M) is added to retard precipitation. The solution should be aquamarine blue.] Solution C (for 1 L): 200 mg/L of $(NH_4)_2SO_4$ in distilled water.

2. All flasks, solutions and other accessories that come into contact with the media solutions must be sterilized or purchased pre-sterilized. For sterilization, a high pressure steam autoclave (AMSCO 3021) was used with a liquid cycle that holds the contents at a temperature of 121°C/14 psi for 25 min.

3. The organism has been deposited with DSM, the German Collection of Microorganisms and Cell Cultures, as # 6414 and also with ATCC, the American Type Culture Collection, as strain number ATCC 700008.

4. Agar plate preparation. Dilute 20 mL of solution A, 10 mL of solution B, 7.5 mL of solution C and 2.5 g of L-arginine to 250 mL in a 1-L flask equipped with a stir bar. In a separate 1-L flask, mix 10 g of Bacto-Agar in 250 mL of distilled H_2O. Sterilize the solutions. A precipitate present in the MSB solution at this point will dissolve with stirring and cooling of the solution. When the MSB and agar solutions are at ~45-50°C, quickly combine, stir and pour the mixed solution into 100-mm diameter plastic Petri dishes (Fisher Scientific Company) (approximately 20 mL of solution per plate). In general, the method used for obtaining single colonies is called "streaking for isolation," shown in Figure 2. Each direction of streaking is to be done

with a flame-sterilized wire loop or a new disposable plastic loop, four loops for the entire procedure. All procedures are performed under sterile conditions in a laminar flow hood.

5. Storage of cells: A culture of cells was grown in sterile Lauria broth (10 g/L of tryptone, 5 g/L of yeast extract, 5 g/L of NaCl) that had been inoculated (transfer of cells as described in part A) with a single colony from an agar plate. These organisms were grown for 24 hr. The above solution (0.8 mL) was transferred to a sterile cryovial, combined with sterile glycerol solution (0.8 mL) and frozen at -78°C. (Glycerol solution: 60 mL of glycerol, 2.5 g of tryptone, 1.25 g of yeast extract, 2.5 g of NaCl, 182 mL of distd water.) The cells are periodically checked for activity (Note 6).

6. Indigo test: Two plates of *Pseudomonas* agar (MSB + 0.2% arginine + 2% Bacto-Agar) were streaked and allowed to grow in the presence of chlorobenzene. In the lid of one dish was placed a vial containing toluene (for use as control) and plugged with cotton. In the lid of the other plate was placed a vial containing chlorobenzene. The plates were placed in an incubator (30°C) overnight. After the plate was grown (~24 hr), it was removed from the incubator and the vial containing the substrate was removed and replaced with several crystals of indole. After the plate was left at room temperature for 30 min, the cells were checked for the presence of a blue color, indicative of indigo formation. (Color should be very apparent in 30 min for toluene-induced cells.) In this case, a blue color was apparent for both plates thus proving that chlorobenzene would induce the production of toluene dioxygenase and that the cells were viable. The same technique can be used for substrates other than chlorobenzene; if a blue color is not observed, the substrate will not induce production of the dioxygenase enzyme during growth of the organism.

7. An aseptic transfer involves first passing the lip of the preculture flask through a flame and repeating the flaming procedure with the lip of the Fernbach flask.

The preculture is then poured into the larger flask carefully, without allowing the flasks to come into contact.

8. To facilitate breaking up any emulsion, the checkers centrifuged (20 min, 7000 rpm) the combined organic layers.

9. The spectral/physical data of 1-chloro-(2S,3S)-dihydroxycyclohexa-4,6-diene are as follows: R_f = 0.32 (1:1 hexane/ethyl acetate) mp 82-84°C; $[\alpha]_D^{25}$ +54° ($CHCl_3$, *c* 0.59); ^1H NMR ($CDCl_3$) δ: 2.63 (d, 1 H, J = 8.4), 2.74 (d, 1 H, J = 7.3), 4.19 (t, 1 H, J = 7.3), 4.48 (m, 1 H), 5.87 (m, 2 H), 6.12 (m, 1 H); ^{13}C NMR ($CDCl_3$) δ: 69.1 (CH), 71.4 (CH), 122.7 (CH), 123.4 (CH), 128.0 (CH), 134.9 (C).

10. The half-life of 1-chloro-(2S,3S)-dihydroxycyclohexa-4,6-diene ($t_{1/2}$) is 4 days in $CDCl_3$ at room temperature. The diol can be stored at -20°C or -80°C for several months without decomposition.

Waste Disposal Information

All toxic materials were disposed of in accordance with "Prudent Practices in the Laboratory"; National Academy Press; Washington, DC, 1995. The bacterial cell mass from fermentation is, after sterilization, judged suitable for disposal to municipal sewers.

3. Discussion

The pioneering work of Gibson[3] on the isolation and mutation of *Pseudomonas* strains that oxidatively degrade aromatic compounds has led, 25 years later, to the application of cyclohexadiene cis-diols in asymmetric synthesis. The first applications of these types of compounds to synthesis were the use of meso-diol derived from benzene for production of polyphenylene[4] by ICI and in the synthesis of racemic pinitol by Ley in 1987.[5] The first use of the title compound in synthesis was in 1990 in the preparation of both enantiomers of erythrose.[6] This field increased greatly in the late 1980s and there are now numerous reports on the use of the title compound as a versatile chiral synthon. Several reviews have appeared.[7]

Over 300 diol metabolites are known.[8] Several diols and some secondary synthons derived from them have recently become commercially available: Eastman Fine Chemicals, Genencor: (1S-cis)-3-Chloro-3,5-cyclohexadiene-1,2-diol, (1S-cis)-3-bromo-3,5-cyclohexadiene-1,2-diol, (1S-cis)-3-iodo-3,5-cyclohexadiene-1,2-diol, (5S-cis)-5,6-dihydroxy-1,3-cyclohexadiene-1-carbonitrile, cis-2R,3S-2,3-dihydroxy-2,3-dihydrobenzonitrile acetonide, (1R-cis)-1,2-dihydro-1,2-naphthalenediol, (1R-cis)-1,2,3,4-tetrahydro-1,2-naphthalenediol, (4S-trans)-4,5-dihydroxy-3-oxo-1-cyclohexene-1-carboxylic acid, furo[3,4-d]-1,3-dioxol-4(3aH)-1-dihydro-6-hydroxy-2,2-dimethyl-[3aR-(3aa,6aa)].

1. (a) Department of Chemistry, Virginia Polytechnic Institute and State University *Address correspondence to this author at: Department of Chemistry, University of Florida, Gainesville, FL 32611; (b) Department of Microbiology, University of Iowa; (c) Genencor International Incorporated, Kimball Way, South San Francisco, CA.

2. Modified from Cohen-Bazire, G.; Sistrom, W. R.; Stanier, *J. Cell. Comp. Phys.* **1957**, *49*, 25; See also, Hudlicky, T.; Boros, C. H.; Boros, E. E. *Synthesis*, **1992**, 174.
3. (a) Gibson, D. T.; Hensley, M.; Yoshioka, H.; Mabry, T. J. *Biochemistry* **1970**, *9*, 1626; (b) Gibson, D. T.; Koch, J. R.; Kallio, R. E. *Biochemistry* **1968**, *7*, 2653.
4. Ballard, D. G. H.; Courtis, A.; Shirley, I. M.; Taylor, S. C. *Macromolecules* **1988**, *21*, 294.
5. Ley, S. V.; Sternfeld, F.; Taylor, S. *Tetrahedron Lett.* **1987**, *28*, 225.

6. Hudlicky, T; Luna, H.; Price, J. D.; Rulin, F. *J. Org. Chem.* **1990**, *55*, 4683.

7. Reviews: (a) Boyd, D. R. *Nat. Prod. Reports* **1998**, 309; (b) Grund, A. D. *SIM News* **1995**, *45*, 59; (c) Hudlicky, T.; Entwistle, D. A.; Pitzer, K. K.; Thorpe, A. J. *Chem. Rev.* **1996**, *96*, 1195; (d) Hudlicky, T. In "Green Chemistry: Designing Chemistry for the Environment"; Anastas, P. T.; Williamson, T., Eds.; ACS Symposium Series; American Chemical Society: Washington, DC, 1996; Vol. 626, p. 180; (e) Sheldrake, G. N. In "Chirality and Industry"; Collins, A. N.; Sheldrake, G. N.; Crosby, J., Eds.; John Wiley and Sons Ltd., 1992; p.127; (f) Hudlicky, T.; Thorpe, A. J. *J. Chem. Soc., Chem. Commun.* **1996**, 1993; (g) Hudlicky, T. *Chem. Rev.* **1996**, *96*, 3; (h) Hudlicky, T.; Reed, J. W. *Adv. Asymmetric Synth.* **1995**, *1*, 271; (i) Illman, D. L. *Chemical and Engineering News*, September 5, 1994, p. 22; (j) Brown, S. M.; Hudlicky, T. *Org. Synth.: Theory Appl.* **1993**, *2*, 113; (k) Carless, H. A. H. *Tetrahedron: Asymm.* **1992**, *3*, 795; (l) Widdowson, D. A.; Ribbons, D. W.; Thomas, S. D. *Janssen Chim. Acta* **1990**, *8*, 3.

8. (a) Stabile, Michele R., Ph.D thesis, Virginia Polytechnic Institute and State University, 1995; (b) McMordie, R. A. S. Ph.D Thesis, School of Chemistry, The Queen's University of Belfast, Belfast BT9 5AG, U. K., 1991; (c) Hudlicky, T.; Gonzalez, D.; Gibson, D. T. *Aldrichimica Acta* **1998**.

Appendix

Chemical Abstracts Nomenclature (Collective Index Number); (Registry Number)

1-Chloro-(2S,3S)-dihydroxycyclohexa-4,6-diene: 3,5-Cyclohexadiene-1,2-diol, 3-chloro-, (1S-cis)- (10); (65986-73-4)

Potassium dihydrogen phosphate: Phosphoric acid, monopotassium salt (9); (7778-77-0)

Disodium hydrogen phosphate: Phosphoric acid, disodium salt (8,9); (7558-79-4)

Nitrilotriacetic acid: CANCER SUSPECT AGENT: Acetic acid, nitrilotri- (8); Glycine, N,N-bis(carboxymethyl)- (9); (139-13-9)

Ammonium molybdate(VI) tetrahydrate: Molybdic acid, hexaammonium salt, tetrahydrate (8,9); (12027-67-7)

Ethylenediaminetetraacetic acid, tetrasodium salt: CANCER SUSPECT AGENT: Glycine, N,N'-1,2-ethanediylbis[N-(carboxymethyl)-, tetrasodium salt, trihydrate (9); (67401-50-7)

Cobalt nitrate hexahydrate: Nitric acid, cobalt (2+ salt), hexahydrate (9); (10026-22-9)

Sodium tetraborate decahydrate: Borax (8,9); (1303-96-4)

L-Arginine hydrochloride: L-Arginine, monohydrochloride (9); (1119-34-2)

Chlorobenzene: Benzene, chloro- (8,9); (108-90-7)

Agar (8,9); (9002-18-0)

Tryptones (Bacteriological): *Chem. Abstr.* See: Peptones, Bacteriological (10); (73049-73-7*)

Indole (8); 1H-Indole (9); (120-72-9)

Figure 1. Shake Flask Apparatus

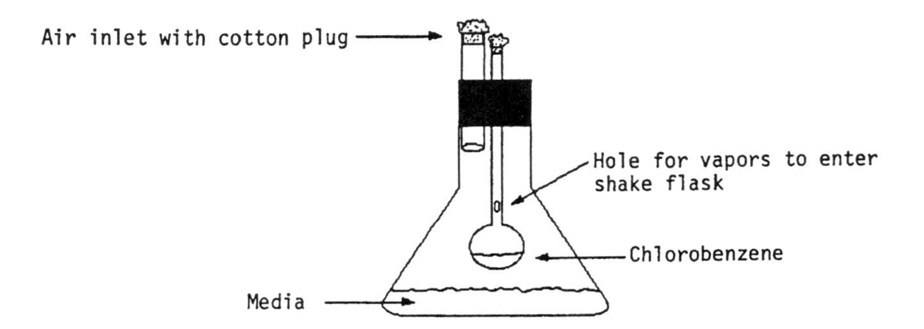

Figure 2. Summary of Steps Used to Streak for Isolation

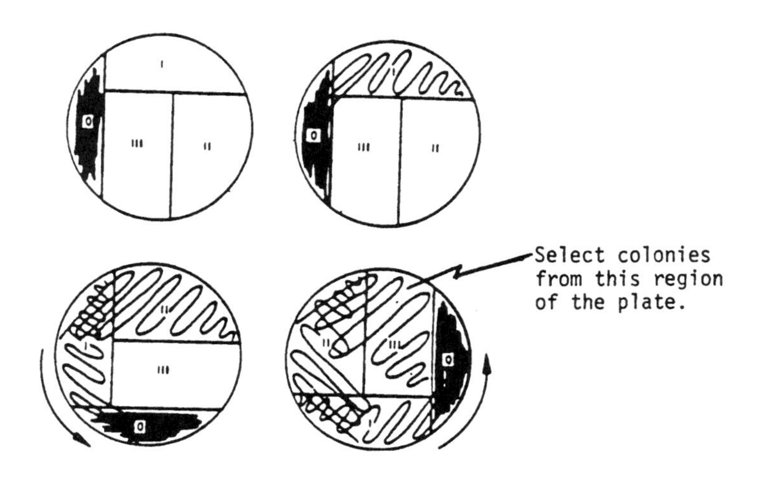

(2S,3S)-(+)-(3-PHENYLCYCLOPROPYL)METHANOL

(Cyclopropanemethanol, 2-phenyl-, (1S-trans)-)

A. BuBr $\xrightarrow{Mg^\circ / \text{Ether}}$ BuMgBr

B. BuMgBr + $B(OMe)_3$ $\xrightarrow[-78^\circ C \text{ to rt}]{\text{Ether}}$ $BuB(OMe)_3MgBr$ $\xrightarrow{10\% \text{ HCl}}$ BuB(OH

C. $BuB(OH)_2$ + HO–CH₂CH₂–NH–CH₂CH₂–OH $\xrightarrow[3Å \text{ molecular sieves}]{\text{Ether} / CH_2Cl_2}$ **1**

D. **1** + **2** $\xrightarrow[NaCl / H_2O]{CH_2Cl_2}$ **3**

E. Ph–CH=CH–CH₂OH $\xrightarrow{Zn(CH_2I)_2 \cdot DME}$ (with dioxaborolane **3**) → (2-phenylcyclopropyl)methanol

Submitted by André B. Charette and Hélène Lebel.[1]

Checked by Kevin Minbiole, Patrick Verhoest, and Amos B. Smith, III.

1. Procedure

A. Butylmagnesium bromide. To a 500-mL, three-necked, round-bottomed flask equipped with an egg-shaped magnetic stirrer, 125-mL pressure-equalizing addition funnel, reflux condenser and a glass stopper (Note 1), is added 17.0 g (0.70 mol) of magnesium turnings (Note 2). Stirring is started, and the system is flame-dried for 2 min. The flask is cooled to room temperature under a flow of argon, and 30 mL of ether (Note 3) is introduced to cover the magnesium. A solution of 24 mL (0.22 mol) of bromobutane (Note 4) in 70 mL of ether is placed in the pressure-equalizing addition funnel. Then, 1 mL (0.01 mol) of bromobutane is added to the suspension of magnesium in ether. The mixture is heated gently to initiate the reaction (Note 5). When the reaction has started, the solution of bromide in ether is added dropwise at a rate sufficient to maintain a gentle reflux (Note 6). After completion of the addition, the funnel is rinsed with 5 mL of ether. The gray solution is stirred for 15 min and then transferred to a dry flask under argon via cannula. The Grignard reagent is titrated with a solution of isopropyl alcohol in benzene using 1,10-phenanthroline as the indicator (Note 7).[2] A 1.90-2.10 M solution of Grignard reagent is obtained.

B. Butylboronic acid. To a 1-L, one-necked, round-bottomed flask equipped with an egg-shaped magnetic stirrer (Note 1) and an internal thermocouple probe (Note 11) is added 220 mL of ether (Note 3), followed by 10 mL (89.2 mmol) of trimethyl borate (Note 12). The clear solution is cooled to -75°C (internal temperature) and stirred vigorously, then 45 mL (87.8 mmol) of a 1.95 M solution of butylmagnesium bromide in ether (Note 13) is added dropwise via cannula at such a rate that the internal temperature does not exceed -65°C (Note 14). After the addition is complete, the resulting white slurry is stirred for an additional 2 hr at -75°C under argon. The cooling bath is removed, and the reaction mixture is allowed to warm to room temperature (Note 15). Hydrolysis is carried out by the dropwise addition of 100 mL of an aqueous 10% solution of hydrochloric acid. The white precipitate is dissolved, and the resulting clear biphasic mixture is stirred for 15 min, at which time the two layers are separated. The aqueous layer is extracted with ether (2 x 50 mL), and the combined extracts are dried over magnesium sulfate. After concentration of the ethereal solution under reduced pressure, the residual white solid is purified by recrystallization as follows: After dissolution in hot water (25 mL), the resulting biphasic solution is cooled to 0°C to induce recrystallization of the boronic acid. The solid is collected on a Büchner funnel, washed with 50 mL of hexanes and placed under vacuum (0.2 mm) for 60 min (Notes 16 and 17). Between 5.0-6.5 g (55-72% yield) of the boronic acid is produced as a white solid (Note 18).

C. [(2-)-N,O,O'[2,2'-Iminobis[ethanolato]]]-2-butylboron, **1**. A 250-mL, one-necked, round-bottomed flask equipped with an egg-shaped magnetic stirrer (Note 1) is charged with 5.15 g (50.5 mmol) of butylboronic acid and 5.31 g (50.5 mmol) of diethanolamine (Note 19). Ether, 100 mL (Note 3) and 50 mL of dichloromethane (Note 20) are added, followed by about 10 g of molecular sieves 3Å (Note 21). The resulting heterogeneous solution (Note 22) is stirred for 2 hr under argon. The solid is triturated with dichloromethane (50 mL), filtered through a medium fritted disk funnel and washed with dichloromethane (2 x 50 mL). The filtrate is concentrated under reduced pressure to produce the crude desired complex. The diethanolamine complex is purified by recrystallization as follows: the white solid is dissolved in hot dichloromethane (20 mL), then ether (50 mL) is added to induce recrystallization of the complex. The mixture is cooled to 0°C and the solid is collected on a Büchner funnel and washed with ether (2 x 30 mL). The product is dried under reduced pressure (0.2 mm) to afford 7.70 g (89%) of the title compound as a white crystalline solid (Note 23).

D. (4R-trans)-2-Butyl-N,N,N',N'-tetramethyl[1,3,2]dioxaborolane-4,5-dicarbox–amide **3**. A 500-mL, one-necked, round-bottomed flask equipped with an egg-shaped magnetic stirrer, under argon is charged with 7.70 g (45.0 mmol) of the butylboronate diethanolamine complex and 11.9 g (58.3 mmol) of (R,R)-(+)-N,N,N',N'-tetramethyltartaric acid diamide (Note 24). The solids dissolve upon the addition of 225 mL of dichloromethane. Brine (70 mL) is added, and the resulting biphasic solution is stirred for 30 min under argon. The two layers are separated, and the aqueous layer is extracted with dichloromethane (50 mL). The combined organic layers are washed with brine (50 mL), dried over magnesium sulfate and filtered. The filtrate is concentrated under reduced pressure and dried under reduced pressure (0.2 mm) to give 11.3 g (93%) of the title compound as a pale yellow oil (Note 25).

E. (2S,3S)-(+)-(3-Phenylcyclopropyl)methanol. A 250-mL, one-necked, round-bottomed flask equipped with an egg-shaped magnetic stirrer (Note 26) and an internal thermocouple probe (Note 11), is charged with 45 mL of dichloromethane (Note 20) and 1.60 mL (14.9 mmol) of 1,2-dimethoxyethane (DME) (Note 27). The solution is cooled to -10°C (internal temperature) with an acetone/ice bath, and 1.50 mL (14.9 mmol) of diethylzinc is added (Note 28). To this stirred solution is added 2.40 mL (29.8 mmol) of diiodomethane (Note 29) over a 15-20 min period while maintaining the internal temperature between -8°C and -12°C. After the addition is complete, the resulting clear solution is stirred for 10 min at -10°C. A solution of 2.41 g (8.94 mmol) of the dioxaborolane ligand in 10 mL of dichloromethane is added via cannula under argon over a 5-6 min period while maintaining the internal temperature below -5°C. A solution of 1.00 g (7.45 mmol) of cinnamyl alcohol (Note 30) in 10 mL of dichloromethane is immediately added via cannula under argon over a 5-6 min period while maintaining the internal temperature under -5°C. The cooling bath is removed, and the reaction mixture is allowed to warm to room temperature and stirred for 8 hr at that temperature (Note 31).

Workup. *Method A.* The reaction is quenched with aqueous saturated ammonium chloride (10 mL) and aqueous 10% hydrochloric acid (40 mL). The mixture is then diluted with ether (60 mL) and transferred to a separatory funnel. The reaction flask is rinsed with ether (15 mL), and aqueous 10% hydrochloric acid (10 mL) and both solutions are transferred to the separatory funnel. The two layers are separated, and the aqueous layer is washed with ether (20 mL). The combined organic layers are transferred to an Erlenmeyer flask, and a solution containing 60 mL of aqueous 2 N sodium hydroxide and 10 mL of aqueous 30% hydrogen peroxide is added in one portion (Note 32). The resulting biphasic solution is stirred vigorously for 5 min. The two layers are separated and the organic layer is washed successively with aqueous 10% hydrochloric acid (50 mL), aqueous saturated sodium sulfite (50 mL), aqueous saturated sodium bicarbonate (50 mL), and brine (50 mL). The organic layer is dried over magnesium sulfate and filtered, and the filtrate is concentrated under reduced pressure. The crude product is left under reduced pressure (0.2 mm) overnight (12-16 hr) to remove the butanol produced in this oxidative work-up. The product is purified by a Kugelrohr distillation (90°C, 0.8 mm) to afford 1.05 g (95%) of (2S,3S)-(+)-(3-phenylcyclopropyl)methanol as a colorless oil (Notes 33 and 34).

Workup. *Method B [with recovery of (R,R)-(+)-N,N,N',N'-tetramethyltartaric acid diamide]*. The mixture is quenched with aqueous saturated ammonium chloride (80 mL), and the resulting biphasic mixture is stirred for 5 min. The two clear layers are separated, and the aqueous layer is washed with dichloromethane (20 mL) (Note 35). The combined organic layers are dried over magnesium sulfate and filtered, and the filtrate is concentrated under reduced pressure. The residual oil is dissolved in ether (75 mL) and water (50 mL). The resulting biphasic mixture is stirred for 1 hr. The layers are separated, and the aqueous layer is washed with ether (20 mL). This aqueous layer is kept for tetramethyltartaric acid diamide recovery (see below). The combined organic layers are treated with 60 mL of aqueous 2 N sodium hydroxide and 10 mL of aqueous 30% hydrogen peroxide (Note 32). The resulting biphasic mixture is stirred for 5 min. The two layers are separated and the organic layer is washed successively with aqueous 10% hydrochloric acid (50 mL), saturated aqueous sodium sulfite (50 mL), saturated aqueous sodium bicarbonate (50 mL), and brine (50 mL). The organic layer is dried over magnesium sulfate and filtered, and the filtrate is concentrated under reduced pressure. The crude product is left under reduced pressure (0.2 mm) overnight (12-16 hr) to remove butanol produced in this oxidative work-up. The product is purified by a Kugelrohr distillation (90°C, 0.8 mm) to afford 1.02 g (93%) of (2*S*,3*S*)-(+)-(3-phenylcyclopropyl)methanol as a colorless oil.

Recovery of (R,R)-(+)-N,N,N',N'-tetramethyltartaric acid diamide. The aqueous layer from the above extraction is concentrated under reduced pressure, and the crude product is recrystallized by an initial dissolution in hot dichloromethane (5 mL) followed by the addition of ethyl acetate (10 mL) to afford between 600 mg to 750 mg (33-41% yield) of the (R,R)-(+)-N,N,N',N'-tetramethyltartaric acid diamide (Notes 36 and 37).

2. Notes

1. All glassware was dried in an oven (110°C) and after assembly was allowed to cool under an atmosphere of argon.

2. Magnesium turnings were purchased from Sigma-Aldrich Fine Chemicals Company Inc. and were used without further purification.

3. Ether was freshly distilled from sodium/benzophenone.

4. Bromobutane was purchased from Fisher Scientific Company and was freshly distilled from phosphorus pentoxide (P_2O_5) (bp 100-104°C).

5. The formation of a gray cloudy suspension indicates that the reaction has started. Furthermore, the reaction is sufficiently exothermic to induce the ether to reflux even when the reaction flask is not heated. If the reaction does not start within 2 to 3 min, repeat the heating procedure.

6. Between 1.5 hr and 2 hr are needed for addition.

7. A dried 10-mL, one-necked, round-bottomed flask is charged with 1 mL of Grignard, some drops of THF (Note 8) and a crystal of 1,10-phenanthroline (Note 9). The slightly pink solution is titrated with a 0.5 M solution of isopropyl alcohol in benzene (Note 10). Between 3.8 and 4.2 mL (±0.2 mL) is obtained to give a clear colorless solution (three titrations).

8. THF was freshly distilled from sodium/benzophenone.

9. 1,10-Phenanthroline was purchased from Sigma-Aldrich Fine Chemicals Company Inc. and was used without further purification.

10. Isopropyl alcohol was freshly distilled from calcium hydride (CaH_2) and benzene was freshly distilled from sodium.

11. A Barnant 100, Type T Thermo-Couple Thermometer was used to monitor the internal temperature of the reaction solution.

12. Anhydrous trimethyl borate (with <5% of methanol) was purchased from Sigma-Aldrich Fine Chemicals Company Inc. and was used without further purification. Alternatively, a non anhydrous reagent can be dried by distillation from calcium hydride (bp 68-69°C).

13. Commercially available (Sigma-Aldrich Fine Chemicals Company Inc.), butylmagnesium chloride, 2.0 M in ether, can be used and a similar yield is observed.

14. Between 20 and 30 min are needed for the addition.

15. Between 1 hr and 1.5 hr are needed.

16. The amount of the boroxine significantly increases if the solid is left under reduced pressure for a longer period of time. The boroxine is always a contaminant of the boronic acid (see discussion).

17. Sometimes a second recrystallization is needed to obtain pure boronic acid by removing by-products resulting from autooxidation.

18. The physical properties are as follows: mp 95-97°C; ^{1}H NMR (400 MHz, DMSO) δ: 0.56 (t, 2 H, J = 7.6), 0.83 (t, 3 H, J = 7.2), 1.31-1.19 (m, 4 H), 7.34 (br s, 2 H);

^{13}C NMR (100 MHz, DMSO) δ: 13.9, 15.3 (br), 25.1, 26.5; ^{11}B NMR (128.4 MHz, DMSO) δ: 32.7.

19. Diethanolamine was purchased from Fisher Scientific Company and was used without further purification.

20. Dichloromethane was freshly distilled from CaH_2.

21. Molecular sieves, 3Å, powder, average particle size 3-5 μ were purchased from Sigma-Aldrich Fine Chemicals Company Inc. and dried under vacuum at 250°C for 24 hr, before using.

22. A few minutes after the addition of the molecular sieves, a white precipitate forms and sometimes maintaining stirring becomes difficult.

23. The physical properties are as follows: mp 145-148°C; ^{1}H NMR (400 MHz, $CDCl_3$) δ: 0.44-0.48 (m, 2 H), 0.88 (t, 3 H, J = 7.1), 1.21-1.37 (m, 4 H), 2.79 (br s, 2 H), 3.26 (br s, 2 H), 3.88 (br s, 2 H), 3.98 (br s, 2 H), 4.80-4.98 (m, 1 H); ^{13}C NMR (100 MHz, $CDCl_3$) δ: 14.1, 18.4 (br), 26.5, 28.1, 51.4, 62.5; ^{11}B NMR (128.4 MHz, $CDCl_3$) δ: 13.1, 32.7. Anal. Calcd for $C_8H_{18}BNO_2$: C, 56.18; H, 10.61; N, 8.19. Found: C, 56.15; H, 10.86; N, 8.07.

24. (R,R)-(+)-N,N,N',N'-Tetramethyltartaric acid diamide was prepared from diethyl tartrate and dimethylamine and was freshly recrystallized with methanol and ethyl acetate.[3] The physical properties are as follows: mp 186-187°C [lit.[2] 189-190°C]; ^{1}H NMR (400 MHz, $CDCl_3$) δ: 3.01 (s, 6 H), 3.13 (s, 6 H), 4.21 (br s, 2 H), 4.65 (s, 2 H); ^{13}C NMR (100 MHz, $CDCl_3$) δ: 36.1, 36.9, 69.8, 170.8; $[\alpha]_D^{20}$ +43° (EtOH, *c* 2.03) [lit.[2] $[\alpha]_D^{20}$ +43° (EtOH, *c* 3.0)]. This product is also commercially available from Sigma-Aldrich Fine Chemicals Company Inc.

25. The physical properties are as follows: ^{1}H NMR (400 MHz, $CDCl_3$) δ: 0.85 (t, 2 H, J = 7.7); 0.87 (t, 3 H, J = 7.2), 1.29-1.41 (m, 4 H), 2.98 (s, 6 H), 3.20 (s, 6 H), 5.53 (s, 2 H); ^{13}C NMR (100 MHz, $CDCl_3$) δ: 9.9 (br), 13.6, 25.0, 25.7, 35.7, 36.9, 75.6, 168.23; ^{11}B NMR (128.4 MHz, $CDCl_3$) δ: 34.2; $[\alpha]_D^{20}$ -104.4° ($CHCl_3$, *c* 1.70). HRMS

Calcd for $C_{12}H_{23}BN_2O_4$: 270.1751. Found: 270.1746. Anal. Calcd for $C_{12}H_{23}BN_2O_4$: C, 53.35; H, 8.58; N, 10.37. Found: C, 53.67; H, 9.07; N, 10.21.

26. All glassware was flame dried, then cooled under a flow of dry argon.

27. DME was freshly distilled from sodium/benzophenone.

28. Diethylzinc is a moisture sensitive and pyrophoric liquid and must be manipulated in an inert atmosphere with gas-tight syringes. Neat diethylzinc was purchased from Akzo Nobel Chemicals Company Inc. and was used without further purification.

29. Diiodomethane was purchased from Acros-Fisher Scientific Company and used without further purification. If necessary, diiodomethane can be purified if it shows any signs of slight decomposition (orange or red color develops over time): diiodomethane is washed with aqueous saturated sodium sulfite, dried over sodium sulfate (Na_2SO_4), and distilled from copper (40°C, 1.0 mm). The pale yellow liquid is collected on copper.

30. Cinnamyl alcohol was purchased from Sigma-Aldrich Fine Chemicals Company Inc. and was freshly purified by Kugelrohr distillation: a first fraction boiling at <70°C (1.0 mm) was discarded, and the alcohol was collected as a white solid at 80°C (1.0 mm).

31. A similar yield is obtained when the submitters stirred this mixture for 14 hr. No noticeable decomposition and side reactions are observed after slightly longer periods of time.

32. Hydrogen peroxide was purchased from ACP Chemicals Company Inc and used without further purification.

33. Alternatively, the product can be purified by flash chromatography on silica gel (78.5 g, 4 cm x 16 cm) using 30% ethyl acetate in hexanes as the mobile phase (800 ml) to afford 1.06 g (96%) of the title compound.

34. The physical properties are as follows: bp 90°C, 0.8 mm; IR (film) cm^{-1}: 3350, 3050, 3000, 2950, 2900, 1600, 1500, 1450, 1100, 1050, 1000, 750, 700; ^{1}H NMR (400 MHz, $CDCl_3$) δ: 0.92-1.01 (m, 2 H), 1.43-1.51 (m, 1 H), 1.75 (br s, 1 H), 1.82-1.86 (m, 1 H), 3.59-3.67 (m, 2 H), 7.07-7.10 (m, 2 H), 7.15-7.20 (m, 1 H), 7.25-7.30 (m, 2 H); ^{13}C NMR (100 MHz, $CDCl_3$) δ: 13.8, 21.2, 25.2, 66.3, 125.6, 125.8, 128.3, 142.5; $[\alpha]_D^{20}$ +82° (EtOH, *c* 1.74) [lit.[4] (2R,3R)-cyclopropylmethanol >99% ee $[\alpha]_D^{20}$ -92° (EtOH, *c* 1.23)]. Anal. Calcd for $C_{10}H_{12}O$: C, 81.04 H, 8.16. Found: C, 81.15; H, 8.30. The enantiomeric excess of the product is determined precisely by GC analylsis of the corresponding trifluoroacetate ester derivative: To a solution of 10 mg of the crude alcohol in 0.75 mL of pyridine is added 0.25 mL of trifluoroacetic anhydride (TFAA). After 30 min at room temperature, an additional 0.25 mL of TFAA is added. After 30 min, the reaction mixture is diluted with 5 mL of ether. This solution was injected directly into the GC (0.5 μL) with the following conditions: Cyclodex G-TA, 0.32 x 30 m; pressure 25 psi; isotherm: 110°C, T_r (minor) 11.5 min, T_r (major) 12.0 min; enantiomeric ratio: 29:1 (93% ee).

35. If the resulting organic layers are not clear, the combined organic layers should be washed with an additional 50 mL of aqueous saturated ammonium chloride.

36. An additional 10-15% of the diol can be recovered from the first aqueous saturated ammonium chloride extract (Note 38): the layer is concentrated on a rotatory evaporator and the white solid is triturated with cold methanol (30 mL), the mixture is filtered on a Büchner funnel, and the solid is washed with cold methanol (20 mL). The filtrate is concentrated to ca. 25 mL and treated with 2.5 g of sodium sulfide (Note 39). The resulting mixture is stirred for 30 min and then filtered on Celite (6 g, 1 cm x 4 cm). The filtrate is concentrated by rotary evaporation, and the residue is purified by flash chromatography on silica gel (75 g, 3.5 cm x 14.5 cm) by dissolving it in 10 mL of 10% methanol in chloroform and eluting with 10% methanol in chloroform. A recrystallization with dichloromethane and ethyl acetate give pure material.

37. The physical properties are identical to those of Note 24.

38. The diol decomposes after a few hours at room temperature in this layer.

39. Sodium sulfide was purchased from Anachemia Science and was used without further purification.

Waste Disposal Information

All toxic materials were disposed of in accordance with "Prudent Practices in the Laboratory"; National Academy Press; Washington, DC, 1995.

3. Discussion

Parts A-D. The preparation of boronic acids by the addition of a Grignard reagent to a trialkyl borate is one of the most convenient and well-established methods involving relatively inexpensive starting materials.[5] The carefully monitored addition of butylmagnesium bromide to trimethyl borate avoids any complications resulting from overaddition, and relatively good yields of the butylboronic acid are obtained after acid hydrolysis. Usually, alkylboronic acids are relatively difficult to characterize and to obtain analytically pure,[6] because they readily tend to form boroxines (anhydrides) under dehydrating conditions (when heated or when left under reduced pressure). The white solid (butylboronic acid) is transformed into a colorless oil (tributylboroxine) if dehydration is pushed to completion. Conversely, butylboronic acid and its boroxine are also readily oxidized by air to generate 1-butanol and boric acid.[7] For these reasons, it is almost impossible to isolate butylboronic acid without any traces of water, its boroxine or oxidation by-products.

$$3\ BuB(OH)_2 \xrightarrow{-3\ H_2O} (BuBO)_3$$

In order to avoid these complications, butylboronic acid is quickly converted to its air-stable and more robust diethanolamine derivative **1**.[8] Complex **1** could be contaminated with some unseparable diethanolamine if a small excess of diethanolamine is used in its preparation. However, this has no effect on the efficiency of the synthesis of the chiral dioxaborolane ligand **3**. Diethanolamine complex **1** reacts quantitatively with a slight excess of tetramethyltartramide **2** under biphasic conditions to produce the desired chiral dioxaborolane ligand **3**. Ligand **3** is relatively stable, and is neither excessively hygroscopic nor oxygen-sensitive. It must be stored under argon for longer periods of time. However, the submitters have shown that the enantioselectivities are directly related to the ligand purity. Consequently, it is generally preferable to use freshly prepared ligand for obtaining optimal results.

Part E. The enantioselective cyclopropanation of allylic alcohols using the chiral dioxaborolane ligand **3** and $Zn(CH_2I)_2$•DME is a powerful tool for synthesizing three-membered rings. This method is much simpler and produces superior enantiomeric excesses compared to those using other stoichiometric chiral ligands.[9] The scope of the reaction is wide and a variety of allylic alcohols have been converted into their cyclopropane derivatives in excellent enantiomeric excesses (88-94%).[10] It was also shown that polyenes can be cyclopropanated at the allylic alcohol position with excellent chemo- and enantioselectivities.[11] Recently, this reaction has been used in the synthesis of cyclopropane containing natural products.[12]

Caution! The previously reported preparation of $Zn(CH_2I)_2$ without a complexing additive[13] *is highly exothermic, and a violent decomposition sometimes occurred. For safety reasons, the use of the $Zn(CH_2I)_2$•DME as reported here is mandatory if this reaction is carried out on a ≥8 mmole scale.*[14] *If the internal* *temperature during the formation of the reagent is carefully monitored, the procedure reported here is extremely safe even on larger scales.*

Note that the structure of $Zn(CH_2I)_2$•DME is derived from the stoichiometry of the reactants (Et_2Zn, CH_2I_2, DME). Substantial quantities of $IZnCH_2I$•DME are necessarily formed at the reaction temperature and as a by-product of the cyclopropanation. Another improvement was made in this procedure: the number of equivalents of the reagent has been decreased to 2.0 equiv (vs 5 equiv in the original paper). However, under these conditions that minimize the amount of Et_2Zn used but require longer reaction times, the yield of the diol recovery dropped to ca. 50%.

The cyclopropanation of cinnamyl alcohol is a good example of the use of dioxaborolane ligand **3** as chiral additive to synthesize chiral cyclopropanes.

1. Departement de Chimie, Université de Montréal, P.O. Box 6128, Station Downtown, Montréal (Québec) Canada, H3C 3J7.
2. Soai, K.; Machida, H.; Yokota, N. *J. Chem. Soc. Perkin Trans. I* **1987**, 1909.
3. Seebach, D.; Kalinowski, H.-O.; Langer, W.; Crass, G.; Wilka, E.-M. *Org. Synth., Coll. Vol. VII* **1990**, 41.
4. Evans, D. A.; Woerpel, K. A.; Hinman, M. M.; Faul, M. M. *J. Am. Chem. Soc.* **1991**, *113*, 726.
5. (a) Srebnik, M.; Cole, T. E.; Ramachandran, V.; Brown, H. C. *J. Org. Chem.* **1989**, *54*, 6085; (b) Washburn, R. M.; Levens, E.; Albright, C. F.; Billig, F. A. *Org. Synth., Coll. Vol. IV* **1963**, 68 and references cited therein.
6. (a) Martichonok, V.; Jones, J. B. *J. Am. Chem. Soc.* **1996**, *118*, 950; (b) Mathre, D. J.; Jones, T. K.; Xavier, L. C.; Blacklock, T. J.; Reamer, R. A.; Mohan, J. J.; Jones, E. T. T.; Hoogsteen, K.; Baum, M. W.; Grabowski, E. J. J. *J. Org. Chem.* **1991**, *56*, 751.

7. (a) Korcek, S.; Watts, G. B.; Ingold, K. U. *J. Chem. Soc., Perkin Trans. 2* **1972**, 242; (b) Johnson, J. R.; Van Campen, Jr., M. G. *J. Am. Chem. Soc.* **1938**, *60*, 121.
8. Brown, H. C.; Vara Prasad, J. V. N. *J. Org. Chem.* **1986**, *51*, 4526.
9. (a) Ukaji, Y.; Nishimura, M.; Fujisawa, T. *Chem. Lett.* **1992**, 61; (b) Ukaji, Y.; Sada, K.; Inomata, K. *Chem. Lett.* **1993**, 1227; (c) Kitajima, H.; Aoki, Y.; Ito, K.; Katsuki, T. *Chem. Lett.* **1995**, 1113.
10. (a) Charette, A. B.; Juteau, H. *J. Am. Chem. Soc.* **1994**, *116*, 2651; (b) Charette, A. B.; Lemay J., *Angew. Chem., Int. Ed. Engl.* **1997**, *36*, 1090.
11. Charette, A. B.; Juteau, H.; Lebel, H.; Deschênes, D. *Tetrahedron Lett.* **1996**, *37*, 7925.
12. For selected examples, see: (a) White, J. D.; Kim, T.-S.; Nambu, M. *J. Am. Chem. Soc.* **1997**, *119*, 103; (b) Barrett, A. G. M.; Kasdorf, K. *J. Chem. Soc., Chem. Comm.* **1996**, 325; (c) Falck, J. R.; Mekonnen, B.; Yu, J.; Lai, J.-Y. *J. Am. Chem. Soc.* **1996**, *118*, 6096; (d) Charette, A. B.; Lebel, H. *J. Am. Chem. Soc.* **1996**, *118*, 10327.
13. Denmark, S. E.; Edwards, J. P. *J. Org. Chem.* **1991**, *56*, 6974.
14. Charette, A. B.; Prescott, S.; Brochu, C. *J. Org. Chem.* **1995**, *60*, 1081.

Appendix

Chemical Abstracts Nomenclature (Collective Index Number); (Registry Number)

(2S,3S)-(+)-(3-Phenycyclopropyl)methanol: Cyclopropanemethanol, 2-phenyl-, (1S-trans)- (12); (110659-58-0)

Butylmagnesium bromide: Magnesium, bromobutyl- (8,9); (693-03-8)

Magnesium (8,9); (7439-95-4)

1-Bromobutane: Butane, 1-bromo- (8,9); (109-65-9)

Isopropyl alcohol: 2-Propanol (8,9); (67-63-0)

1,10-Phenanthroline (8,9); (66-71-7)

Butylboronic acid: 1-Butaneboronic acid (8); Boronic acid, butyl- (9); (4426-47-5)

Trimethyl borate: Boric acid, trimethyl ester (8,9); (121-43-7)

Diethanolamine: Ethanol, 2,2'-iminodi- (8); Ethanol, 2,2'-iminobis- (9): (111-42-2)

(4R-trans)-2-Butyl-N,N,N',N'-tetramethyl[1,3,2]dioxaborolane-4,5-dicarboxamide: 1,3,2-Dioxaborolane-4,5-dicarboxamide, 2-butyl-N,N,N'N'-tetramethyl-, (4R-trans)- (13); (161344-85-0)

(R,R)-(+)-N,N,N',N'-Tetramethyltartaric acid diamide: Tartramide, N,N,N'N'-tetramethyl-, (+)- (8); Butanediamide, 2,3-dihydroxy-N,N,N'N'-tetramethyl-, [R-(R*,R*)]- (9); (26549-65-5)

Dimethoxyethane: Ethane, 1,2-dimethoxy- (8,9); (110-71-4)

Diethylzinc: Zinc, diethyl- (8,9); (557-20-0)

Diiodomethane: Methane,diiodo- (8,9); (75-11-6)

Cinnamyl alcohol (8); 2-Propen-1-ol, 3-phenyl- (9); (104-54-1)

Hydrogen peroxide (8,9); (7722-84-1)

Sodium sulfite: Sulfurous acid, disodium salt (8,9); (7757-83-7)

Copper (8,9); (7440-50-8)

Trifluoroacetic anhydride: Acetic acid, trifluoro-, anhydride (8,9); (407-25-0)

Sodium sulfide (8,9); (1313-82-2)

SYNTHESIS OF CHIRAL NON-RACEMIC DIOLS FROM (S,S)-1,2,3,4-DIEPOXYBUTANE: (2S,3S)-DIHYDROXY-1,4-DIPHENYLBUTANE

(2,2'-Bioxirane, [S-(R*,R*)]- and 2,3-Butanediol, 1,4-diphenyl-, [S-(R*,R*)]-)

A. 1. CH_3SO_2Cl, pyr; 2. CH_3SO_3H, EtOH, H_2O

B. KOH, EtOH–H_2O

C. 1. PhMgBr, CuI; 2. NH_4^+ Cl^-

Submitted by Michael A. Robbins,[1a] Paul N. Devine,[1a] and Taeboem Oh.[1b]
Checked by Marc Renard and Léon Ghosez.

1. Procedure

A. L-Threitol 1,4-bismethanesulfonate. A dry, 1-L, round-bottomed flask equipped with a magnetic stirring bar, vacuum adapter, rubber septum, and a nitrogen line (Note 1) is charged with 2,3-O-isopropylidene-L-threitol (25.0 g, 0.154 mol) (Note 2), pumped under high vacuum for 10 min, and a nitrogen atmosphere is introduced.

Methylene chloride (308 mL) (Note 3) and pyridine (37.4 mL, 0.462 mol) (Note 4) are added, and the stirred solution is cooled to 0°C with an ice-water bath. Methanesulfonyl chloride (29.8 mL, 0.385 mol) (Note 5) is added dropwise via a 50-mL glass syringe over a period of 10 min. After an additional 30 min, the ice-water bath is removed, and the stirred solution is allowed to warm to room temperature. After an additional 6 hr a precipitate forms (pyridinium chloride); 300 mL of an aqueous saturated solution of sodium bicarbonate ($NaHCO_3$) is added slowly, dissolving the precipitate. The solution is stirred for a further 30 min and then transferred to a 1-L separatory funnel. The layers are separated, and the aqueous layer is extracted with methylene chloride (3 x 100 mL). The organic layers are combined, dried with anhydrous sodium sulfate, and the drying agent is removed by filtration. The solvent is removed by rotary evaporation to give a tan solid that can be used as such or recrystallized (Note 6) from 1:1 chloroform-diethyl ether to give the product as a crystalline white solid, mp 77-78°C, 43 g (88% yield) (Note 7).

A 1-L, one-necked, round-bottomed flask equipped with a heating mantle, magnetic stirring bar, and a reflux condenser is charged with 2,3-O-isopropylidene-L-threitol 1,4-bismethanesulfonate (40 g, 0.126 mol), 95% ethanol (250 mL) (Note 8) and methanesulfonic acid (0.204 mL, 3.14 mmol) (Note 9), and brought to a gentle reflux. The solution is refluxed for 10 hr and then cooled to 0°C with an ice-water bath resulting in the formation of crystals. The crystals are collected by suction filtration, washed with cold ethanol (2 x 50 mL) and diethyl ether (2 x 50 mL), and dried in a vacuum desiccator at 60°C under full vacuum for 4 hr to give the product as white crystals, mp 101-102°C, 29.6 g, (84% yield) (Note 10).

B. (S,S)-1,2,3,4-Diepoxybutane. A 250-mL, two-necked, round-bottomed flask equipped with a magnetic stirring bar, nitrogen line, and a 125-mL pressure-equalizing addition funnel is charged with L-threitol 1,4-bismethanesulfonate (25.0 g, 0.0898 moles) and diethyl ether (180 mL). The mixture is stirred vigorously to form a suspension, and a solution of potassium hydroxide (11.6 g, 0.207 mol) (Note 11) in water (35 mL) is added dropwise via the addition funnel over a period of 15 min. The clear mixture is stirred for an additional 45 min at room temperature, and the ether layer is decanted. The aqueous layer is transferred to a 500-mL separatory funnel and extracted with methylene chloride (3 x 50 mL). The combined ether and methylene chloride extracts are dried with anhydrous sodium sulfate, the sodium sulfate is removed by filtration, and the solution is concentrated to approximately 50 mL total volume by rotary evaporation. The concentrate is fractionally distilled through a 13-cm Vigreux distillation column at atmospheric pressure to give the product as a clear oil, bp 138-140°C, 6.3 g (81%) (Notes 12 and 13).

C. (2S,3S)-Dihydroxy-1,4-diphenylbutane. A 500-mL, one-necked, round-bottomed flask is equipped with a vacuum adapter, septum, magnetic stirring bar and an argon line (Note 14). The flask is flame-dried under reduced pressure, an argon atmosphere is reintroduced, and copper iodide (CuI) (1.89 g, 9.9 mmol) (Note 15) is added. The flask with CuI is pumped under reduced pressure for 10 min, and the argon atmosphere is reintroduced. Then 30 mL of tetrahydrofuran (THF) (Note 16) is added to the flask, stirring is initiated, and the slurry is cooled to -30°C with a dry ice-bromobenzene bath. Phenylmagnesium bromide in THF (18.0 g, 0.099 mol, 99 mL of a 1 M solution in THF) (Note 17) is added via a 50-mL syringe over a period of 10 min, and the stirred slurry is aged a further 10 min. (S,S)-1,2,3,4-Diepoxybutane (2.84 g, 0.033 mol, 11 mL of a 3 M solution in THF) is added dropwise via syringe over a period of 10 min. After an additional 10 min at -30°C, the reaction mixture is warmed to 0°C with an ice-water bath. After an additional 2.5 hr at 0°C, the reaction is quenched by the slow addition of 200 mL of an aqueous saturated ammonium chloride solution, the ice-water bath is removed, and the quenched reaction is stirred for 15 min yielding a deep blue color. The reaction mixture is transferred to a 1-L separatory funnel, and the layers are separated. The aqueous layer is extracted with methylene chloride (4 x 50 mL), the combined organic layers are dried with magnesium sulfate the drying agent is removed by filtration, and solvent is removed by rotary evaporation to give 7.1 g of a light yellow solid. The solid is dissolved in 225 mL of toluene with heat (heating mantle) then and cooled to 0°C with an ice-water bath resulting in the formation of fine needles. The needles are collected via suction filtration and dried in a vacuum desiccator at 40°C under full vacuum for 6 hr to yield 6.0 g of a white solid. The filtrate is evaporated to dryness via rotary evaporation to yield 1.1 g of a yellow solid. The yellow solid is flash-chromatographed on 30 g of Merck 230-400 mesh silica gel with 1:1 hexanes-ethyl acetate to give 1.0 g of a white solid (Note 18). The combined recrystallized and chromatographed product yields a total of 7.0 g of diol (88%) (Note 19).

2. Notes

1. The reaction was maintained under a positive pressure of dry nitrogen except during the quenching process. All glassware was dried in a 150°C oven for 30 min and cooled in a desiccator prior to use.

2. The submitters used 2,3-O-isopropylidene-L-threitol prepared according to the procedure in *Org. Synth., Coll. Vol. VIII* **1993**, 155-161. The diol was recrystallized from 1:1 hexanes-diethyl ether prior to use. Alternatively, 2,3-O-isopropylidene-L-threitol can be purchased from Aldrich Chemical Company, Inc.

3. Methylene chloride was distilled from calcium hydride under nitrogen immediately prior to use. The methylene chloride was transferred from the still to the reaction flask in two portions via a dry 250-mL pressure-equalizing addition funnel.

4. Pyridine was distilled from calcium hydride and stored over KOH until needed. The pyridine was added via a 50-mL glass syringe.

5. Methanesulfonyl chloride was obtained from Aldrich Chemical Company, Inc., distilled under reduced pressure and stored under nitrogen at 0°C (freezer) until needed.

6. The tan solid is recrystallized as follows. It is dissolved in approximately 250 mL of 1:1 chloroform-diethyl ether with heat (heating mantle). Upon cooling to 0°C with an ice-water bath, a solid precipitates that is collected by suction filtration. The solid is washed twice with cold 50-mL portions of diethyl ether and dried in a vacuum desiccator at 40°C under full vacuum to give the product as a white crystalline solid.

7. This product was found to be greater than 98% pure by ^1H and ^{13}C NMR. Physical properties and spectral data are as follows: $[\alpha]_D$ -22.1 (acetone, *c* 1.9,), lit.[2] $[\alpha]_D$ -21.3 (acetone, *c* 2.0); ^1H NMR ($CDCl_3$) δ: 1.49 (s, 6 H), 3.10 (s, 6 H), 4.20 (q, 2 H), 4.39 (q, 4 H); ^{13}C NMR ($CDCl_3$) δ: 27.1, 37.9, 67.9, 75.7, 111.4.

8. 95% ethanol from Pharmco Products Inc. was used as obtained.

9. Methanesulfonic acid, from Aldrich Chemical Company, Inc., was used as obtained. Methanesulfonic acid was measured and added with a 1-mL syringe.

10. This product was found to be greater than 98% pure by ^1H and ^{13}C NMR. Physical properties and spectral data are as follows: $[\alpha]_D$ -5.2 (acetone, *c* 1.85), lit.[2] $[\alpha]_D$ -5.5 (acetone, *c* 2); ^1H NMR ($CDCl_3$ + DMSO-d_6) δ: 2.80 (s, 6 H), 3.62 (m, 2 H), 3.79 (br s, 2 H), 4.00 (m, 4 H); ^{13}C NMR ($CDCl_3$ + DMSO) δ: 36.5, 67.5, 70.1.

11. Potassium hydroxide, from J.T. Baker Inc., was used as obtained.

12. This product was found to be greater than 98% pure by ^1H and ^{13}C NMR. Physical properties and spectral data are as follows $[\alpha]_D$ +25.7 ($CHCl_3$, *c* 2.1), lit.[3] $[\alpha]_D$ +23.6 ($CHCl_3$, *c* 2); ^1H NMR ($CDCl_3$) δ: 2.56 (m, 2 H), 2.68 (m, 4 H); ^{13}C NMR ($CDCl_3$) δ: 44.4, 51.0.

13. The product was stored in a dry, round-bottomed flask under argon in a freezer until needed.

14. The reaction was maintained under a positive pressure of dry argon except during the quenching process. All glassware was dried in a 150°C oven for 30 min and cooled in a desiccator prior to use.

15. Copper iodide, from Aldrich Chemical Company, Inc., was purified by Soxhlet extraction in THF followed by drying under full vacuum overnight.

16. THF was distilled from potassium under nitrogen immediately prior to use.

17. Phenylmagnesium bromide was prepared immediately prior to use according to established procedures. Phenylmagnesium bromide, 1.0 M in THF, can be purchased from Aldrich Chemical Company, Inc.

18. The diol elutes at an $R_f = 0.64$, (SiO_2, 1:1, hexanes-ethyl acetate).

19. This product was found to be greater than 98% pure by ^1H and ^{13}C NMR. Physical properties and spectral data are as follows: $[\alpha]_D$ +6.2 (chloroform, *c* 1.7); ^1H NMR ($CDCl_3$) δ: 2.07 (bd, 2 H), 2.88 (m, 4 H), 3.75 (m, 2 H), 7.23 (m, 6 H), 7.31 (m, 4 H); ^{13}C NMR ($CDCl_3$) δ: 40.3, 74.0, 126.6, 128.6, 129.4, 138.0. The checkers recrystallized the product in hexane-ethyl acetate (1:1).

Waste Disposal Information

All toxic materials were disposed of in accordance with "Prudent Practices in the Laboratory"; National Academy Press; Washington, DC, 1995.

3. Discussion

The present approach provides a general method for the preparation of optically active diols. The Table illustrates the preparation of several diols.[4] The procedures are simple, and there is no possibility of racemization in any of the steps. This method does not require any difficult-to-handle or disposal of metals, which makes it a good alternative to osmium-catalyzed dihydroxylation.[5] A variety of diols can be prepared from the one diepoxide substrate. The optically active epoxides and diols are highly useful in organic synthesis.[6] A variety of diols have been useful as chiral ligands in asymmetric catalyst.[7]

TABLE[2]

OPENING OF THE DIEPOXIDE

Nucleophile	Product	Yield(%)	$[\alpha]_D^{23}$
BnMgBr/CuI	$BnCH_2$(OH)CH–CH(OH)CH_2Bn	80	-35.0° ($CHCl_3$, *c* 2.2)
MeMgBr/CuI	Et(OH)CH–CH(OH)Et	89	-23° ($CHCl_3$, *c* 5.0)
i-PrMgCl/CuI	i-$PrCH_2$(OH)CH–CH(OH)CH_2-i-Pr	66	-53.2° ($CHCl_3$, *c* 3.8)
t-BuMgCl/CuI	t-$BuCH_2$(OH)CH–CH(OH)CH_2-t-Bu	31	-43.0° ($CHCl_3$, *c* 2.3)

1. (a) Department of Chemistry, Binghamton University, Binghamton, NY 13902; Present address, Merck Sharp & Dohme Research Laboratories, P.O. Box 2000, Rahway, NJ; (b) Department of Chemistry, California State University, Northridge, CA 91330-8200.
2. Feit, P. W. *J. Med. Chem.* **1964**, *7*, 14.
3. Feit, P. W. *Chem. Ber.* **1960**, *93*, 116.
4. (a) The data for this table was taken from Ph. D. Dissertation, Paul N. Devine, Binghamton University; (b) see also, Devine, P. N.; Oh, T. *Tetrahedron Lett.* **1991**, *32*, 883.
5. For a review of osmium-catalyzed dihydroxylation, see: Kolb, H. C.; VanNieuwenhze, M. S.; Sharpless, K. B. *Chem. Rev.* **1994**, *94*, 2483.
6. (a) Behrens, C. H.; Sharpless, K. B. *Aldrichimica Acta* **1983**, *16*, 67; (b) Kishi, Y. *Aldrichimica Acta* **1980**, *13*, 23; (c) Wallace, T.W.; Wardell, I.; Li, K.-D.; Leeming, P.; Redhouse, A.D.; Challand, S. R. *J. Chem. Soc., Perkin Trans. 1* **1995**, 2293.
7. Whitesell, J. K. *Chem. Rev.* **1989**, *89*, 1581.

Appendix

Chemical Abstracts Nomenclature (Collective Index Number); (Registry Number)

(S,S)-1,2,3,4-Diepoxybutane: Butane, 1,2:3,4-diepoxy-, (2S,3S)- (8); 2,2'-Bioxirane, [S-(R*,R*)]- (9); (30031-64-2)

L-Threitol 1,4-methanesulfonate: 1,2,3,4-Butanetetrol, 1,4-dimethanesulfonate, [S-(R*,R*)]- (9); (299-75-2)

2,3-Di-O-isopropylidene-L-threitol: Aldrich: (+)-2,3-O-Isopropylidene-L-threitol: 1,3-Dioxolane-4,5-dimethanol, 2,2-dimethyl-, (4S-trans)- (9); (50622-09-8)

Pyridine (8,9); (110-86-1)

Methanesulfonyl chloride (8,9); (124-63-0)

Pyridinium chloride: Pyridinium hydrochloride (8,9); (628-13-7)

Methanesulfonic acid (8,9); (75-75-2)

(2S,3S)-Dihydroxy-1,4-diphenylbutane: 2,3-Butanediol, 1,4-diphenyl-, [S-(R*,R*)]- (12); (133644-99-2)

Copper iodide (8,9); (7681-65-4)

Phenylmagnesium bromide: Magnesium, bromophenyl- (8,9); (100-58-3)

PREPARATION OF ENANTIOMERICALLY PURE α-N,N-DIBENZYLAMINO ALDEHYDES: S-2-(N,N-DIBENZYLAMINO)-3-PHENYLPROPANAL

(Benzenepropanal, α-[bis(phenylmethyl)amino]-, (S)-)

A. H_2N–CH($PhCH_2$)–CO_2H —(BnBr; NaOH/K_2CO_3)→ Bn_2N–CH($PhCH_2$)–CO_2Bn

B. Bn_2N–CH($PhCH_2$)–CO_2Bn —($LiAlH_4$)→ Bn_2N–CH($PhCH_2$)–CH_2OH

C. Bn_2N–CH($PhCH_2$)–CH_2OH —($(COCl)_2$; DMSO)→ Bn_2N–CH($PhCH_2$)–CHO

Submitted by Manfred T. Reetz,[1] Mark W. Drewes,[2] and Renate Schwickardi.[1]
Checked by Yan Dong, Anthony Laurenzano, and Steven Wolff.

1. Procedure

Caution! Lithium aluminum hydride is a flammable solid, that upon contact with moisture, forms hydrogen, a highly flammable and potentially explosive gas.

A. *Benzyl (S)-2-(N,N-dibenzylamino)-3-phenylpropanoate.* A 250-mL, three-necked, round-bottomed flask, equipped with a magnetic stirring bar, a reflux condenser, and a dropping funnel, is charged with a solution of 16.6 g (120 mmol) of potassium carbonate and 4.8 g (120 mmol) of sodium hydroxide in 100 mL of water. Following the addition of 9.9 g (60 mmol) of S-phenylalanine (Note 1), the stirred suspension is heated at reflux to form a clear solution. To this refluxing solution is added dropwise 31.0 g (181 mmol) of distilled benzyl bromide. The mixture is heated at reflux for an additional 1 hr and then cooled to room temperature. The organic phase is separated, and the aqueous phase is extracted twice with 75 mL of diethyl ether. The combined organic phases are washed with about 75 mL of a saturated aqueous solution of sodium chloride (NaCl) and dried over magnesium sulfate ($MgSO_4$). Following the removal of the solvent under reduced pressure, the crude product is purified using flash chromatography (250 g of 230-400 mesh silica gel; hexane/ethyl acetate 10:1, v/v) to provide 15.14-18.0 g (58-69%) of benzyl (S)-2-(N,N-dibenzylamino)-3-phenylpropanoate (Note 2).

B. (S)-2-(N,N-Dibenzylamino)-3-phenyl-1-propanol. A dry, 100-mL, three-necked, round-bottomed flask, equipped with a magnetic stirring bar and a dropping funnel, is charged with 60 mL of dry diethyl ether and 1.13 g (30 mmol) of lithium aluminum hydride (Note 3) under an atmosphere of argon (Note 4). The suspension is cooled to 0°C, and 10.9 g (25 mmol) of benzyl (S)-2-(N,N-dibenzylamino)-3-phenylpropanoate in 10 mL of diethyl ether are added dropwise. The mixture is stirred for 16 hr at room temperature and cooled again to 0°C. The mixture is carefully worked up by the *dropwise* and sequential addition of 1.1 mL of water, 1.1 mL of a 15% aqueous sodium hydroxide solution and an additional 3.4 mL of water. The reaction mixture is filtered through a coarse filtration frit to remove aluminum salts, and the latter are washed four times with 8-mL portions of diethyl ether. The combined filtrates and washings are dried over magnesium sulfate and concentrated under reduced pressure. Most of the benzyl alcohol is removed under high vacuum (10^{-5} mbar, 0.75 x 10^{-5} mm) at a bath temperature of 50°-60°C. The crude product is then purified by flash chromatography (100 g of 230-400 mesh silica gel; hexane/ethyl acetate 95:5, v:v) or recrystallized from hexane to afford 6.2-7.2 g (75-87% yield) of a white solid (Note 5) having a melting point of 67°C and an optical rotation of

$[\alpha]_D^{20}$ +35.6° (CH_2Cl_2, *c* 1.91). The enantiomeric purity as measured by HPLC [Varian 5560; chiral stationary phase: 250 mm Chiralcel OD-H, 4.6 mm i. d.; mobile phase: heptane : 2-propanol = 90 : 10; T/p/F: 308 K/3.2 Mpa/0.5 mL/min; sample volume: 5 μL (2.3 mg in 0.5 mL of heptane, 0.05 mL of 2-propanol); detector: UV 200, 254 nm, TC = 0.05 sec. E = 0.1] is >99% (Note 6).

C. (S)-2-(N,N-Dibenzylamino)-3-phenylpropanal. This procedure is based on the Swern oxidation[3] (Note 7). A dry, 250-mL, three-necked, round-bottomed flask equipped with a magnetic stirrer and a dropping funnel is charged with 100 mL of dry dichloromethane under an atmosphere of argon (Note 3). After cooling to -78°C, 1.52 g (12 mmol) of oxalyl chloride (Note 4) and 1.56 g (20 mmol) of dry dimethyl sulfoxide are added dropwise to the stirred solution. After 5 min, 3.31 g (10 mmol) of (S)-2-(N,N-dibenzylamino)-3-phenyl-1-propanol in 5 mL of dichloromethane are added with stirring. The mixture is stirred for an additional 0.5 hr at -78°C, and 4.05 g (40 mmol) of freshly distilled triethylamine is added. The mixture is then allowed to warm to room temperature over 0.5 hr, whereupon 50 mL of water is added. The phases are separated, and the aqueous phase is extracted three times with 50 mL of diethyl ether or dichloromethane. The combined organic phases are washed successively with 10 mL of aqueous 1% hydrochloric acid, 10 mL of water, 10 mL of aqueous 5% sodium bicarbonate, and 10 mL of saturated aqueous NaCl. The organic layer is dried over $MgSO_4$, and the solvent is removed under reduced pressure, to provide 3.22 g (95-98% yield) of the aldehyde as a crude product having a purity of ≥95% (Notes 8 and 9).

2. Notes

1. S-Phenylalanine was purchased from Aldrich Chemical Company, Inc., and used without further purification.

2. The product has the following spectral properties: ^1H NMR (400 MHz, $CDCl_3$) δ: 3.00 and 3.14 (ABX System, 2 H, J_{AB} = 14.1, J_{AX} = 7.3 and J_{BX} = 5.9, $C\underline{H}_2$-CH-N), 3.54 and 3.92 (AB System, 4 H, J_{AB} = 13.9, N-$C\underline{H}_2$), 3.71 (t, 1 H, J = 7.6, CH_2-C$\underline{H}$-N), 5.11 and 5.23 (AB System, 2 H, J_{AB} = 12.3, O-$C\underline{H}_2$-C_6H_5) and 7.18 (m, 20 H, O-CH_2-$C_6\underline{H}_5$, N-CH_2-$C_6\underline{H}_5$ and $C_6\underline{H}_5$-CH_2); ^{13}C NMR (100 MHz, $CDCl_3$) δ: 35.8 (t, C_6H_5-$C\underline{H}_2$), 54.5 (t, N-$C\underline{H}_2$), 62.5 (d, CH_2-C$\underline{H}$-N), 66.0 (t, O-$C\underline{H}_2$-C_6H_5), 126.3-139.3 (m, O-CH_2-$C_6\underline{H}_5$, N-CH_2-$C_6\underline{H}_5$ and $C_6\underline{H}_5$-CH_2) and 172.1 (s, COO-CH_2). Anal. Calcd for $C_{30}H_{29}NO_2$: C, 82.72; H, 6.72; N, 3.22. Found: C, 82.86; H, 6.73; N, 3.50.

3. Nitrogen can also be used as the inert gas.

4. Oxalyl chloride was purchased from Merck, Darmstadt/Germany or Aldrich Chemical Company, Inc. and used as received.

5. The product has the following spectral properties: ^1H NMR (400 MHz, $CDCl_3$) δ: 2.36-2.50 (m, 1 H, C$\underline{H}$-C_6H_5), 3.30-3.58 (m, 2 H), 3.29-3.38 (m, 1 H), 3.48 and 3.92 (AB System, 4 H, J_{AB} = 13.3, N-$C\underline{H}_2$), 3.45-3.52 (m, 1 H) and 7.06-7.36 (m, 15 H, N-CH_2-$C_6\underline{H}_5$ and $C_6\underline{H}_5$-CH_2); ^{13}C NMR (100 MHz, $CDCl_3$) δ: 31.9 (t, C_6H_5-$C\underline{H}_2$), 53.4 (t, N-$C\underline{H}_2$), 60.6 (d, CH_2-C$\underline{H}$-N), 61.1 (t, $C\underline{H}_2$-OH) and 126.2-134.3 (m, N-CH_2-$C_6\underline{H}_5$ and $C_6\underline{H}_5$-CH_2). The signal for the hydroxy hydrogen is centered at δ = 2.97 and is very broad. Anal. Calcd for $C_{23}H_{25}NO$: C, 83.35; H, 7.60; N, 4.23. Found: C, 83.25; H, 7.65; N, 4.20.

6. An alternative synthesis of this compound which avoids $LiAlH_4$ involves N-benzylation of commercially available (e. g., Aldrich Chemical Company, Inc.) (S)-2-amino-3-phenylpropanol as follows: A mixture of 3.78 g (25 mmol) of (S)-2-amino-3-phenyl-1-propanol and 6.91 g (50 mmol) of potassium carbonate in 50 mL of 96% ethanol and 10 mL of water is brought to reflux temperature. To this stirred, two-phase mixture is added dropwise 10.69 g (62.5 mmol) of benzyl bromide. The vigorously stirred mixture is heated at reflux for an additional 0.5 hr and cooled to room temperature, and 30 mL of water are added. The product is extracted three times with

100 mL of ether, and the combined organic phases are washed with a saturated NaCl solution and dried over $MgSO_4$. Following removal of the solvents using a rotary evaporator, the crude product is recrystallized from 20 mL of boiling hexane to provide 7.92 g (96% yield) of white crystals having a melting point of 68-69°C. The ^{1}H and ^{13}C NMR spectra and enantiomeric purity are identical to those previously recorded (Note 5).

7. The oxidation can also be performed efficiently with pyridine/sulfur trioxide, but not with Jones reagent; pyridinium dichromate affords poor yields (40-50%).[4]

8. Thin layer chromatography of the crude product shows one spot with R_f = 0.64. The material can be purified by column chromatography (silica gel; petroleum ether/ethyl acetate 95:5, v/v) to provide a 92% yield of analytically pure aldehyde.[4] However, in further reactions of the aldehyde it is best to use the crude material as it is. The spectral properties are as follows: ^{1}H NMR (400 MHz, $CDCl_3$) δ: 2.94 and 3.15 (ABX System, 2 H, J_{AB} = 13.9, J_{AX} = 7.3 and J_{BX} = 6.2, C_6H_5-$C\underline{H}_2$), 3.56 (t, 1 H, 7.1, CH_2-C$\underline{H}$-N), 3.69 and 3.82 (AB System, 4 H, J_{AB} = 13.7, N-$C\underline{H}_2$), 7.12-7.30 (m, 15 H, N-CH_2-$C_6\underline{H}_5$ and $C_6\underline{H}_5$-CH_2) and 9.72 (s, 1 H, C-C$\underline{H}$O); ^{13}C NMR (100 MHz, $CDCl_3$) δ: 29.2 (t, C_6H_5-$C\underline{H}_2$), 53.8 (t, N-$C\underline{H}_2$), 67.5 (d, CH_2-C$\underline{H}$-N), 125.2-138.1 (M, N-CH_2-$C_6\underline{H}_5$ and $C_6\underline{H}_5$-CH_2) and 200.9 (d, C-C$\underline{H}$O). Anal. Calcd for $C_{23}H_{23}NO$: C, 83.85; H, 7.04; N, 4.25. Found: C, 84.02; H, 7.13; N, 4.23.

9. The procedure may be conducted on a larger scale. For example, the submitters used 40 mmol of (S)-2-(N,N-dibenzylamino)-3-phenyl-1-propanol, and obtained 12.9 g (98%) of the aldehyde.

Waste Disposal Information

All toxic materials were disposed of in accordance with "Prudent Practices in the Laboratory"; National Academy Press; Washington, DC, 1995.

3. Discussion

The procedure described here provides a simple way to prepare S-configured 2-(N,N-dibenzylamino)-3-phenylpropanal from naturally occurring phenylalanine using protocol that is quite general for the conversion of naturally occurring α-amino acids into the corresponding N,N-dibenzyl-protected α-amino aldehydes.[5,6] The amino acids that have been used so far include alanine,[5,6] phenylalanine,[5,6] valine,[5,6] leucine,[5,6] isoleucine,[6] lysine,[6] serine,[6] threonine,[6] ornithine,[6] and tryptophan.[7] Upon using the enantiomers of the natural α-amino acids, R-configured N,N-dibenzyl-protected α-amino aldehydes may be prepared.[6,8]

N,N-Dibenzylamino aldehydes are useful building blocks in a wide variety of diastereoselective C-C bond forming reactions. These include: Grignard-type reactions[5,6,9] of RMgX, RLi, R_2CuLi, or $RTi(OiPr)_3$; aldol additions involving lithium (Li) enolates,[5,6,8] zinc reagents,[10] enolsilanes,[6,8,11] or enolboranes;[12] trimethylsilyl cyanide (Me_3SiCN) additions catalyzed by ZnX_2;[6,13] sulfur ylide additions;[6,8,14] and hetero Diels-Alder reactions[6,8] (Scheme 1). In almost all cases a high degree of non-chelation control (diastereoselectivity 90-99%) has been observed, an unusual phenomenon that can be explained either by applying the Felkin-Anh model or by invoking ground state effects.[6,15] If highly Lewis acidic conditions are used, reversal of diastereoselectivity may result, but chelation control is not general.[5,6]

A number of research groups have used N,N-dibenzylamino aldehydes in these and in other C-C bond forming reactions.[6-14,16] In most cases the products were shown to be enantiomerically pure (ee > 98%), which demonstrates the absence of any appreciable racemization during the formation or reaction of these aldehydes. Nevertheless, it is best to use freshly prepared samples in crude form as soon as possible. During isolation of the N,N-dibenzylamino aldehydes it is not necessary to

Scheme 1

Typical Non-chelation-Controlled Reactions of N,N-Dibenzylamino Aldehydes

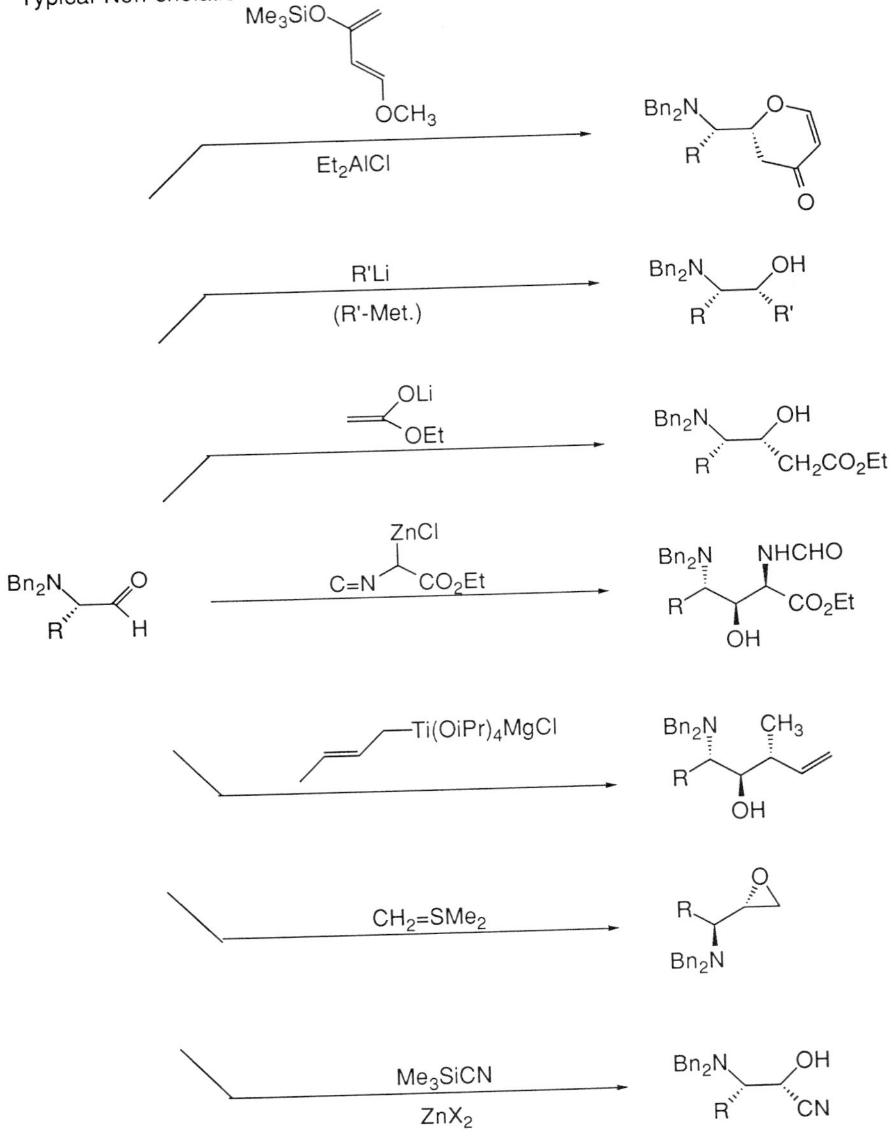

Me_3SiO / OCH_3 / Et_2AlCl

$R'Li$ / (R'-Met.)

OLi / OEt

ZnCl / C=N / CO_2Et

$Ti(OiPr)_4MgCl$

$CH_2=SMe_2$

Me_3SiCN / ZnX_2

Bn_2N, R, O, H

Bn_2N, R, O, O

Bn_2N, OH, R, R'

Bn_2N, OH, R, CH_2CO_2Et

Bn_2N, NHCHO, R, CO_2Et, OH

Bn_2N, CH_3, R, OH

R, O, Bn_2N

Bn_2N, OH, R, CN

work at low temperatures. This is in contrast to the configurationally much more labile N-tert-butoxycarbonyl(Boc)-protected analogs, that have to be handled in cold ether.[6,17] The N-Boc-protected α-amino aldehydes generally react more or less stereo-randomly with such reagents as RLi, RMgX, Li-enolates, $NaCN/NaHSO_3$, or sulfur ylides,[6,17] the so-called Garner aldehyde derived from serine being a notable exception.[6,18] Under special Lewis acidic conditions N-Boc-protected α-amino aldehydes afford the chelation controlled products,[17-19] which means that such processes are stereochemically complementary to the reactions of N,N-dibenzylamino aldehydes. Deprotection of the reaction products, i.e., removal of the two benzyl groups at nitrogen, is best achieved using the Pearlman catalyst according to the method of Yoshida.[6,20]

N,N-Dibenzylamino aldehydes are also useful building blocks in the preparation of other classes of amino compounds as summarized in a review article covering the literature up to mid 1998.[6] These include α,β-unsaturated ester or ketone derivatives prepared by Wittig or Wittig-Horner reactions[21] as well as α-amino aldimines prepared by condensation reactions of the aldehydes.[22] Such compounds are in themselves synthetically interesting building blocks in a variety of C-C bond forming processes, Michael additions and hydrogenation reactions.[21,22] The industrial synthesis and use of the title compound has been described on a 190 kg (576 mol) scale.[23] Finally, N,N-dibenzylamino aldehydes and ketones have been prepared in enantiomerically pure form from precursors other than α-amino acids, adding to the synthetic versatility of this class of compounds.[24]

1. Max-Planck-Institut für Kohlenforschung, Kaiser-Wilhelm-Platz 1, D-45470 Mülheim/Ruhr, Germany.
2. Present address: Bayer AG, Pflanzenschutzzentrum, D-40789 Monheim, Germany.
3. Mancuso, A. J.; Huang, S.-L.; Swern, D. *J. Org. Chem.* **1978**, *43*, 2480.
4. Drewes, M. W. Dissertation, Universität Marburg 1988.
5. Reetz, M. T.; Drewes, M. W.; Schmitz, A. *Angew. Chem.* **1987**, *99*, 1186; *Angew. Chem., Int. Ed. Engl.* **1987**, *26*, 1141.
6. For reviews of N,N-dibenzylamino aldehydes and related compounds see: Reetz, M. T. *Angew. Chem.* **1991**, *103*, 1559; *Angew. Chem., Int. Ed. Engl.* **1991**, *30*, 1531; Reetz, M. T. *Pure Appl. Chem.* **1992**, *64*, 351; Reetz, M. T. *Chem. Rev.*, in press.
7. Kano, S.; Yokomatsu, T.; Shibuya, S. *Tetrahedron Lett.* **1991**, *32*, 233.
8. Reetz, M. T.; Drewes, M. W.; Schmitz, A.; Holdgrün, X.; Wünsch, T.; Binder, J. *Philos. Trans. R. Soc. London A* **1988**, *326*, 573; Reetz, M. T. *Pure Appl. Chem.* **1988**, *60*, 1607.
9. Reetz, M. T.; Reif, W.; Holdgrün, X. *Heterocycles* **1989**, *28*, 707.
10. Reetz, M. T.; Wünsch, T.; Harms, K. *Tetrahedron: Asymmetry* **1990**, *1*, 371.
11. Reetz, M. T.; Schmitz, A.; Holdgrün, X. *Tetrahedron Lett.* **1989**, *30*, 5421; Reetz, M. T.; Fox, D. N. A. *Tetrahedron Lett.* **1993**, *34*, 1119.
12. Reetz, M. T.; Rivadeneira, E.; Niemeyer, C. *Tetrahedron Lett.* **1990**, *31*, 3863.
13. Reetz, M. T.; Drewes, M. W.; Harms, K.; Reif, W. *Tetrahedron Lett.* **1988**, *29*, 3295.
14. Reetz, M. T.; Binder, J. *Tetrahedron Lett.* **1989**, *30*, 5425.
15. Frenking, G.; Köhler, K. F.; Reetz, M. T. *Tetrahedron* **1993**, *49*, 3983.

16. For further examples of the use of N,N-dibenzylamino aldehydes see: Midland, M. M.; Afonso, M. M. *J. Am. Chem. Soc.* **1989**, *111*, 4368; Jurczak, J.; Golebiowski, A.; Raczko, J. *J. Org. Chem.* **1989**, *54*, 2495; Kano, S.; Yokomatsu, T.; Shibuya, S. *Tetrahedron Lett.* **1991**, *32*, 233; Hormuth, S.; Reissig, H. U. *Synlett* **1991**, 179; Ina, H.; Kibayashi, C. *Tetrahedron Lett.* **1991**, *32*, 4147; Klute, W.; Dress, R.; Hoffmann, R. W. *J. Chem. Soc., Perkin Trans. 2* **1993**, 1409; Yokomatsu, T.; Yamagishi, T.; Shibuya, S. *Tetrahedron: Asymmetry* **1993**, *4*, 1401; Rehders, F.; Hoppe, D. *Synthesis* **1992**, 859; Nagai, M.; Gaudino, J. J.; Wilcox, C. S. *Synthesis* **1992**, 163; DeCamp, A. E.; Kawaguchi, A. T.; Volante, R. P.; Shinkai, I. *Tetrahedron Lett.* **1991**, *32*, 1867; Mikami, K.; Kaneko, M.; Loh, T.-P.; Terada, M.; Nakai, T. *Tetrahedron Lett.* **1990**, *31*, 3909; Priepke, H.; Brückner, R.; Harms, K. *Chem. Ber.* **1990**, *123*, 555; Hormuth, S.; Reissig, H.-U.; Dorsch, D. *Angew. Chem.* **1993**, *105*, 1513; *Angew. Chem., Int. Ed. Engl.* **1993**, *32*, 1449; Cooke, J. W. B.; Davies, S. G.; Naylor, A. *Tetrahedron* **1993**, *49*, 7955; Grieco, P. A.; Moher, E. D. *Tetrahedron Lett.* **1993**, *34*, 5567; Stanway, S. J.; Thomas, E. J. *J. Chem. Soc., Chem. Commun.* **1994**, 285; Ipaktschi, J.; Heydari, A. *Chem. Ber.* **1993**, *126*, 1905; Ng, J. S.; Przybyla, C. A.; Liu, C.; Yen, J. C.; Muellner, F. W.; Weyker, C. L. *Tetrahedron* **1995**, *51*, 6397; Gennari, C.; Pain, G.; Moresca, D. *J. Org. Chem.* **1995**, *60*, 6248; Beaulieu, P. L.; Wernic, D.; Duceppe, J.-S.; Guindon, Y. *Tetrahedron Lett.* **1995**, *36*, 3317; Furuta, T.; Iwamura, M. *J. Chem. Soc., Chem. Commun.* **1994**, 2167; Gmeiner, P.; Kärtner, A. *Synthesis* **1995**, 83; Hanessian, S.; Devasthale, P. V. *Tetrahedron Lett.* **1996**, *37*, 987; Barluenga, J.; Baragaña, B.; Concellón, J. M. *J. Org. Chem.* **1995**, *60*, 6696; Hoffmann, R. W.; Klute, W. *Chem.-Eur. J.* **1996**, *2*, 694; O'Brien, P.; Warren, S. *Tetrahedron Lett.* **1996**, *37*, 4271; Beaulieu, P. L.; Wernic, D. *J. Org. Chem.* **1996**, *61*, 3635; Arrastia, I.; Lecea, B.; Cossìo, F. P. *Tetrahedron Lett.* **1996**, *37*, 245; Laib, T.;

Chastanet, J.; Zhu, J. *J. Org. Chem.* **1998**, *63,* 1709; Paquette, L. A.; Mitzel, T. M.; Isaac, M. B.; Crasto, C. F. Schomer, W. W. *J. Org. Chem.* **1997**, *62*, 4293; Hanessian, S.; Park, H.; Yang, R.-Y. *Synlett* **1997**, 351; Shibata, N.; Katoh, T.; Terashima, S. *Tetrahedron Lett.* **1997**, *38*, 619; Gennari, C.; Moresca, D.; Vulpetti, A.; Pain, G. *Tetrahedron* **1997**, *53*, 5593; O'Brien, P.; Powell, H. R.; Raithby, P. R.; Warren, S. *J. Chem. Soc., Perkin Trans. 1* **1997**, 1031; Andrés, J. M.; Barrio, R.; Martìnez, M. A.; Pedrosa, R.; Pérez-Encabo, A. J. *Org. Chem.* **1996**, *61*, 4210; Aggarwal, V. K.; Ali, A.; Coogan, M. P. *J. Org. Chem.* **1997**, *62*, 8628; Concellón, J. M.; Bernad, P. L.; Pérez-Andrés, J. A. *J. Org. Chem.* **1997**, *62*, 8902.

17. Jurczak, J.; Golebiowski, A. *Chem. Rev.* **1989**, *89*, 149.

18. Garner, P.; Park, J. M. *J. Org. Chem.* **1987**, *52*, 2361; Herold, P. *Helv. Chim. Acta* **1988**, *71*, 354; Casiraghi, G.; Colombo, L.; Rassu, G.; Spanu, P. *J. Chem. Soc., Chem. Commun.* **1991**, 603.

19. Vara Prasad, J. V. N.; Rich, D. H. *Tetrahedron Lett.* **1990**, *31*, 1803; Takemoto, Y.; Matsumoto, T.; Ito, Y.; Terashima, S. *Tetrahedron Lett.* **1990**, *31*, 217; Reetz, M. T.; Rölfing, K.; Griebenow, N. *Tetrahedron Lett.* **1994**, *35*, 1969.

20. Yoshida, K.; Nakajima, S.; Wakamatsu, T.; Ban, Y.; Shibasaki, M. *Heterocycles* **1988**, *27*, 1167.

21. Reetz, M. T.; Röhrig, D. *Angew. Chem.* **1989**, *101*, 1732; *Angew. Chem., Int. Ed. Engl.* **1989**, *28,* 1706; Reetz, M. T.; Wang, F.; Harms, K. *J. Chem. Soc., Chem. Commun.* **1991**, 1309; Reetz, M. T.; Lauterbach, E. H. *Tetrahedron Lett.* **1991**, *32*, 4477; Reetz, M. T.; Kayser, F. *Tetrahedron: Asymmetry* **1992**, *3*, 1377; Reetz, M. T.; Kayser, F.; Harms, K. *Tetrahedron Lett.* **1992**, *33*, 3453; Reetz, M. T.; Röhrig, D.; Harms, K.; Frenking, G. *Tetrahedron Lett.* **1994**, *35*, 8765.

22. Reetz, M. T.; Jaeger, R.; Drewlies, R.; Hübel, M. *Angew. Chem.* **1991**, *103,* 76; *Angew. Chem., Int. Ed. Engl.* **1991**, *30*, 103; Reetz, M. T.; Hübel, M.; Jaeger, R.; Schwickardi, R.; Goddard, R. *Synthesis* **1994**, 733.

23. Liu, C.; Ng, J. S.; Behling, J. R.; Yen, C. H.; Campbell, A. L.; Fuzail, K. S.; Yonan, E. E.; Mehrotra, D. V. *Org. Process Res. Dev.* **1997**, *1*, 45; see also Beaulieu, P. L.; LavallÈe, P.; Abraham, A.; Anderson, P. C.; Beucher, C.; Bousquet, Y.; Duceppe, J.-S.; Gillard, J.; Gorys, V.; Grand-MaÓtre, C.; Grenier, L.; Guindon, Y.; Guse, I.; Plamondon, L.; Soucy, F.; Valois, S.; Wernic, D.; Yoakim, C. *J. Org. Chem.* **1997**, *62*, 3440.

24. Banwell, M.; De Savi, C.; Hockless, D.; Watson, *K. Chem. Commun. (Cambridge)* **1998**, 645; Ina, H.; Kibayashi, C. *Tetrahedron Lett.* **1991**, *32*, 4147; Hoffmann, R. V.; Tao, J. *J. Org. Chem.* **1997**, *62*, 2292.

Appendix

Chemical Abstracts Nomenclature (Collective Index Number); (Registry Number)

S-2-(N,N-Dibenzylamino)-3-phenylpropanal: Benzenepropanal, α-[bis(phenylmethyl)amino]-, (S)- (12); (111060-64-1)

Lithium aluminum hydride: Aluminate (1-), tetrahydro-, lithium (8); Aluminate (1-) tetrahydro-, lithium, (I-4)- (9); (16853-85-3)

Benzyl (S)-2-(N,N-dibenzylamino)-3-phenylpropanoate: L-Phenylalanine, N,N-bis(phenylmethyl)-, phenylmethyl ester (12); (111138-83-1)

(S)-Phenylalanine: Alanine, phenyl-, L- (8); L -Phenylalanine (9); (63-91-2)

Benzyl bromide: Toluene, α-bromo- (8); Benzene, (bromomethyl)- (9); (100-39-0)

(S)-2-(N,N-Dibenzylamino)-3-phenyl-1-propanol: Benzenepropanol, β-[bis(phenylmethyl)amino]-, (S)- (12); (111060-52-7)

Benzyl alcohol (8); Benzenemethanol (9); (100-51-6)

Oxalyl chloride (8); Ethanedioyl dichloride (9); (79-37-8)

Dimethyl sulfoxide: Methyl sulfoxide (8); Methane, sulfinylbis- (9); (67-68-5)

Triethylamine (8); Ethanamine, N,N-diethyl- (9); (121-44-8)

(S)-2-Amino-3-phenyl-1-propanol: 1-Propanol, 2-amino-3-phenyl-, L- (8); Benzenepropanol, 2-amino-, (S)- (9); (3182-95-4)

METHYL (S)-2-PHTHALIMIDO-4-OXOBUTANOATE

(2H-Isoindole-2-acetic acid, 1,3-dihydro-1,3-dioxo-α-(2-oxoethyl)-, methyl ester, (S)-)

A. CH_3S–CH_2CH_2–CH($NH_3^+Cl^-$)–CO_2CH_3 → (phthalic anhydride; NEt_3, toluene) → CH_3S–CH_2CH_2–CH(NPhth)–CO_2CH_3

B. CH_3S–CH_2CH_2–CH(NPhth)–CO_2CH_3 → (1. NCS, CCl_4; 2. H_2O) → H–C(=O)–CH_2–CH(NPhth)–CO_2CH_3

Submitted by Patrick Meffre, Philippe Durand, and François Le Goffic.[1]
Checked by David A. Ellis, Yvan Mermet-Bouvier, and David J. Hart.

1. Procedure

A. Methyl (S)-2-phthalimido-4-methylthiobutanoate. Into a 2-L, round-bottomed flask fitted with a Dean-Stark apparatus, reflux condenser, and drying tube containing calcium chloride are placed L-methionine methyl ester hydrochloride (50.0 g, 0.25 mol, Notes 1 and 2), phthalic anhydride (37.1 g, 0.25 mol), triethylamine (100 mL, 0.72 mol), and toluene (1 L). The mixture is magnetically stirred and heated under reflux for 4.5 hr at which point approximately 4.5 mL of water has separated. The reaction mixture is allowed to cool to room temperature and the precipitated triethylamine hydrochloride (34 g) is collected by suction filtration. The filtrate is washed with four 300-mL portions of 1 N aqueous hydrochloric acid followed by three 300-mL portions of water. The organic layer is dried over magnesium sulfate, filtered, and the filtrate is concentrated under reduced pressure using a rotary evaporator. The residual oil is placed under reduced pressure for 12 hr at 0.1-0.5 mm, followed by trituration with 200 mL of pentane to give 59 g (80%) of product as a white solid after collection and drying at room temperature under reduced pressure (mp 37-40°C) (Note 3).

B. Methyl (S)-2-phthalimido-4-oxobutanoate. Caution! This reaction must be performed under a well-ventilated hood. Into a 500-mL, two-necked, round-bottomed flask fitted with a gas inlet and a drying tube containing calcium chloride are placed methyl (S)-2-phthalimido-4-methylthiobutanoate (15.45 g, 52.7 mmol) and carbon tetrachloride (CCl_4) (160 mL) (Note 4). The mixture is magnetically stirred under nitrogen at room temperature, N-chlorosuccinimide (7.04 g, 52.7 mmol) is added in one portion, and the reaction mixture is stirred for 2 hr at room temperature. The resulting mixture is filtered through a sintered glass funnel with suction into a 1-L, three-necked, round-bottomed flask, and the collected succinimide (5.2 g) is rinsed with 100 mL of carbon tetrachloride. The three-necked flask containing the filtrate and rinse is equipped with a gas inlet (Note 5), a stopper, and a cold water condenser. The condenser is equipped with a gas outlet that in turn is connected to a 500-mL Erlenmeyer flask containing 200 mL of sodium hypochlorite solution (commercial bleach) to scavenge volatile sulfur-containing by-products. Water (320 mL) is added in one portion, and nitrogen is bubbled through the solution at a rate of 2.4-3.0 mL/sec for a period of 20 hr at room temperature. The resulting phases are separated using a separatory funnel, and the acidic aqueous phase is extracted with dichloromethane (2 x 50 mL). The combined organic phases are washed with saturated aqueous sodium bicarbonate (100 mL), and water (100 mL) and then dried over sodium sulfate. The solution is filtered, and the filtrate is concentrated at reduced pressure using a rotary evaporator to give 14.0 g of crude product as a pale yellow oil (Note 6). Chromatography of this oil over silica gel (Note 7) gives a white solid that is triturated at -15°C (ice-salt bath) using 20 mL of diethyl ether to provide 7.50-9.19 g (54-67%) of analytically pure aldehyde as a white solid (mp 52-103°C) (Notes 8, 9 and 10).

2. Notes

1. Unless stated otherwise, all solvents and reagents were used as purchased (reagent grade) without further purification.

2. Commercial L-methionine methyl ester hydrochloride (from Aldrich Chemical Company, Inc.) is used or can be easily prepared using a literature procedure.[2]

3. The properties of methyl (S)-2-phthalimido-4-methylthiobutanoate follow: mp 37-40°C, lit.[3] mp 33-34°C; $[\alpha]_D^{20}$ -41.6° ($CHCl_3$, *c* 1.49), lit.[3] $[\alpha]_D^{20}$ -46.2° ($CHCl_3$, *c* 1); ^1H NMR (200 MHz, $CDCl_3$) δ: 2.02 (s, 3 H), 2.45-2.60 (m, 4 H), 3.69 (s, 3 H), 5.05 (m, 1 H), 7.73 (m, 2 H), 7.81 (m, 2 H); ^{13}C NMR (50.3 MHz, $CDCl_3$) δ: 15.2, 28.0, 30.7, 50.7, 52.8, 123.5, 131.7, 134.2, 167.5, 169.5; MS (CI, DCI-NH_3): 294 $(M+H)^+$, 311 $(M+NH_4)^+$. Anal. Calcd. for $C_{14}H_{15}NO_4S$: C, 57.33; H, 5.12; N, 4.78. Found: C, 57.29; H, 5.19; N, 4.72. This material was shown to have 96-98% ee based on HPLC analysis over a 25-cm x 0.46-cm Chiracel OJ column using hexane-isopropyl alcohol (98:2) as eluant and a flow rate of approximately 1 mL min^{-1}. The S-isomer elutes with a retention time of approximately 47 min, while the R-isomer has a retention time of approximately 66 min.

4. Carbon tetrachloride is stored over 4 Å molecular sieves before use. The chlorination is performed using oven-dried glassware quickly assembled before use.

5. The gas inlet consists of an 18-gauge needle passed through a rubber septum reaching to the bottom of the 1-L reaction vessel.

6. TLC analysis using silica gel 60-F254 (0.2-mm thick on aluminum sheets) and visualized under UV lamp or with ammonium molybdate-based reagent [36 mL of H_2O, 4 mL of concd H_2SO_4, 1 g of $(NH_4)_6Mo_7O_{24}\cdot 4H_2O$, and 0.4 g of $Ce(SO_4)_2$]

shows the presence of two spots with $R_f = 0.3$ and $R_f = 0.15$ (cyclohexane/ethyl acetate, 2:1).

7. The oil is dissolved in 20 mL of cyclohexane-ethyl acetate (3:1) and 10 mL of dichloromethane, applied to a 25-cm high x 8-cm wide column of Merck silica gel 60 (230-400 mesh), and eluted using flash chromatography.[4] The column is eluted with 4 L of cyclohexane-ethyl acetate (3:1) followed by 3 L of cyclohexane-ethyl acetate (1:1). After a forerun of 2.3 L is collected, early fractions containing the higher R_f material (Note 6) are obtained (2.2 g). This material is a mixture of starting material, alkenes **A** and **B**, thioacetal **C**, and other material. Elution of the aldehyde begins after about 4 L of eluant have been passed through the column.

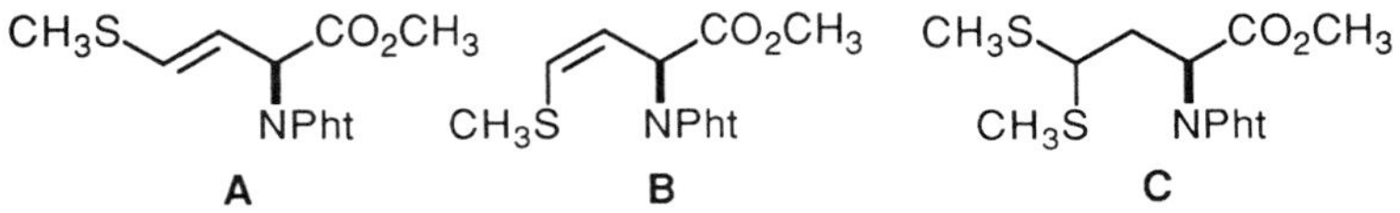

CH_3S, CO_2CH_3, NPht — **A**; CO_2CH_3, CH_3S, NPht — **B**; CH_3S, CO_2CH_3, CH_3S, NPht — **C**

8. The properties of methyl (S)-2-phthalimido-4-oxobutanoate follow: $[\alpha]_D^{20}$ -44.4° ($CHCl_3$, *c* 1.49); ^1H NMR (200 MHz, $CDCl_3$) δ: 3.26 (ddd, 1 H, J = 18.4, 7.7, 0.8), 3.55 (ddd, 1 H, J = 18.4, 6.0, 0.8), 3.74 (s, 3 H), 5.51 (dd, 1 H, J = 7.7, 6.0), 7.77 (m, 2 H), 7.90 (m, 2 H), 9.75 (t, 1 H, J = 0.8); ^{13}C NMR (50.3 MHz, $CDCl_3$) δ: 42.8, 45.9, 53.1, 123.6, 131.5, 134.3, 167.1, 168.8, 197.4 (CHO); MS (CI, DCI-NH_3): 262 $(M+H)^+$, 279 $(M+NH_4)^+$; IR (neat) cm^{-1}: 1720, 1750, 1780, 2720, 2820. Anal. Calcd. for $C_{13}H_{11}NO_5$: C, 59.77; H, 4.24; N, 5.36. Found: C, 59.47; H, 4.49; N, 5.32.

9. The (R)-aldehyde could be prepared from the hydrochloride salt of D-methionine methyl ester using a similar two-step procedure.

10. This material was analytically pure, but it melted over a broad range because of the presence of a mixture of enantiomers. This material was shown to have approximately 90-92% ee using the following procedure:[5] To a stirred solution

of (1R,2R)-(+)-N,N'-dimethyl-1,2-bis(3-trifluoromethyl)phenyl-1,2-ethanediamine (**D**) (50 mg, 0.13 mmol) in 10 mL of ether containing 1 g of 4 Å molecular sieves at room temperature under an argon atmosphere is added a solution of the (L)-aldehyde (35 mg, 0.13 mmol) in 10 mL of ether. The mixture is stirred at room temperture for 2 hr, whereupon TLC analysis (silica gel, cyclohexane-ethyl acetate, 2:1) indicates that the reaction is complete. The solution is filtered, and the filtrate is concentrated to give a mixture of diastereomeric imidazolidines (**E** and **F**) whose ratio is best determined by integration of the methine signals (^{1}H NMR in benzene-d_6 at 500 MHz) that appear at δ 3.68 in diastereomer **E** and at δ 3.94 in diastereomer **F**.

CF_3 CF_3 NHMe NHMe **D** —L-aldehyde→ CF_3 CF_3 Me N N Me CO_2Me R_1 R_2

E R_1=NHPhth; R_2=H
F R_1=H;R_2=NHPhth

Care should be taken to take NMR spectra with a pulse delay long enough to ensure that relaxation of all nuclei occurs. The checkers found that integration of other signals in the ^{1}H and ^{19}F spectra that were less well resolved gave lower % ee values. They also found that integration of the aforementioned signals on lower field instruments (200 and 300 MHz) gave lower % ee values because of partial overlap of signals.

Waste Disposal Information

All toxic materials were disposed of in accordance with "Prudent Practices in the Laboratory"; National Academy Press; Washington, DC, 1995.

3. Discussion

The synthetic utility of amino acids for the generation of chiral reagents, intermediates, and final products including amino acids is well documented.[6] Derivatives of (S)-2-amino-4-oxobutyric acid (aspartic acid β-semialdehyde, 3-formyl-alanine) are useful chiral intermediates for the synthesis of biologically relevant molecules of wide interest including: nicotinamine and analogues, iron chelating agents,[7] naturally occurring unusual α-amino acids,[8] serine-phosphate peptide isosteres,[9] penicillin or cephalosporin analogues,[10] and compounds used in some enzyme studies.[11] These compounds have been synthesized from the expensive allylglycine via ozonolysis or oxidative cleavage,[7a,7b,11a,11b] from aspartic acid via reduction of the acid side chain to give homoserine derivatives,[7c,8a,8b,9] from methionine via a homoserine derivative[8c] and subsequent oxidation to the aldehyde, or directly from the expensive homoserine.[7d,7e,8d,8e,10,11c] Direct reduction of reactive aspartic acid derivatives also led to the aldehyde.[8f,8g,8h] These strategies often suffer from low yields because of the number of steps or from the use of expensive or impractical starting materials or reagents. The synthesis of ethyl (S)-2-phthalimido-4-oxobutanoate has been described[12] from L-methionine using corrosive sulfuryl chloride as the chlorinating agent followed by a time-consuming hydrolysis, but partially racemized compounds are obtained, probably because of the harsh experimental conditions. The procedure described here provides material of greater than 90% ee using inexpensive and easily handled reagents.

In the procedure presented here, the phthaloyl group is chosen as an amino protecting group to avoid internal amine participation.[13] Amine diprotection is necessary since the same reaction conducted on the N-benzyloxycarbonyl (NHCbz) and the N-tert-butyloxycarbonyl (NHBoc) analogues did not led to clean α-chlorination but rather to unidentified products. This protection was performed using a standard procedure.[14]

The second step is a Pummerer-like reaction using easily handled N-chlorosuccinimide (NCS)[15,16] in carbon tetrachloride. α-Chlorination was chosen because of its regioselectivity towards an α-methylene compared with an α-methyl group.[17] NMR analysis of the crude mixture of chlorinated intermediates (mixture of isomers)[12,15] indicates that chlorination occurs mainly at the methylene group, although some vinyl sulfide products are also present at this stage of the reaction. Succinimide formed in the reaction is not soluble in CCl_4, and floats on the surface (unlike N-chlorosuccinimide) indicating completion of the reaction; it is easily separated by filtration. Hydrolysis of the transient chlorinated intermediates is then immediately performed using water and bubbling nitrogen to help remove methanethiol from the reaction mixture. Hydrolysis using $HgCl_2/CdCO_3/H_2O$,[18] $NCS/AgNO_3$/acetonitrile,[19] or $CuCl_2 \cdot 2H_2O/CuO/H_2O$/acetone[20] leads to lower yields.

1. Laboratoire de Bioorganique et Biotechnologies, Associé au C.N.R.S., ENSCP, 11, rue Pierre et Marie Curie, 75231 PARIS Cédex 05, FRANCE

2. Rachele, J. R. *J. Org. Chem.* **1963**, *28*, 2898.
3. Sato, Y.; Nakai, H.; Mizoguchi, T.; Kawanishi, M.; Hatanaka, Y.; Kanaoka, Y. *Chem. Pharm. Bull.* **1982**, *30*, 1263.
4. Still, W. C.; Kahn, M.; Mitra, A. *J. Org. Chem.* **1978**, *43*, 2923.
5. Cuvinot, D.; Mangeney, P.; Alexakis, A.; Normant, J.-F.; Lellouche, J.-P. *J. Org. Chem.* **1989**, *54*, 2420.

6. (a) Williams, R. M. "Synthesis of Optically Active α-Amino Acids"; Pergamon Press: Oxford, 1989; (b) Coppola, G. M.; Schuster, H. F. "Asymmetric Synthesis: Construction of Chiral Molecules Using Amino Acids"; Wiley-Interscience: New York, 1987; (c) Duthaler, R. O. *Tetrahedron* **1994**, *50*, 1539.

7. (a) Fushiya, S.; Sato, Y.; Nakatsuyama, S.; Kanuma, N.; Nozoe, S. *Chem. Lett.* **1981**, 909; (b) Fushiya, S.; Nakatsuyama, S.; Sato, Y.; Nozoe, S. *Heterocycles* **1981**, *15*, 819; (c) Faust, J., Schreiber, K., Ripperger, H. *Z. Chem.* **1984**, *24*, 330; (d) Oida, F.; Ota, N.; Mino, Y.; Nomoto, K.; Sugiura, Y. *J. Am. Chem. Soc.* **1989**, *111*, 3436; (e) Ohfune, Y.; Tomita, M., Nomoto, K. *J. Am. Chem. Soc.* **1981**, *103*, 2409.

8. (a) Ramsamy, K.; Olsen, R. K.; Emery, T. *Synthesis* **1982**, 42; (b) Knapp, S.; Hale, J. J.; Bastos, M.; Molina, A.; Chen, K. Y. *J. Org. Chem.* **1992**, *57*, 6239; (c) Baldwin, J. E.; Flinn, A. *Tetrahedron Lett.* **1987**, *28,* 3605; (d) Fushiya, S.; Maeda, K.; Funayama, T.; Nozoe, S. *J. Med. Chem.* **1988**, *31*, 480; (e) Keith, D. D.; Tortora, J. A.; Ineichen, K.; Leimgruber, W. *Tetrahedron* **1975**, *31*, 2633; (f) Hoffmann, M. G.; Zeiss, H.-J. *Tetrahedron Lett.* **1992**, *33*, 2669; (g) Wernic, D.; DiMaio, J.; Adams, J. *J. Org. Chem.* **1989**, *54*, 4224; (h) Ornstein, P. L.; Melikian, A.; Martinelli, M. J. *Tetrahedron Lett.* **1994**, *35*, 5759.

9. (a) Tong, G.; Perich, J. W.; Johns, R. B. *Tetrahedron Lett.* **1990**, *31*, 3759; (b) Valerio, R. M.; Alewood, P. F.; Johns, R. B. *Synthesis* **1988**, 786; (c) Perich, J. W. *Synlett* **1992**, 595.

10. (a) Baldwin, J. E.; Lowe, C.; Schofield, C. J.; Lee, E. *Tetrahedron Lett.* **1986**, *27,* 3461; (b) Saito, T.; Nishihata, K., Fukatsu, S. *J. Chem. Soc., Perkin Trans. I* **1981**, 1058.

11. (a) Tudor, D. W.; Lewis, T.; Robins, D. J. *Synthesis* **1993**, 1061; (b) Turner, N. J.; Whitesides, G. M. *J. Am. Chem. Soc.* **1989**, *111*, 624; (c) Chang, C.-D.; Coward, J. K. *J. Med. Chem.* **1976**, *19*, 684.

12. Dehmlow, E. V., Westerheide, R. *Synthesis* **1993**, 1225.

13. Lambeth, D. O.; Swank, D. W. *J. Org. Chem.* **1979**, *44*, 2632.

14. Bose, A. K.; Greer, F.; Price, C. C. *J. Org. Chem.* **1958**, *23*, 1335.

15. Meffre, P.; Lhermitte, H.; Vo-Quang, L.; Vo-Quang, Y.; Le Goffic, F. *Tetrahedron Lett.* **1991**, *32*, 4717.

16. (a) Wilson, Jr., G. E. *Tetrahedron* **1982**, *38*, 2597; (b) Dilworth, B. M.; McKervey, M. A. *Tetrahedron* **1986**, *42*, 3731.

17. (a) Tuleen, D. L.; Stephens, T. B. *J. Org. Chem.* **1969**, *34*, 31; (b) Wilson, Jr, G. E.; Albert, R. *J. Org. Chem.* **1973**, *38*, 2156 and 2160.

18. Paquette, L. A.; Klobucar, W. D.; Snow, R. A. *Synth. Commun.* **1976**, *6*, 575.

19. Corey, E. J.; Erickson, B. W. *J. Org. Chem.* **1971**, *36*, 3553.

20. Bakuzis, P.; Bakuzis, M. L. F.; Fortes, C. C.; Santos, R. *J. Org. Chem.* **1976**, *41*, 2769.

Appendix

Chemical Abstracts Nomenclature (Collective Index Number); (Registry Number)

Methyl (S)-2-phthalimido-4-oxobutanoate: 2H-Isoindole-2-acetic acid, 1,3-dihydro-1,3-dioxo-α-(2-oxoethyl)-, methyl ester, (S)- (12); (137278-36-5)

Methyl (S)-2-phthalimido-4-methylthiobutanoate: 2H-Isoindole-2-acetic acid, 1,3-dihydro-α-[2-(methylthio)ethyl]-1,3-dioxo-, methyl ester, (S)- (9); (39739-05-4)

L-Methionine methyl ester hydrochloride: Methionine, methyl ester, hydrochloride, L- (8); L-Methionine, methyl ester, hydrochloride (9); (2491-18-1)

Phthalic anhydride (8); 1,3-Isobenzofurandione (9); (85-44-9)

Triethylamine (8); Ethanamine, N,N-diethyl- (9); (121-44-8)

Toluene (8); Benzene, methyl- (9); (108-88-3)

Carbon tetrachloride: CANCER SUSPECT AGENT (8); Methane, tetrachloro- (9); (56-23-5)

N-Chlorosuccinimide: Succinimide, N-chloro- (8); 2,5-Pyrrolidinedione, 1-chloro- (9); (128-09-6)

Sodium hypochlorite solution: Hypochlorous acid, sodium salt (8,9); (7681-52-9)

CONVERSION OF NITRILES INTO TERTIARY AMINES: N,N-DIMETHYLHOMOVERATRYLAMINE

(Benzeneethanamine, 3,4-dimethoxy-N,N-dimethyl-)

A. (3,4-dimethoxyphenyl)acetonitrile + $(CH_3)_2NH$ —CuCl→ 2-(3,4-dimethoxyphenyl)-N,N-dimethylacetamidine

B. 2-(3,4-dimethoxyphenyl)-N,N-dimethylacetamidine —$NaBH_4$→ N,N-dimethylhomoveratrylamine

Submitted by Guilhem Rousselet,[1] Patrice Capdevielle,[2] and Michel Maumy.[2]
Checked by Sam Derrer and Andrew B. Holmes.

1. Procedure

A. 2-(3,4-Dimethoxyphenyl)-N,N-dimethylacetamidine. An oven-dried, 250-mL, single-necked, round-bottomed flask containing a 3-cm magnetic stirring bar is charged with 7.96 g (0.045 mol) of (3,4-dimethoxyphenyl)acetonitrile (Note 1) and 4.95 g (0.050 mol) of copper(I) chloride (Note 2), fitted with a septum, flushed with argon, and maintained under a static pressure of argon using a gas bubbler. Using a 20-mL gas-tight syringe, 45 mL (0.067 mol) of a 1.5 M ethanolic solution of dimethylamine (Note 3) is added successively in three 15-mL portions at room temperature with vigorous stirring. The heterogeneous pale brown mixture is then heated at 70°C for 24 hr, during which time it becomes brown-red. The mixture is cooled to room temperature and poured with vigorous stirring into a 250-mL Erlenmeyer flask containing 70 mL of aqueous 30% sodium hydroxide and 100 mL of diethyl ether (Note 4); the mixture was stirred vigorously for 3 min. The organic layer is separated, and the aqueous layer is extracted with three 75-mL portions of diethyl ether (Note 5). The combined organic extracts are dried over sodium sulfate and filtered through a sintered glass funnel layered with 2 cm of Celite. The solid residue is washed with diethyl ether (20 mL). The combined filtrate and washings are concentrated by rotary evaporation followed by drying under reduced pressure (0.1 mm) for 1.5 hr to provide 9.1-9.4 g (93-96% yield) of a dark brown oil (Notes 6,7). The compound is stored under argon at -18°C until used in step B.

B. N,N-Dimethylhomoveratrylamine. An oven-dried, 100-mL, Erlenmeyer flask is equipped with a magnetic stir bar, and charged with 9.05 g (0.041 mol) of 2-(3,4-dimethoxyphenyl)-N,N-dimethylacetamidine and 37 mL of methanol (Note 8). The solution is cooled to 0°C in an ice bath and 1.87 g (0.049 mol, 1.2 equiv) of sodium borohydride (Note 9) is added with stirring in portions of ca. 0.1 g. The solution is allowed to stand at room temperature for 4 hr, then poured with vigorous stirring into a 250-mL Erlenmeyer flask containing 50 mL of aqueous 30% sodium hydroxide and 100 mL of diethyl ether. The organic layer is separated, and the aqueous layer is extracted with three 75-mL portions of diethyl ether. The combined organic extracts are dried over sodium sulfate and filtered through a sintered glass funnel. The solid residue is washed with 20 mL of diethyl ether. The combined filtrate and washings are concentrated by rotary evaporation followed by drying under reduced pressure (0.3 mm) for 1 hr to provide 7.9-8.1 g (93-95% yield) of N,N-dimethylhomoveratrylamine as a yellow oil. NMR and GC analysis indicate a purity greater than 95% (Note 10).

2. Notes

1. (3,4-Dimethoxyphenyl)acetonitrile was obtained from Aldrich Chemical Company, Inc., and used without further purification.

2. Copper(I) chloride was obtained as a light green powder from Aldrich Chemical Company, Inc. (ref. Aldrich 21,294-6). The submitters obtained material with the same catalog number as sticks, which were carefully ground immediately before use.

3. The 1.5 M solution of dimethylamine in ethanol was prepared as follows: Aqueous 40% dimethylamine is heated at 65°C under a flow of argon. The gas is passed through potassium hydroxide pellets and blown across a known quantity (100 mL) of absolute ethanol (analytical grade used without purification), then cooled to 0°C in an ice bath, with continuous stirring, for 3 hr. The ethanolic solution is weighed and diluted to 150 mL with absolute alcohol. The ethanolic dimethylamine solution (ca. 1.5 M) is carefully capped and kept under an inert atmosphere at -18°C. The concentration may be measured by pouring a 1-mL aliquot of the ethanolic dimethylamine solution into aqueous 1 N hydrogen chloride. Water and ethanol are removed by rotary evaporation, and the residue is dried over potassium hydroxide pellets under vacuum (0.3 mbar, 2.5 mm) for one night. The weight of residue allows the exact concentration of the ethanolic dimethylamine solution to be measured.

With other less volatile amines (see Table I), the mixture of Cu(I)Cl, nitrile, and amine is simply refluxed in ethanol. In such a case, 1.1 equiv only of amine can be used.

4. Reagent grade diethyl ether may be used without further purification.

5. The phases were separated with difficulty because of the presence of copper salts. Some product adheres to the flask and separatory funnel.

6. The submitters carried out the drying at 0.3 mbar (2.5 mm) (1 hr) and reported 98% yield. Under these conditions the checkers observed residual ethanol (δ 1.13, t, 3 H J = 7; 3.58, q, 2 H, J = 7) in the ^1H NMR spectrum of the product. Most of the ethanol was removed after drying for the longer time at higher vacuum.

7. The product exhibits the following spectral properties : IR (film) ν cm^{-1}: 3380 (ν_{N-H}), 1590 ($\nu_{C=N}$); ^1H NMR (300 MHz, $CDCl_3$) δ: 2.96 (s, 6 H, N-C$\underline{H}_3$), 3.59 (s, 2 H), 3.84 (s, 3 H, 4-OC$\underline{H}_3$), 3.87 (s, 3 H, 3-OC$\underline{H}_3$), 5.50 (br s, 1 H, N$\underline{H}$), 6.78-6.87 (m, 3 H); ^{13}C NMR (75 MHz, $CDCl_3$) δ: 37.8 (2 x CH_3), 40.9 (CH_2), 55.7 (2 x CH_3), 111.2 (CH), 112.3 (CH), 121.6 (CH), 127.8 (Cq), 147.9 (Cq), 148.9 (Cq), 167.2 (Cq); Mass (EI) m/z 222 ($M^{+\bullet}$, 75%), 207 (50%), 177 (45%), 152 (50%), 151 (50%), 71 (100%). HRMS Calcd. for $C_{12}H_{18}N_2O_2$: 222.136826. Found: 222.136823. The level of ethanol contamination still remaining was evident from the elemental analysis: Calcd. for $C_{12}H_{18}N_2O_2$: C, 64.8; H, 8.2; N, 12.6. Found: C, 63.75; H, 8.2; N, 12.0 %.

8. Methanol was distilled from magnesium methoxide.

9. Sodium borohydride was obtained from Aldrich Chemical Company, Inc. and used without further purification.

10. The product exhibits the following spectral properties, in agreement with literature data:[3] IR (film) ν cm^{-1}: 2810 (ν_{N-Me}), 2750 (ν_{N-Me}); ^1H NMR (300 MHz, $CDCl_3$) δ: 2.28 (s, 6 H, N-C$\underline{H}_3$), 2.46-2.58 (m, 2 H, Ar-C$\underline{H}_2$), 2.70-2.82 (m, 2 H, N-C$\underline{H}_2$), 3.85 (s, 3 H, 4-OC$\underline{H}_3$), 3.89 (s, 3 H, 3-OC$\underline{H}_3$), 6.72-6.82 (m, 3 H); ^{13}C NMR (75 MHz, $CDCl_3$) δ: 34.0 (CH_2), 45.5 (2 x CH_3), 55.8 (CH_3), 55.9 (CH_3), 61.7 (CH_2), 111.3 (CH), 111.9 (CH), 120.5 (CH), 133.0 (Cq), 147.3 (Cq), 148.8 (Cq); Mass (EI) m/z 209 ($M^{+\bullet}$, 6%), 151 (8%), 58 (100%). HRMS. Calcd. for $C_{12}H_{19}NO_2$: 209.1417. Found: 209.1419. The checkers could not obtain satisfactory elemental analyses. The freshly prepared sample was >95% pure as shown by NMR. Further distillation (through a 5"-Vigreux column or Kugelrohr apparatus) afforded material, bp 96-98°C/0.1 mm, of >98% purity [Hewlett-Packard 5890 series II GC, capillary column SGE 25QC3BP5-

0.5, 5% phenylpolysiloxane, temperature gradient 100°C (1 min), 20°C/min, 250°C (5 min)] with 85% mass recovery. After 3 months storage the purity of material dropped to ca. 90%. The hydrochloride has mp 195-196°C (submitters reported 197°C) (from ethanol/diethyl ether) (lit.,[3] 196-197°C).

Waste Disposal Information

All toxic materials were disposed of in accordance with "Prudent Practices in the Laboratory"; National Academic Press; Washington, DC, 1995.

3. Discussion

The procedure described above illustrates a general, two-step method for the preparation of secondary or tertiary amines. It can be considered as a reductive N-alkylation of a nitrile or an N-monoalkylation of a primary or secondary amine. The first step in the procedure involves direct addition of an aliphatic amine to a nitrile promoted by a stoichiometric amount of cuprous chloride, as fully described recently.[4] This method may be used with a large variety of nitriles and primary or secondary aliphatic amines. The nitrile itself may be used as solvent (acetonitrile, benzonitrile). In the case of a primary amine, substrate stoichiometry must be adapted to obtain selectively either the N-monosubstituted amidine [1 eq amine, 1.2 eq Cu(I)Cl in acetonitrile] or the N,N-disubstituted amidine [4 eq amine, 1 eq Cu(I)Cl, 1 eq acetonitrile in alcohol or DMSO].[4]

As summarized in Table I, this strategy is applicable to the synthesis of many secondary or tertiary amines. It must, however, be noted that some arylamines are not sufficiently nucleophilic to be used in this synthesis. For example, anilines do not react under these conditions, although 2-aminopyridine does. The functionality in some amines is incompatible with copper; thus those amines that chelate with copper do not react with nitriles.

The amidine reduction by sodium borohydride is efficient, although longer reaction times (up to 15 hr) may be required depending on the amine used. The mode of decomposition of the intermediate geminal diamine is known to favor preferential expulsion of the less basic amine leaving group.[5] In fact, the alkylated amine was obtained with a selectivity higher than 80% (up to 99%) with all the compounds the submitters tested (see Table I).

The only comparable transformation of nitriles available in the literature is the reduction over palladium with excess dimethylamine in methanol. This reaction requires a large amount of catalyst (one-quarter or more of the weight of nitrile).[6]

The common synthetic route[3] to N,N-dimethylhomoveratrylamine involves acyl chloride formation from (3,4-dimethoxyphenyl)acetic acid with thionyl chloride (84%),[7] followed by amide formation with dimethylamine (99%),[8] and reduction with lithium aluminum hydride (71%).[9] The procedure provides N,N-dimethylhomoveratrylamine in 59% overall yield, requires three steps and more expensive substrates and reagents.

Another classical alternative is the Eschweiler-Clarke procedure, that involves drastic experimental conditions in which a solution of 88% aqueous formic acid, 35% aqueous formaldehyde and the corresponding primary amine in dimethylformamide is refluxed for 5 hr.[10] Moreover, only methylated amines can be prepared.

Compared with these methods, the reductive N-alkylation of nitriles is much more efficient and practical. The starting materials and reagents are cheaper, and only two steps are involved that proceed in a higher overall yield. Reductive N-alkylation affords, without any chromatographic separation, a product of high purity.

1. Guilhem Rousselet is a graduate student of the "Institut de Formation Supérieure Biomédicale" (IFSBM), Institut Gustave Roussy (Villejuif, France).
2. Laboratoire de Chimie Organique, URA CNRS 476, ESPCI, 10 rue Vauquelin, F-75231 Paris Cedex 05.
3. Bather, P. A.; Smith, L. J. R.; Norman, R. O. C. *J. Chem. Soc. (C)* **1971**, 3060-3068.
4. Rousselet, G.; Capdevielle, P.; Maumy, M. *Tetrahedron Lett.* **1993**, *34*, 6395-6398.
5. Moad, G.; Benkovic, S. J. *J. Am. Chem. Soc.* **1978**, *100*, 5495-5499.
6. Buck, J. S.; Baltzly, R.; Ide, W. S. *J. Am. Chem. Soc.* **1938**, *60*, 1789-1792.
7. Law, K.-Y.; Bailey, F. C. *J. Org. Chem.* **1992**, *57*, 3278-3286.
8. Stansbury, Jr., H. A.; Cantrell, R. F. *J. Org. Chem.* **1967**, *32*, 824-826.
9. Gallagher, T.; Magnus, P.; Huffman, J. C. *J. Am. Chem. Soc.* **1983**, *105*, 4750-4757.
10. Cherayil, G. D. *J. Pharm. Sci.* **1973**, *62*, 2054-2055.

TABLE I

SECONDARY AND TERTIARY AMINE SYNTHESES BY REDUCTIVE N-ALKYLATION OF NITRILE

Nitrile	Starting Amine	Final Amine	Selectivity of Step B	Overall Yield
CH_3CN	hexylamine	$CH_3CH_2NH(CH_2)_5CH_3$	95%	75%
CH_3CN	benzylamine	$CH_3CH_2NHCH_2C_6H_5$	85%	70%
CH_3CN	piperidine	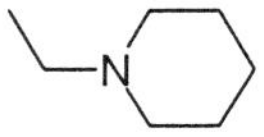 N	95%	90%
CH_3CN	morpholine	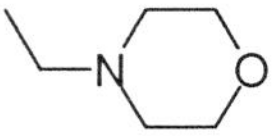 N O	95%	80%
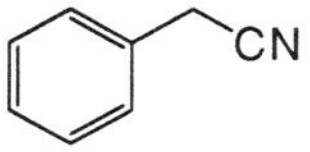 CN	dimethyl-amine	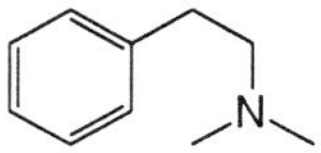 N	> 99%	95%
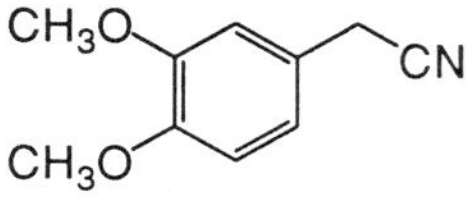 CH_3O CH_3O CN	dimethyl-amine	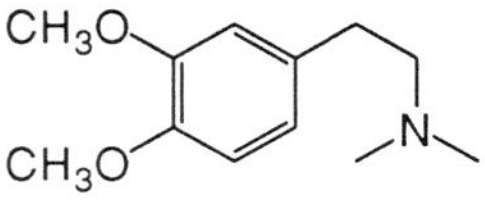 CH_3O CH_3O N	> 99%	95%
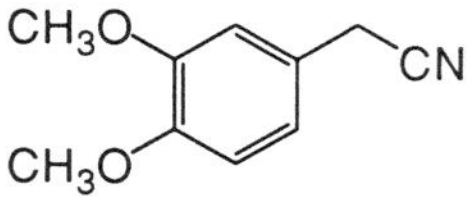 CH_3O CH_3O CN	piperidine	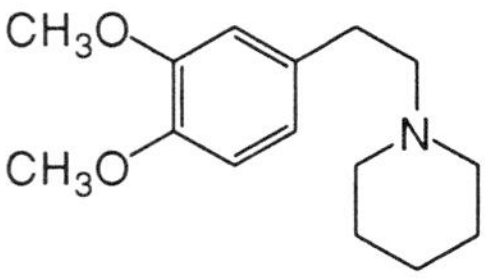 CH_3O CH_3O N	> 99%	93%
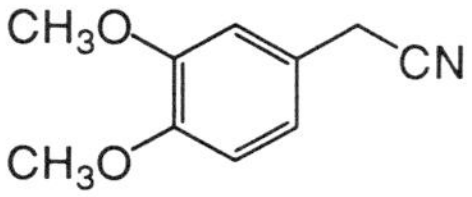 CH_3O CH_3O CN	morpholine	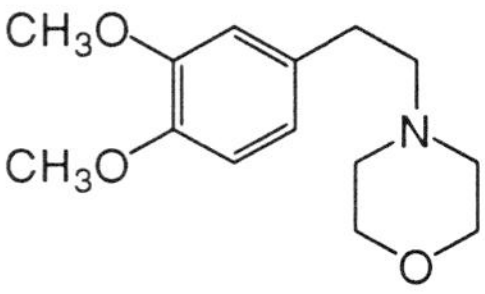 CH_3O CH_3O N O	80%	75%

Appendix

Chemical Abstracts Nomenclature (Collective Index Number); (Registry Number)

N,N-Dimethylhomoveratrylamine: Benzeneethanamine, 3,4-dimethoxy-N,N-dimethyl- (9); (3490-05-9)

(3,4-Dimethoxyphenyl)acetonitrile: Acetonitrile, (3,4-dimethoxyphenyl)- (8); Benzeneacetonitrile, 3,4-dimethoxy- (9); (93-17-4)

Copper(I) chloride: Copper chloride (8,9); (7758-89-6)

Dimethylamine (8); Methanamine, N-methyl- (9); (124-40-3)

Sodium borohydride: Borate (1-), tetrahydro-, sodium- (8,9); (16940-66-2)

ETHYL 5-CHLORO-3-PHENYLINDOLE-2-CARBOXYLATE

(1H-Indole-2-carboxylic acid, 5-chloro-3-phenyl-, ethyl ester)

A. 1 → 2 (ClCOCOOEt, Pyridine, CH_2Cl_2)

B. 2 → 3 ($TiCl_3$, Zn, DME)

Submitted by Alois Fürstner, Achim Hupperts, and Günter Seidel.[1]

Checked by Michael C. Hillier and Stephen F. Martin.

1. Procedure

Ethyl oxalyl chloride is a corrosive lacrymator and the reaction ***should be*** *carried out in a well-ventilated hood.*

A. N-(2-Benzoyl-4-chlorophenyl)oxanilic acid ethyl ester. A two-necked, round bottomed flask (500 mL) equipped with a Teflon-coated magnetic stirring bar, a gas inlet, and a dropping funnel is purged with argon. The flask is charged with 2-amino 5-chlorobenzophenone (13.9 g, 60 mmol, Note 1), dichloromethane (100 mL), and pyridine (20 mL, Note 2). A solution of ethyl oxalyl chloride (9.6 g, 70.3 mmol, Note 1 in dichloromethane (20 mL) is added dropwise through the addition funnel over a period of 45 min at 0°C (ice bath), and the resulting suspension is stirred for another 1.5 hr at ambient temperature. An aqueous saturated solution of sodium bicarbonate (40 mL) is added dropwise, and the biphasic system is stirred for 1.5 hr until evolution of gas ceases. The layers are separated, and the aqueous layer is extracted with dichloromethane (3 x 50 mL). The combined organic phases are washed with water (50 mL), dried (Na_2SO_4), filtered, and evaporated. The residual pyridine is removed by azeotropic distillation with toluene (3 x 100 mL) under reduced pressure on a rotary evaporator, and the residue is dried under vacuum (10^{-3} mm) to afford N-(2-benzoyl-4-chlorophenyl)oxanilic acid ethyl ester (19.6 g, 98%), that is directly used in the next step (Notes 3, 4).

B. Ethyl 5-chloro-3-phenylindole-2-carboxylate. An oven-dried, two-necked, round-bottomed, 500-mL flask equipped with a Teflon-coated magnetic stirring bar, a glass stopper, and a reflux condenser connected to the argon line is flushed with argon. The flask is charged with N-(2-benzoyl-4-chlorophenyl)oxanilic acid ethyl ester (13.27 g, 40 mmol), titanium(III) chloride ($TiCl_3$) (12.34 g, 80 mmol), zinc dust (10.45 g, 160 mmol, Note 5), and ethylene glycol dimethyl ether (DME) (250 mL, Note 6). The resulting suspension is heated at reflux for 2 hr with stirring, during which time a characteristic color change from violet ($TiCl_3$) to blue to black occurs (Note 7). The mixture is allowed to cool to ambient temperature and then slowly filtered through a short pad of silica on a sintered glass funnel. The inorganic residues are thoroughly washed with ethyl acetate (5 x 50 mL, Note 8), and the combined filtrates are concentrated to dryness on a rotary evaporator. For purification, the crude product is refluxed in toluene (120 mL, Note 9), and the resulting yellow solution is decanted from the oily residues. The product crystallizes upon standing at ambient temperature. The precipitated needles are collected on a funnel, washed with cold hexane (3 x 10 mL), and dried under reduced pressure to afford a first crop of ethyl 5-chloro-3-phenylindole-2-carboxylate as pale-yellow needles (9.3-9.7 g, 78-81%). Evaporation of the filtrate and recrystallization of the residue from toluene (20 mL) as described above gives a second fraction of product (0.8-1.4 g, 7-12%) (Note 10).

2. Notes

1. 2-Amino-5-chlorobenzophenone (98%) and ethyl oxalyl chloride (98%) were purchased from Aldrich Chemical Company, Inc., and used as received.

2. Dichloromethane (99.6%) was freshly distilled under argon from calcium hydride. Pyridine (99%+) was dried over activated molecular sieves (4 Å) and distilled under reduced pressure prior to use.

3. The crude product is ≥ 98% pure by GC. The checkers used the following conditions for GLC analysis: injector: 250°C, column: 90°C to 300°C, rate; 12°C/min.

4. The product has the following properties: mp 132-134°C (uncorrected), 135.9°C (differential scanning calorimetry, DSC), ^1H NMR (300 MHz, $CDCl_3$) δ: 1.42 (t, 3 H, J = 7.2), 4.42 (q, 2 H, J = 7.2), 7.45-7.76 (m, 7 H), 8.67 (d, 1 H, J = 9.6), 12.04 (br s, 1 H, -NH); ^{13}C NMR (75 MHz, $CDCl_3$) δ: 13.9, 63.6, 122.8, 125.5, 128.5, 129.0, 129.9, 132.7, 133.0, 133.8, 137.0, 137.4, 155.0, 160.1, 197.6.

5. $TiCl_3$ (Aldrich Chemical Company, Inc., 99%) and zinc dust (Aldrich Chemical Company, Inc., 98+%, <10 micron) were used as received. A substrate : $TiCl_3$ ratio of 1 : 2 is usually required to ensure quantitative conversions.

6. DME (Merck-Schuchardt, 99%+) was freshly distilled under argon from either sodium-potassium alloy or potassium/benzophenone prior to use.

7. Thorough mixing of the black suspension during reflux was neccessary to obtain optimal yields.

8. Occasionally the mixture must be passed through silica twice to obtain a clear filtrate.

9. Toluene (99%+) was distilled prior to use. Alternatively, the product can be recrystallized from ethyl acetate (50 mL) / hexane (350 mL).

10. The compound has the following properties: pale-yellow crystals, purity: $\geq$ 98% (GC); mp 169-172°C (uncorrected), 178°C (DSC); ^{1}H NMR (300 MHz, DMSO-d_6) δ: 1.21 (t, 3 H, J = 7.0), 4.28 (q, 2 H, J = 7.0), 7.34-7.61 (m, 8 H), 12.20 (br s, 1 H, -NH); ^{13}C NMR (75 MHz, DMSO-d_6) δ: 14.1, 60.7, 114.7, 119.6, 122.1, 124.5, 125.4, 127.3, 128.1, 130.6, 133.2, 134.8, 161.2.

Waste Disposal Information

All toxic materials were disposed of in accordance with "Prudent Practices in the Laboratory"; National Academy Press; Washington, DC, 1995.

3. Discussion

Low-valent titanium, formed from $TiCl_x$ (x = 3, 4) and various reducing agents, xhibits a high oxophilicity and a strong reducing ability. This particular combination f properties provides the driving force for the reductive coupling of carbonyl ompounds to alkenes. Generally referred to as the "McMurry olefin synthesis",[2] two nportant extensions to this reaction have recently been found:

(1) As far as starting materials are concerned, its scope has been significantly xpanded beyond aldehydes and ketones. Thus, a new approach to aromatic eterocycles such as furans, benzo[b]furans, pyrroles, and indoles has been devised, ased on the reductive cyclization of oxo-ester- or oxo-amide derivatives,[3-7] although mides have previously been considered inert towards activated titanium. This new ethod has already found applications in the syntheses of alkaloids as well as of harmaceutically active target molecules.[6]

(2) The experimental set-up of titanium-induced reductions has been considerably simplified. McMurry reactions were generally performed in two consecutive steps, consisting of the preparation of the active titanium slurry by reduction of $TiCl_x$ ($x = 3, 4$) with strong and potentially hazardous reducing agents (e.g., K, Li, Na, C_8K, $LiAlH_4$), followed by the addition of the respective carbonyl compound. The submitters have devised a shorter method, which relies upon the preparation of the active species *in the presence of the substrate* and the use of commercial zinc dust as the reducing agent.[5] Precoordination of the $TiCl_3$ to the carbonyl group thereby ensures a "site selective" formation of the coupling agent in this one-pot procedure.

The reductive cyclization of substrate **2** to ethyl 5-chloro-3-phenylindole-2-carboxylate, **3**, which is a known precursor for diazepam (Valium),[8] nicely illustrates these advancements. Although compound **2** bears four different reducible groups (amide, aryl chloride, ester, and ketone), the desired indole **3** is formed in a completely chemo- and regioselective way. This strong bias for a low-valent, titanium-promoted oxo-amide coupling is quite representative and renders the method compatible with an array of different functional groups including acetals, alkenes, alkyl chlorides, (remote) amides, aryl halides, (remote) esters, ethers, nitriles, cyclopropyl-, furanyl-, pyridyl-, thiazolyl-, trifluoromethyl-, and N-tosyl groups. Even free carboxylic acids and unprotected, remote ketone groups may be resistant to the reaction conditions.[4-7]

This one-pot procedure for titanium-induced reactions is also applicable to the synthesis of crowded products,[5] to completely chemo- and regioselective "zipper-type" polycyclizations,[7] to bimolecular reductions of alkynes,[5] and to conventional McMurry reactions of aldehydes or ketones.[5] Some representative examples are compiled in the Table.

1. Max-Planck-Institut für Kohlenforschung, Kaiser-Wilhelm-Platz 1, D-45470 Mülheim a. d. Ruhr, Germany.
2. (a) McMurry, J. E. *Chem. Rev.* **1989**, *89*, 1513; (b) Fürstner, A.; Bogdanovic', B. *Angew. Chem., Int. Ed. Engl.* **1996**, *35*, 2443.
3. Fürstner, A.; Jumbam, D. N. *Tetrahedron* **1992**, *48*, 5991.
4. Fürstner, A.; Jumbam, D. N. *J. Chem. Soc., Chem. Commun.* **1993**, 211.
5. Fürstner, A.; Hupperts, A.; Ptock, A.; Janssen, E. *J. Org. Chem.* **1994**, *59*, 5215.
6. (a) Fürstner, A.; Jumbam, D. N.; Seidel, G. *Chem. Ber.* **1994**, *127* 1125; (b) Fürstner, A.; Ernst, A. *Tetrahedron* **1995**, *51*, 773; (c) Fürstner, A.; Weintritt, H.; Hupperts, A. *J. Org. Chem.* **1995**, *60*, 6637; (d) Fürstner, A.; Ernst, A.; Krause, H.; Ptock, A. *Tetrahedron* **1996**, *52*, 7329.
7. Fürstner, A.; Ptock, A.; Weintritt, H.; Goddard, R.; Krüger, C. *Angew. Chem., Int. Ed. Engl.* **1995**, *34*, 678.
8. Yamamoto, H.; Inaba, S.; Hirohashi, T.; Ishizumi, K. *Chem. Ber.* **1968**, *101*, 4245.

Appendix

Chemical Abstracts Nomenclature (Collective Index Number); (Registry Number)

:thyl 5-chloro-3-phenylindole-2-carboxylate: Indole-2-carboxylic acid, 5-chloro-3-)henyl-, ethyl ester (8); 1H-Indole-2-carboxylic acid, 5-chloro-3-phenyl-, ethyl ester (9); 21139-32-2)

:thyl oxalyl chloride: Glyoxylic acid, chloro-, ethyl ester (8); Acetic acid, chlorooxo-, thyl ester (9); (4755-77-5)

N-(2-Benzoyl-4-chlorophenyl)oxanilic acid ethyl ester: Oxanilic acid, 2'-benzoyl-4'-chloro-, ethyl ester (8); Acetic acid, [(2-benzoyl-4-chlorophenyl)amino]oxo-, ethyl ester (9); (19144-20-8)

2-Amino-5-chlorobenzophenone: Benzophenone, 2-amino-5-chloro- (8); Methanone, (2-amino-5-chlorophenyl)phenyl- (9); (719-59-5)

Pyridine (8,9); (110-86-1)

Titanium(III) chloride (8,9); (7705-07-9)

Zinc (8,9); (7440-66-6)

Ethylene glycol dimethyl ether: Ethane, 1,2-dimethoxy- (8,9); (110-71-4)

TABLE

REDUCTIVE CYCLIZATIONS WITH LOW-VALENT TITANIUM REAGENTS

Substrate	Product	Yield
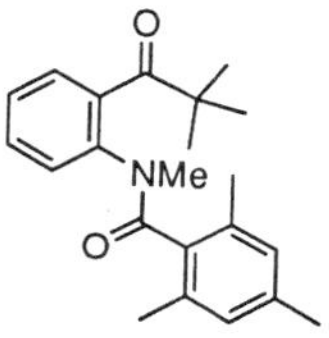 O; NMe; O	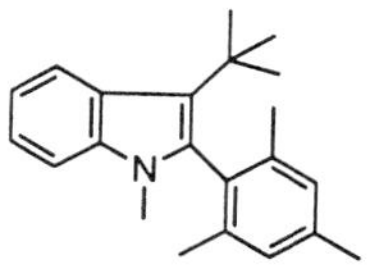 N	88%
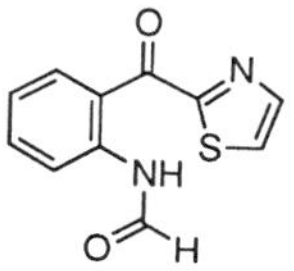 O; N; S; NH; O; H	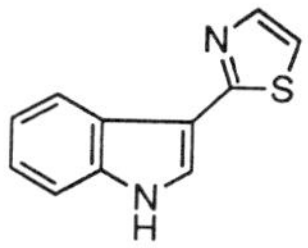 N; S; N; H	71%[a]
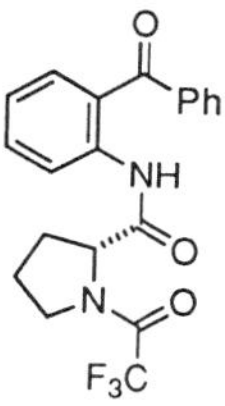 O; Ph; NH; O; N; O; F_3C	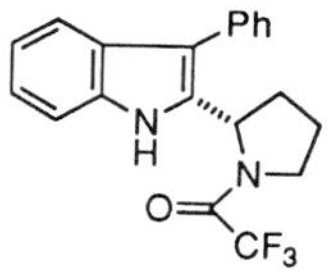 Ph; N; H; N; O; CF_3	90%[b]
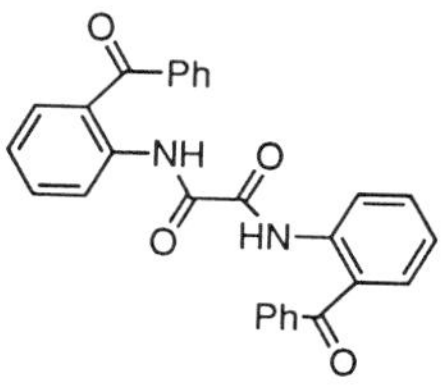 O; Ph; NH; O; O; HN; Ph; O	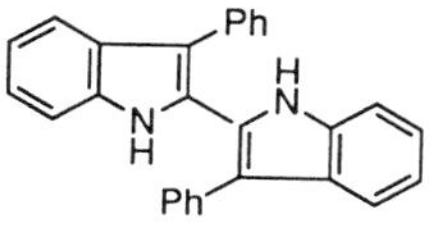 Ph; N; H; H; N; Ph	81%
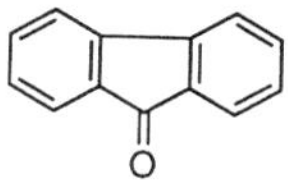 O	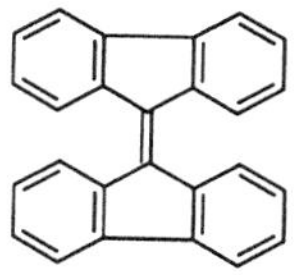	92%

TABLE contd.

REDUCTIVE CYCLIZATIONS WITH LOW-VALENT TITANIUM REAGENTS

Substrate	Product	Yield
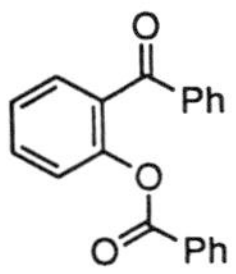	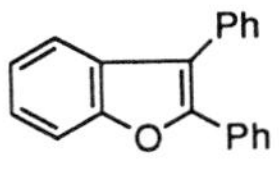	76%
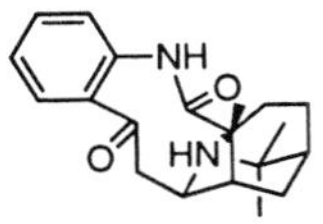	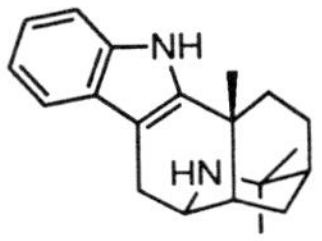	75%[a]

[a]After work-up with aq. EDTA; [b] ee (substrate) = ee (product) = 93%.

GENERATION AND USE OF LITHIUM PENTAFLUOROPROPEN-2-OLATE: 4-HYDROXY-1,1,1,3,3-PENTAFLUORO-2-HEXANONE HYDRATE

(2,2,4-Hexanetriol, 1,1,1,3,3-pentafluoro- from 1-Propen-2-ol, 1,1,3,3,3-pentafluoro-, lithium salt)

Submitted by Cheng-Ping Qian, Yu-Zhong Liu, Katsuhiko Tomooka, and Takeshi Nakai.[1]

Checked by Christine Kim, Peter Belica, and Steven Wolff.

1. Procedure

An oven-dried (Note 1), 300-mL, three-necked, round-bottomed flask containing a magnetic stirring bar, and fitted with a rubber septum inlet, a thermometer and a balloon full of nitrogen is charged with 100 mL of dry tetrahydrofuran (Note 2) and 6.3 mL (60 mmol) of hexafluoroisopropyl alcohol (Note 3). The flask and its contents are cooled to -70°C with a dry ice/acetone bath. After the apparatus has cooled, 76.9 mL of a 1.6 M hexane solution of butyllithium (123 mmol) is added dropwise using a 100-mL syringe over a period of 15 min. The reaction temperature is kept below -40°C (Note 4). After the addition is complete, the reaction mixture is further stirred for 10 min at -70°C and warmed to 0°C with an ice bath for 1 hr to afford lithium pentafluoropropen-2-olate, **1**, (Note 5). To the pale yellow solution is added dropwise

4.8 mL (66 mmol) of propionaldehyde (Note 6) via a syringe over 30 min, and the reaction mixture is stirred for 1 hr at 0°C. The flask is opened to the atmosphere, and about 65 mL of 1.0 M hydrochloric acid is added for neutralization. The quenched reaction solution is stirred for 10 min at room temperature and then transferred to a 500-mL separatory funnel. The aqueous layer is separated. The organic layer is diluted with 100 mL of ethyl acetate and washed with 150 mL of saturated brine. The combined aqueous material is extracted with two 75-mL portions of ethyl acetate. The organic extracts are combined, dried over anhydrous magnesium sulfate, filtered and evaporated. The yellow oily residue is further dried under reduced pressure at room temperature for several hours until crude crystals (about 12-14 g) are formed (Note 7). To the crude crystals is added 20 mL of hexane, and the resulting material is left at 5°C overnight. After the supernatant liquid is decanted, the wet crystals are washed further with two 20-mL portions of hexane followed by decantation and dried under reduced pressure to give a first crop of pure crystalline product. The hexane washings are concentrated and treated as described for the oily residue, giving a second crop of pure crystalline product (Note 8). The total yield is 8.64 g (64%) of 4-hydroxy-1,1,1,3,3-pentafluoro-2-hexanone hydrate (mp 74-78°C) (Note 9).

2. Notes

1. All the glassware was washed with acetone and oven-dried for a least 4 hr at 80°C, assembled hot, and allowed to cool under a nitrogen atmosphere.

2. Tetrahydrofuran was distilled under a nitrogen atmosphere from sodium/benzophenone prior to use.

3. Hexafluoroisopropyl alcohol was obtained from Central Glass Company Ltd., and used without additional purification. It is advisable to keep this reagent under a nitrogen atmosphere because it is highly hygroscopic.

4. During the addition of butyllithium, a temperature higher than -40°C must be avoided to prevent the formation of a by-product arising from further reaction of the lithium perfluoroenolate with butyllithium.

5. This perfluoro enolate displays the following spectroscopic data: ^{19}F NMR (56.45 MHz, THF, $CFCl_3$) δ: -68.5 (dd, 3 F, J = 28.6, 9.4, $C\underline{F}_3$), -109.8 (br d, 1 F, J = 84.7, $C\underline{F}_2$), -117.8 (br d, 1 F, J = 84.7, $C\underline{F}_2$).

6. Propionaldehyde was distilled prior to use.

7. Crystals are difficult to form unless the solvent is removed as completely as possible.

8. The ratio of the first and second crops, and the total amount of crystallized product depend mainly on whether the ethyl acetate is removed completely. If necessary, the second hexane washings can be treated again to give a third crop of product.

9. The spectral properties of the product are as follows: ^{1}H NMR (300 MHz, DMSO-d_6) δ: 0.94 (t, 3 H, J = 7.5, $C\underline{H}_3$), 1.45 (ddq, 1 H, J = 14.0, 9.9, 7.2, $C\underline{H}_2$), 1.76 (dddq, 1 H, J = 14.0, 2.7, 2.3, 7.6, $C\underline{H}_2$), 3.85-4.05 (m, 1 H, HOC$\underline{H}$), 5.96 (d, 1 H, J = 6.5, $\underline{H}$OCH), 7.54 [s, 1 H, C(O$\underline{H}$)$_2$], 7.92 [s, 1 H, C(O$\underline{H}$)$_2$]; ^{1}H NMR (300 MHz, $CDCl_3$) δ: 1.08 (t, 3 H, J = 7.4, $C\underline{H}_3$), 1.60-1.80 (m, 1 H, $C\underline{H}_2$), 1.80-2.05 (m, 1 H, $C\underline{H}_2$), 2.2-2.9 (br s, 1 H, O$\underline{H}$), 3.8-4.4 (br s, 1 H, O$\underline{H}$), 4.27 (dddd, 1 H, J = 21.7, 9.9, 3.1, 1.4, HOC$\underline{H}$), 5.4-6.3 (br s, 1 H, O$\underline{H}$); ^{13}C NMR (75 MHz, DMSO-d_6) δ: 9.9 ($\underline{C}H_3$), 22.4 ($\underline{C}H_2$), 71.0 (dd, J = 27.7, 22.7, HO$\underline{C}$H), 92.6 [m, $\underline{C}$(OH)$_2$], 118.5 (t, J = 256.5, $\underline{C}F_2$), 122.6 (q, J = 289.3, $\underline{C}F_3$); ^{19}F NMR (56.45 MHz, AcOEt, $CFCl_3$) δ: -80.0 (dd, 3 F, J = 13.5, 11.7, $C\underline{F}_3$), -120.0 (ddq, 1 F, $J_{F_A-F_B}$ = 275.3, J_{F_A-H} = 2.0, $J_{F_A-CF_3}$ = 13.5, $C\underline{F_A}F_B$), -132.0 (ddq, 1 F, $J_{F_B-F_A}$ = 275.3, J_{F_B-H} = 21.9, $J_{F_B-CF_3}$ = 11.7, $CF_A\underline{F_B}$); IR (neat film/NaCl plate) cm^{-1}: 3400, 2980, 1440, 1200, 1100, 1070, 980, 920, 850, 770, 750. Anal. Cacld for $C_6H_9F_5O_3$: C, 32.15; H, 4.05; F, 42.38. Found: C, 32.09; H, 3.96; F. 42.04.

Waste Disposal Information

All toxic materials were disposed of in accordance with "Prudent Practices in the Laboratory"; National Academy Press; Washington, DC, 1995.

3. Discussion

The present procedure describes an extremely convenient preparation of lithium perfluoropropenolate (**1**) and illustrates its use in organofluorine synthesis. The following scheme shows the wide spectrum of reactivity observed with lithium F-enolate **1**.[2-4] Several features of the F-enolate reactivity have been demonstrated: (a) The F-enolate exhibits the usual enolate reactivities, i.e., O- and C-nucleophilicity, to afford various classes of polyfluorinated compounds of types **2**~**4** that are otherwise difficult to obtain. (b) The F-enolate is capable of undergoing aldol reactions with various carbonyl compounds to afford adducts as hydrates **2**. (c) In reactions with reagents bearing an active hydrogen, the initial protonation occurs at the β-carbon, not at the oxygen, to yield products of type **4**. (d) Most significantly, the F-enolate exhibits a rather unusual electrophilic reactivity toward organometallic reagents to generate a geometric mixture of the β-alkyl F-enolates **5** via an addition-elimination process. This means that the F-enolate behaves as a kind of perfluoroolefin that is well-known to undergo a similar type of substitution reaction. Some examples of these reactions are shown in Tables I and II.

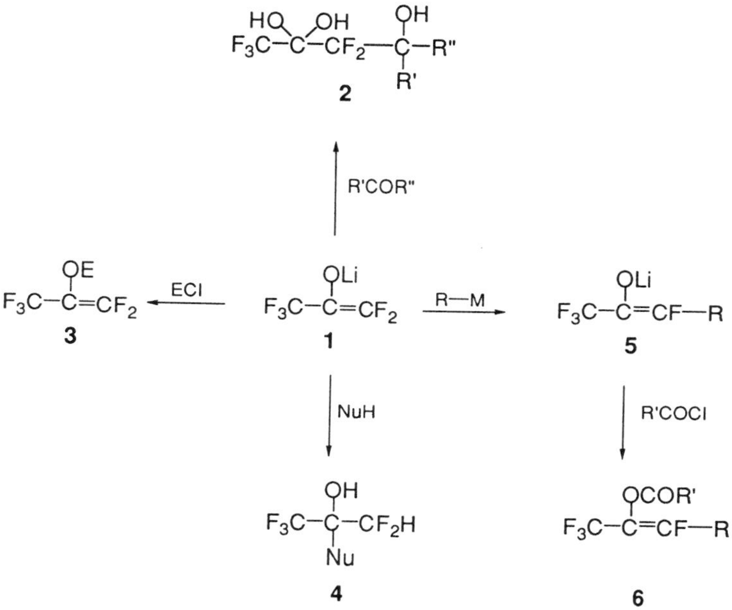

$F_3C-C(OH)_2-CF_2-C(OH)(R')-R''$

2

↑ R'COR''

$F_3C-C(OE)=CF_2$ ← ECl — $F_3C-C(OLi)=CF_2$ — R—M → $F_3C-C(OLi)=CF-R$

3 **1** **5**

1 ↓ NuH; 5 ↓ R'COCl

$F_3C-C(OH)(Nu)-CF_2H$ $F_3C-C(OCOR')=CF-R$

4 **6**

Finally, it should be noted that the present procedure is applicable to preparations of various lithium F-enolates such as **7** and **8** that exhibit different reactivity from that of **1**.[3,4]

$CF_3\sim CF=C(OLi)-CF_2CF_3$ **7**

$CF_3\sim CF=C(OLi)-CF_3$ **8**

1. Department of Chemical Technology, Tokyo Institute of Technology, Meguro-ku Tokyo, Japan.
2. Qian, C.-P.; Nakai, T. *Tetrahedron Lett.* **1988**, *29*, 4119.
3. Qian, C.-P.; Nakai, T.; Dixon, D. A.; Smart, B. E. *J. Am. Chem. Soc.* **1990**, *112* 4602.
4. Qian, C.-P.; Nakai, T. In "Selective Fluorination in Organic and Bioorgani Chemistry"; Welch, J. T., Ed.; American Chemical Society: Washington, DC 1991; ACS Symposium Series 456; p. 82.

Appendix
Chemical Abstracts Nomenclature (Collective Index Number); (Registry Number)

Lithium pentafluoropropen-2-olate: 1-Propen-2-ol, 1,1,3,3,3-pentafluoro-, lithium salt (12); (116019-90-0)

4-Hydroxy-1,1,1,3,3-pentafluoro-2-hexanone hydrate: 2,2,4-Hexanetriol, 1,1,1,3,3-pentafluoro- (12); (119333-90-3)

Hexafluoroisopropanol alcohol: 2-Propanol, 1,1,1,3,3,3-hexafluoro- (8,9); (920-66-1)

Butyllithium: Lithium, butyl- (8,9); (109-72-8)

Propionaldehyde (8); Propanal (9); (123-38-6)

TABLE I

REACTION OF LITHIUM PENTAFLUOROPROPEN-2-OLATE WITH ELECTROPHILES

Electrophile	Product (% Yield)	Electrophile	Product (% Yield)
EtCHO	$F_3C-C(OH)_2-CF_2-CH(OH)-C_2H_5$ (71)	Me_3SiCl	$F_3C-C(OSiMe_3)=CF_2$ (80)
PhCHO	$F_3C-C(OH)_2-CF_2-CH(OH)-C_6H_5$ (72)	$(MeO)_2SO_2$	$F_3C-C(OCH_3)=CF_2$ (72)
PhCOMe	$F_3C-C(OH)_2-CF_2-C(OH)(CH_3)-C_6H_5$ (82)	H_2O	$F_3C-C(OH)_2-CF_2H$ (86)
CF_3CHO	$F_3C-C(OH)_2-CF_2-CH(OH)-CF_3$ (87)	$PhCH_2OH$	$F_3C-C(OH)(OCH_2C_6H_5)-CF_2H$ (71)
$(C_2F_5)_2CO$	$F_3C-C(OH)_2-CF_2-C(OH)(C_2F_5)-C_2F_5$ (74)	$PhCONH_2$	$F_3C-C(OH)(NHCOC_6H_5)-CF_2H$ (74)
PhCOCl	$F_3C-C(OCOC_6H_5)=CF_2$ (99)	$CHF(CO_2Et)_2$	$F_3C-C(OH)(CF(CO_2Et)_2)-CF_2H$ (61)

TABLE II

REACTION OF LITHIUM PENTAFLUOROPROPEN-2-OLATE WITH NUCLEOPHILES

Nucleophile	Product[a]	Yield (%)
BuLi	$F_3C-C(OCOCF_3)=CF-C_4H_9$ (E/Z = 72/28)	71
PhLi	$F_3C-C(OCOCH_3)=CF-C_6H_5$ (E/Z = 77/23)	64
PhMgBr	$F_3C-C(OCOCH_3)=CF-C_6H_5$ (E/Z = 83/17)	48
Red-Al	$F_3C-C(OCOC_6H_5)=CF-H$ (E/Z = 43/57)	55

[a]Metal enolates **5** thus formed were isolated after acylation with an acyl chloride.

BROMOFLUORINATION OF ALKENES: 1-BROMO-2-FLUORO-2-PHENYLPROPANE

(Benzene, (2-bromo-1-fluoro-1-methylethyl))

NBS, $Et_3N \cdot 3\,HF$

CH_2Cl_2, 0→20°C

Submitted by Günter Haufe,[1] Gerard Alvernhe,[2] André Laurent,[2] Thomas Ernet,[1] Olav Goj,[1] Stefan Kröger,[1] and Andreas Sattler.[1]

Checked by Dudley W. Smith and Stephen F. Martin.

1. Procedure

1-Bromo-2-fluoro-2-phenylpropane (Note 1). A magnetically stirred mixture of α-methylstyrene (7.1 g, 60 mmol) (Note 2), triethylamine trihydrofluoride (Notes 3 and 4) (14.7 mL, 90 mmol) and dichloromethane (Note 5) (60 mL) contained in a 250 mL, single-necked, round-bottomed flask is treated with N-bromosuccinimide (11.8 g, 66 mmol) (Note 6) at 0°C. After 15 min, the bath is removed, and stirring is continued at room temperature for 5 hr (Note 7). The reaction mixture is poured into ice water (1000 mL), made slightly basic with aqueous 28% ammonia (Note 8), and extracted with dichloromethane (4 x 150 mL). The combined extracts are washed with 0.1 N hydrochloric acid (2 x 150 mL) and 5% sodium hydrogen carbonate solution (2 x 150 mL) and then dried over magnesium sulfate. After removal of the solvent by rotary evaporation, the crude product is distilled (Note 9) to give the product: 11.6 g (89%); bp 50-52°C (0.15 mm), n_D^{20} 1.5370 (Note 10).

2. Notes

1. Other 1-bromo-2-fluoro compounds that may be prepared following this procedure are listed in the Table.

2. The submitters have scaled this procedure up to 100 mmol for several alkenes in the Table and to a 400-mmol scale for 1-pentene.

3. Triethylamine trihydrofluoride is less corrosive than Olah's reagent[3] or anhydrous hydrogen fluoride itself, but all contact with the skin must still be avoided. The reagent has been tested for laboratory use only. The experiments should be done under an efficient hood.

4. Triethylamine trihydrofluoride[4] is an oily liquid that does not attack borosilicate glassware. The checkers purchased it from Aldrich Chemical Company Inc., but it is also available from Fluka Chemical Corp. and other suppliers.

5. Dichloromethane was dried over calcium hydride and distilled.

6. N-Bromosuccinimide was purchased from Aldrich Chemical Company, Inc. and used without purification; the purity of the compound is about 90%.

7. The reaction times for other olefins are given in the Table.

8. About 25-30 mL of aqueous 28% ammonia is necessary to make the solution slightly basic. If the aqueous layer is not made basic (pH 9-10), decomposition of the product is observed during distillation.

9. The distillation was performed using a 5-cm Vigreux column; there was no forerun. The product is somewhat sensitive to light and temperature.

10. Spectral data for the product were: ^{1}H NMR (300 MHz, $CDCl_3$) δ: 1.74 (3 H, d, $^{3}J_{HF}$ = 21.9, CH_3), 3.57 (1 H, $^{2}J_{AB}$ = 11.4, $^{3}J_{HF}$ = 22.8, CH_2Br), 3.61 (1 H, $^{2}J_{AB}$ = 11.4, $^{3}J_{HF}$ = 15.8, CH_2Br), 7.27 (m, 5 H, arom. H); ^{13}C NMR (75.5. MHz, $CDCl_3$) δ: 25.3 (d, $^{2}J_{CF}$ = 24.3, CH_3), 40.3 (d, $^{2}J_{CF}$ = 28.3, CH_2Br), 94.3 (d, $^{1}J_{CF}$ = 178.7, CF), 124.1 (d, $^{3}J_{CF}$ = 9.2, o-C), 128.0 (p-C), 128.3 (d, $^{4}J_{CF}$ = 1.1, m-C), 141.3 (d, $^{2}J_{CF}$

21.6, ipso-C); ^{19}F NMR (188 MHz, $CDCl_3$, $CFCl_3$) δ: -147.5 (m); (GC-MS (70 eV): m/z (%): 216/218 (9) [M^+], 196/198 (1) [M^+-HF], 123 (100) [M^+-CH_2Br]; HRMS, 217.0039 (Calcd for $C_9H_{10}BrF$, 217.0028).

Waste Disposal Information

All toxic materials were disposed of in accordance with "Prudent Practices in the Laboratory"; National Academy Press; Washington, DC, 1995.

3. Discussion

Fluorinated organic compounds are receiving increased interest because of their biological activity. One of the most useful methods for introducing single fluorine substituents into a molecule is by the halofluorination of unsaturated substrates. Although a number of reagents are available for effecting the bromofluorination of alkenes,[3] each suffers some disadvantage(s). The combination of N-bromosuccinimide, which is a source of electropositive bromine, and $Et_3N \cdot 3HF$ is a very convenient reagent for effecting the efficient bromofluorinations of various alkenes.[5]

The formal electrophilic addition of "BrF" to a double bond proceeds stereospecifically in an anti-sense as evidenced by the formation in high yields of trans-1-bromo-2-fluorocycloalkanes from cis-cycloalkenes or of cis-1-bromo-2-fluorocyclododecane from trans-cyclododecene, respectively.[5] The addition is regioselective with the observed regiochemistry being in accordance with the Markovnikov rule. For example, the bromofluorinations of α-substituted styrenes give the 1-bromo-2-fluoro-2-phenylalkanes with virtually complete regioselectivity; only traces (<1%) of the regioisomeric adducts were detectable by ^{19}F NMR spectroscopy of the crude reaction mixtures. In the bromofluorinations of other simple α-olefins such as 1-alkenes or allylbenzene, the Markovnikov products also predominate (9:1 to 19:1) over the corresponding anti-Markovnikov compounds. Very high regioselectivity has also been found for the bromofluorination of methallyl chloride and methallylphenyl ether,[6] whereas with methallylphenylthio ether and ω-unsaturated fatty acids,[7] the selectivity is about 9:1 favoring the Markovnikov product.

In unsaturated hydrocarbon systems where Wagner-Meerwein-type rearrangements, transannular hydrogen shifts or transannular π-participations are possible, such reactions do occur.[8] Moreover, in the bromofluorination of 9-oxabicyclo[6.1.0]non-4-ene, a transannular reaction involving oxygen participation has been observed.[9]

This method for bromofluorination of ethylenic compounds has been extended by others to symmetrical alkenes,[10] terminal allylic alcohols,[11] vinyl oxiranes,[12] enol esters,[13] and vinyl fluorides.[14]

Bromofluoro compounds are themselves useful starting materials for the preparation of monofluorinated compounds. For example, reduction of 1-bromo-2-fluorocyclododecanes with tributyltin hydride gives fluorocyclododecane, while other methods to produce this compound have been unsuccessful.[15] The elimination of hydrogen bromide from vicinal bromofluorides also gives vinyl fluorides in good yields,[10,16] as is illustrated by the recent synthesis of several substituted α-fluorostyrenes for studies related to [4+2]-cycloadditions.[17]

In other applications, the cyclizations of ω-bromo-(ω-1)-fluorocarboxylic acids by the intramolecular nucleophilic displacement of bromide ion by a carboxylate may be used for the syntheses of monofluorinated, medium-sized and large ring lactones.[7] 1-Acetoxy-2-fluoro-2-phenylalkanes, which were prepared by treating several of the 1-bromo-2-fluoro-2-phenylalkanes shown in the Table with acetate, have been used to prepare 2-fluoro-2-phenylalkanoic acids,[18] including several 2-fluorinated analogs of the "profen-family" of anti-inflammatory drugs.[19] Finally, several γ-fluoro-α-amino acids have been prepared in racemic[20] or optically active[21] form using 1-bromo-2-fluoroalkanes as alkylating agents.

1. Organisch-Chemisches Institut, Universität Münster, Corrensstraße 40, D-48149, Münster, Germany.
2. Laboratoire de Chimie Organique 3, Université Claude Bernard, Lyon I, 43 Bd du 11 Novembre 1918, F-69622 Villeurbanne, France.
3. (a) Olah, G. A.; Li, X. Y. In "Synthetic Fluorine Chemistry"; Olah, G. A.; Chambers, R. D.; Prakhash, G. K. S., Eds.; Wiley: New York, 1992, p. 163; (b) Sharts, C. M.; Sheppard, W. A. *Org. React.* **1974**, *21*, 125.
4. Franz, R. *J. Fluorine Chem.* **1980**, *15*, 423.
5. Alvernhe, G.; Laurent, A.; Haufe, G. *Synthesis* **1987**, 562.
6. Haufe, G.; Weßel, U.; Schulze, K.; Alvernhe, G. *J. Fluorine Chem.* **1995**, *74*, 283.
7. Sattler A.; Haufe G. *Tetrahedron* **1996**, *52*, 5469.
8. (a) Haufe, G.; Alvernhe, G.; Laurent, A. *Tetrahedron Lett.* **1986**, *27*, 4449; (b) Alvernhe, G.; Anker, D.; Laurent, A.; Haufe, G.; Beguin, C. *Tetrahedron* **1988**, *44*, 3551.
9. Haufe, G.; Alvernhe, G.; Laurent, A. *J. Fluorine Chem.* **1990**, *46*, 83.
10. Suga, H.; Hamatani, T.; Guggisberg, Y.; Schlosser, M. *Tetrahedron* **1990**, *46*, 4255.
11. Chehidi, I.; Chaabouni, M. M.; Baklouti, A. *Tetrahedron Lett.* **1989**, *30*, 3167.
12. Hedhli, A.; Baklouti, A. *J. Org. Chem.* **1994**, *59*, 5277.
13. Limat, D.; Guggisberg, Y.; Schlosser, M. *Liebigs Ann. Chem.* **1995**, 849.
14. Kremlev, M. M.; Haufe, G. *J. Fluorine Chem.* **1998**, *90*, 121.
15. Matsubara, S.; Matsuda, H.; Hamatani, T.; Schlosser, M. *Tetrahedron* **1988**, *44*, 2855.
16. Eckes, L.; Hanack, M. *Synthesis* **1978**, 217.
17. Ernet, T.; Haufe, G. *Tetrahedron Lett.* **1996**, *37*, 7251.

18. Goj, O.; Haufe, G. *Liebigs Ann. Chem.* **1996**, 1289.

19. Goj, O.; Kotila, S.; Haufe, G. *Tetrahedron* **1996**, *52*, 12761.

20. Kröger, S.; Haufe, G. *Amino Acids* **1997**, *12*, 363.

21. Kröger, S.; Haufe, G. *Liebigs Ann. Chem.* **1997**, 1201.

Appendix

Chemical Abstracts Nomenclature (Collective Index Number); (Registry Number)

1-Bromo-2-fluoro-2-phenylpropane: Benzene, (2-bromo-1-fluoro-1-methylethyl)- (9); (59974-27-5)

α-Methylstyrene: Styrene, α-methyl- (8); Benzene, (1-methylethenyl)- (9); (98-83-9)

Triethylamine trihydrofluoride: Ethanamine, N,N-diethyl-, trishydrofluoride (10); (73602-61-6)

N-Bromosuccinimide: Succinimide, N-bromo- (8); 2,5-Pyrrolidinedione, 1-bromo- (9); (128-08-5)

TABLE

BROMOFLUORINATION OF ALKENES WITH THE REAGENT COMBINATION N-BROMOSUCCINIMIDE/TRIETHYLAMINE TRISHYDROFLUORIDE

Substrate	Product (ratio of regioisomers)[1]	Reaction time	Tempe-rature	B.p. (mm)	Isolated yield (%)
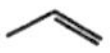	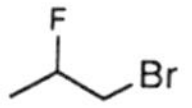 F Br (94:6)	15 hr	-78°C	72°C	70
	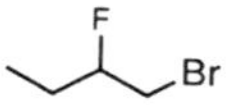 F Br (94:6)	15 hr	-20°C	75-77°C	73
	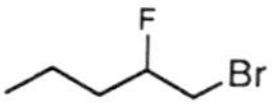 F Br (92:8)	15 hr	r.t.	50°C (18)	78
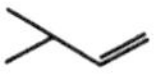	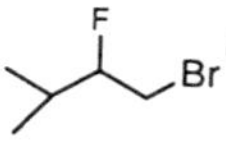 F Br (80:20)	15 hr	r.t.	42°C (18)	60
	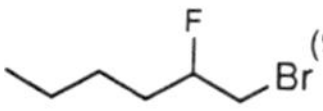 F Br (90:10)	15 hr	r.t.	55°C (18)	66
	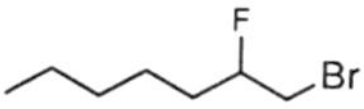 F Br (86:14)	15 hr	r.t.	70°C (18)	65
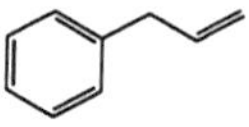	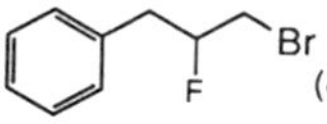 Br F (89:11)	5 hr	r.t.	69-70°C (0.8)	83
 HOOC ( )6	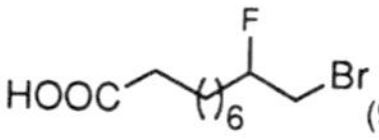 HOOC ( )6 F Br (90:10)	5 hr	r.t.	M.p. 49°C	91
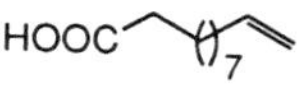 HOOC ( )7	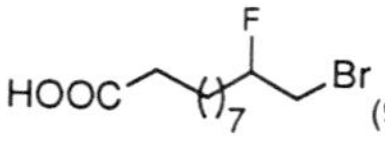 HOOC ( )7 F Br (92:8)	5 hr	r.t.	M.p. 59°C	95
	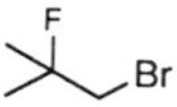 F Br	15 hr	r.t.	101-102°C	53

TABLE (contd.)

Substrate	Product	Reaction time	Tempe-rature	B.p. (mm)	Isolated yield (%)
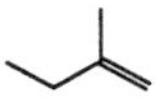	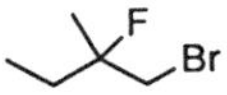 F, Br	15 hr	r.t.	90°C (18)	68
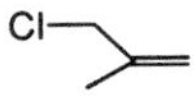 Cl	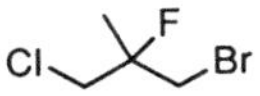 Cl, F, Br	18 hr	r.t.	57°C (12)	92
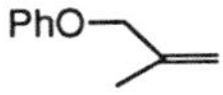 PhO	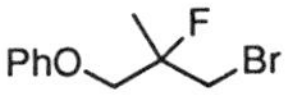 PhO, F, Br	12 hr	r.t.	n.d.[2]	88 (crude)
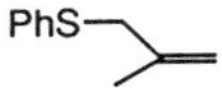 PhS	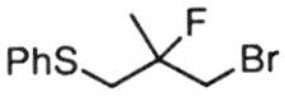 PhS, F, Br	12 hr	r.t.	n.d.[2]	61
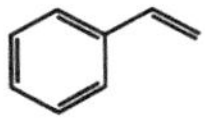	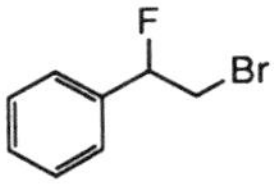 F, Br	5 hr	r.t.	n.d.[2]	85
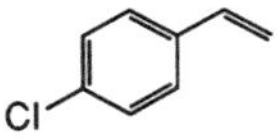 Cl	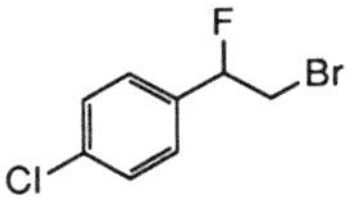 F, Br, Cl	18 hr	r.t.	n.d.[2]	63
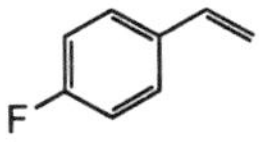 F	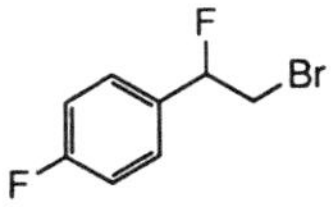 F, Br, F	18 hr	r.t.	n.d.[2]	63
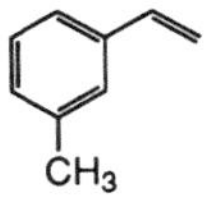 CH_3	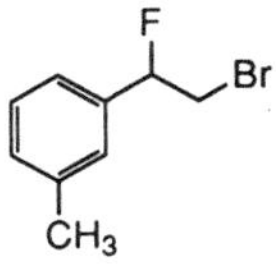 F, Br, CH_3	5 hr	r.t.	n.d.[2]	81
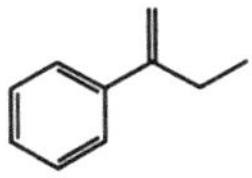	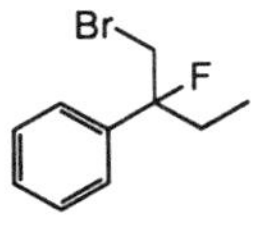 Br, F	14 hr	r.t.	n.d.[2]	94
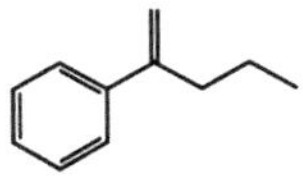	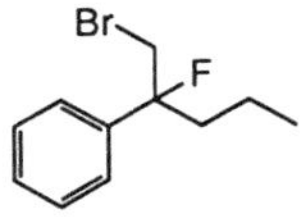 Br, F	14 hr	r.t.	M.p. 22°C	95
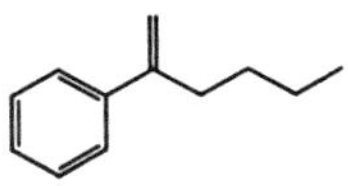	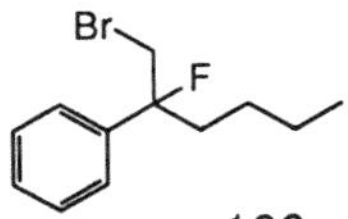 Br, F	14 hr	r.t.	n.d.[2]	93

TABLE (contd.)

Substrate	Product (ratio of regioisomers)[1]	Reaction time	Tempe-rature	B.p. (mm)	Isolated yield (%)
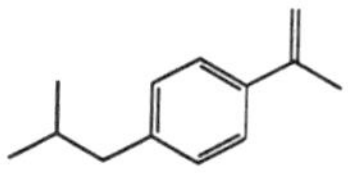	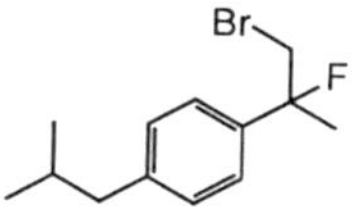 Br, F	14 hr	r.t.	n.d.[2]	80
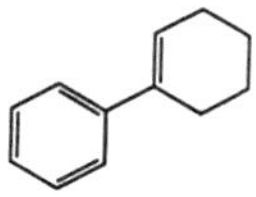	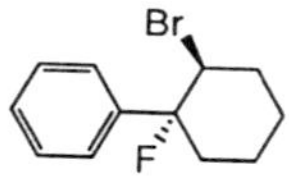 Br, F	15 hr	r.t.	M.p. 66°C	44
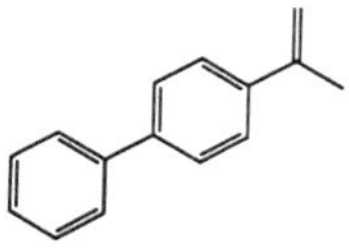	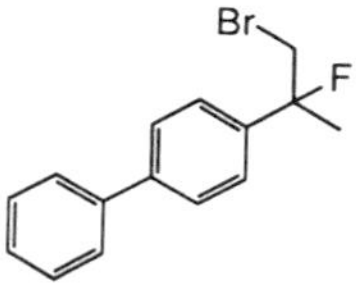 Br, F	14 hr	r.t.	M.p. 52°C	90
	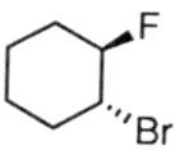 F, Br	5 hr	r.t.	76-77°C (18)	88
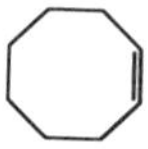	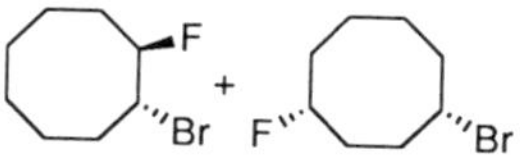 F, Br + F, Br (21 : 79)	5 hr	r.t.	49°C (0.09)	13 and 42[2]
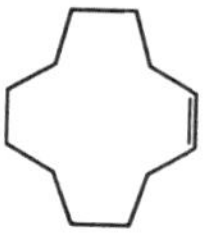	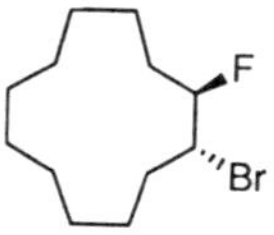 F, Br	5 hr	r.t.	79-80°C (0.1)	95
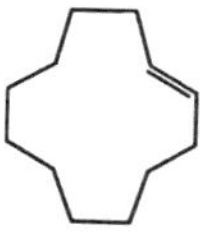	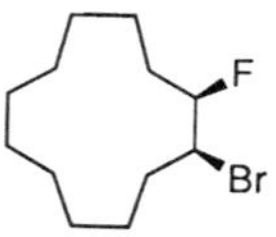 F, Br	5 hr	r.t.	82-83°C (0.1)	91

TABLE (contd.)

Substrate	Product	Reaction time	Tempe-rature	B.p. (mm)	Isolated yield (%)
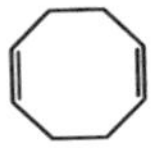	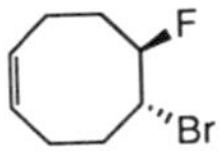	5 hr	r.t.	43°C (0.06)	71[4]
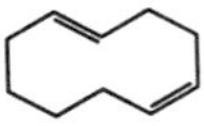	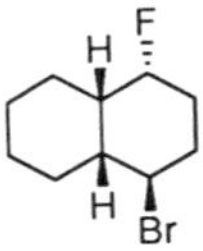	5 hr	r.t.	65-66°C (0.15)	78 (GC)[5]

[1]Determined by ^{19}F NMR; other regioisomers 2-bromo-1-fluoroalkanes.

[2]Isolated by column chromatography. Chromatography was done in a 20-cm glass column of 2-cm diameter with 25 g silica gel (70-260 mesh, Merck) per g of the bromofluoride using about 500 mL of cyclohexane/ethylacetate (9:1).

[3]In addition 18% of 5-bromocyclooctene was isolated.

[4]In addition two isomeric 2-bromo-6-fluoro-cis-bicyclo[3.3.0]octanes were formed (together 8%)

[5]Plus additional isomers (Ref. 8a).

MONO-C-METHYLATION OF ARYLACETONITRILES AND METHYL ARYLACETATES BY DIMETHYL CARBONATE: A GENERAL METHOD FOR THE SYNTHESIS OF PURE 2-ARYLPROPIONIC ACIDS. 2-PHENYLPROPIONIC ACID

(Benzeneacetic acid, α-methyl-)

CN
1 + $CH_3OCOOCH_3$ —(K_2CO_3, 180 °C)→ 2 + CH_3OH + CO_2
CN

2 —(1. 10% aq NaOH, reflux; 2. aq HCl)→ 3
COOH

Submitted by Pietro Tundo, Maurizio Selva, and Andrea Bomben.[1]

Checked by Peter Belica, Robert Koehler, and Steven Wolff.

1. Procedure

Caution! The reaction must be carried out in a pressure vessel because it occurs at temperatures >160°C, and dimethyl carbonate (DMC) boils at 90°C. An autoclave is recommended.

A stainless-steel (AISI 316) autoclave (internal volume 500 mL) equipped with a purging valve (Note 1), a pressure gauge, a thermocouple, and a magnetic stir bar (Note 2), is charged with a mixture of phenylacetonitrile (12.0 g, 0.10 mol), dimethyl carbonate (DMC) (147.8 g, 1.64 mol), and potassium carbonate (K_2CO_3) (28.4 g, 0.21 mol) (Note 3), and is heated in an electrical oven at 180°C. The reaction mixture is magnetically stirred (900 rpm) for 18 hr (Note 4) in the autoclave at 180°C.

The autoclave is removed from the oven and cooled to room temperature. After the coproduct, carbon dioxide, is released by the purging valve, the autoclave is opened, and the pale-yellow suspension is transferred to a 500-mL separatory funnel (Note 5). Water (120 mL) is added, and the mixture is extracted with diethyl ether (3 x 60 mL). The combined organic extracts are dried over sodium sulfate (25 g), which is then filtered off and washed with two 50-mL portions of diethyl ether.

The solvent and excess dimethyl carbonate are removed by rotary evaporation, and the remaining yellow liquid (13.2 g of **2**) is transferred to a 250-mL, round-bottomed flask equipped with a condenser. An aqueous solution of sodium hydroxide (10%, 60 mL) is added to the flask, and the mixture is heated with magnetic stirring in an oil bath at reflux temperature for 4.5 hr (Note 6). The course of the reaction is monitored by gas-chromatographic analysis (Note 7). After the solution is cooled to room temperature, it is extracted with diethyl ether to remove nonacidic material (primarily traces of amide). An aqueous solution of hydrochloric acid (15%, 50 mL) is added portionwise (Note 8). The suspension that forms is poured into a 250-mL separatory funnel and extracted with diethyl ether (3 x 50 mL). The combined organic extracts are washed with water (60 mL) and dried over sodium sulfate (25 g). After filtration, the solvent is removed by rotary evaporation, and the yellow liquid residue is distilled from a 25-mL flask, under reduced pressure, in a Claisen apparatus equipped with a 2-cm Vigreux column and a fused-on Liebig condenser. Distillation affords 2-phenylpropionic acid (**3**) (14.3 g, 93%) as a pale-yellow liquid, bp 93-94°C/0.9 mm (lit.[3] bp 147°C/11 mm), with > 98% purity by gas chromatography (Note 9).

2. Notes

1. At room temperature, a stream of nitrogen (about 400 mL/min for 3 min) is admitted through the purging valve to remove air before the reaction. The autoclave used by the checkers was charged, then pressurized with nitrogen and vented three times before heating was initiated.

2. The thermocouple is inserted into a 1/8-in stainless-steel pipe (fixed on the autoclave head) that dips into the reaction mixture. The magnetic stir bar is 60 x 6 mm (length x diameter). The checkers used a 500-mL, electrically heated, stainless steel autoclave equipped with a mechanical stirrer, a thermocouple, and a sampling tube.

3. DMC is used in a large excess (15 molar excess with respect to the substrate) acting both as the methylating agent and the solvent. Previous investigations have been shown that DMC as solvent provides a suitable polar-aprotic environment for the reaction and that it may be totally recovered (by distillation) after the reaction.[2] Potassium carbonate may also be used in catalytic amounts (5% molar with respect to the substrate), but longer reaction times result.[2]

4. Autogenic pressure reaches 12 bar (900 mm). Because of the presence of solid K_2CO_3, vigorous stirring is needed. Actually, the time required for complete substrate conversion is highly dependent on the rate of stirring of the mixture which, in turn, depends on the shape of the reaction vessel. Reaction times for the checkers' runs varied between 5 and 6.5 hr. If the reaction is not monitored by gas chromatography, a substantial amount of the undesired dimethylated product may form by further reaction of 2-phenylpropionitrile (**2**).

5. A sample of a few drops of the mixture is diluted with diethyl ether (2 mL), washed with water (2 mL), and analyzed by GC and GC/MS (DB5 capillary column, 30-m x 0.25-mm i. d., 0.25 μm film thickness). At a conversion of up to 99%, the yield

of monomethyl derivative **2** (2-phenylpropionitrile) is 98.5% while the dimethylated product is 0.15%.

6. The temperature of the oil bath is 130°C. The initially observed biphasic (aqueous-organic) reaction system gradually turns to a homogeneous solution as the reaction proceeds.

7. At intervals (1 hr), small aliquots (0.2-0.3 mL) of the reaction mixture are withdrawn, acidified with a few drops of concd HCl (35%), and extracted with diethyl ether (2 mL). Then they are analyzed by GC (DB5 capillary column); both the amide and the acid derived from the reacting nitrile are present. After 4.5 hr, the product is 99% acid **3** (2-phenylpropionic acid). The checkers followed the hydrolysis by TLC analysis: SiO_2 plates; 4:1 hexane:ethyl acetate; short-wave UV detection; R_f (acid) = 0.43, R_f (amide) = 0.61.

8. Aliquots (5-7 mL) of the HCl solution are added in 10 min. The final pH of the mixture is about 2.

9. The product shows the following spectroscopic properties: ^1H NMR (400 MHz, $CDCl_3$, TMS) δ: 1.51 (d, 3 H, J = 7.2, CH_3), 3.73 (q, 1 H, J = 7.2, CH), 7.31-7.33 (m, 4 H, Ph), 11.5-11.7 (br s, 1 H, OH); ^{13}C NMR (100 MHz, $CDCl_3$, TMS) δ: 18.02, 45.36, 127.24, 127.56, 128.64, 139.68, 181.12; mass spectrum (70 eV) m/z (relative intensity): 150 (M^+, 29), 106 (11), 105 (100), 104 (5),103 (12), 79 (14), 78 (6), 77 (17), 51 (7). Anal. Calcd for $C_9H_{10}O_2$: C, 71.98; H, 6.71. Found: C, 71.77; H, 6.69.

Waste Disposal Information

All toxic materials were disposed of in accordance with "Prudent Practices in the Laboratory"; National Academy Press; Washington, DC, 1995.

3. Discussion

2-Phenylpropionic acid is the simplest member of the 2-arylpropionic acids class to which a number of widely used antiinflammatory drugs belongs. Important examples of non-steroidal analgesics are 2-(p-isobutylphenyl)-, 2-(m-benzoylphenyl)- and 2-(6-methoxy-2-naphthyl)propionic acids well known commercially as Ibuprofen, Ketoprofen, and Naproxen, respectively.

Among the different synthetic procedures available for the preparation of hydratropic acids (e.g., indirect methylation of arylacetic acids,[4] asymmetric hydroformylation of styrenes,[5] rearrangements of α-bromoalkyl aryl ketals,[6] etc.), direct methylation of arylacetic acid derivatives seems the most attractive from both economical and synthetic aspects: the reagents are easily accessible and a one-pot reaction is involved. Nevertheless, this procedure is seldom used since the yields of the monomethyl derivatives are severely limited by the low selectivity of the reaction. Sizeable amounts of dimethylated by-products form.[2] Even under phase-transfer catalysis conditions, high selectivity in monomethylation is elusive.[7-8]

The procedure described here overcomes this difficulty by using dimethyl carbonate as a methylating agent.[9] In this case, although high reaction temperatures (> 180°C) are required, the methylation of CH_2-acidic compounds by DMC proceeds with a selectivity up to 99% toward the monomethyl derivatives, at complete conversion. The present reaction has been successfully carried out on a number of arylacetonitriles and methyl arylacetates[10] both on the multigram- (Table) and the gram-scale [2-(o-, m- and p-methoxyphenyl)-, 2-(o- and p-methylphenyl)-, 2-(p-fluorophenyl)-, 2-(p-chlorophenyl)-, 2-(m-carboxymethylphenyl)-propionitriles; methyl 2-phenylpropionate].[2]

Mechanistic investigations of the reaction reported here suggest that the observed high selectivity may occur as a result of the double reactivity that DMC may exhibit.[2] It acts first as a methoxy carbonylating reagent (via a $B_{Ac}2$ mechanism) and then as a methylating agent (via a $B_{Al}2$ mechanism), as shown in Scheme 1.

Scheme 1

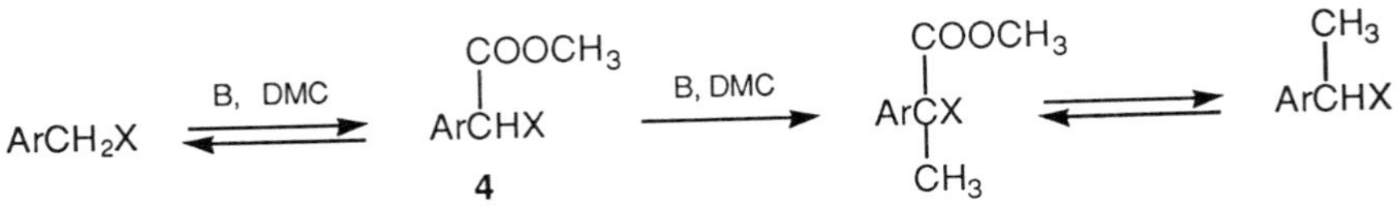

$$ArCH_2X \underset{}{\overset{B,\ DMC}{\rightleftharpoons}} \underset{\mathbf{4}}{ArCH(COOCH_3)X} \xrightarrow{B,\ DMC} ArC(COOCH_3)(CH_3)X \rightleftharpoons ArCH(CH_3)X$$

X = CN, $COOCH_3$; B = base

Selectivity arises from the fact that the reaction occurs only via the methoxy carbonylated intermediate (**4**) and no direct methylation takes place on the $ArCH_2X$.

Beside the synthetic benefits of the procedure, the method also represents a true example of the *Green Chemistry* concept intended as a new approach to synthesis, processing, and use of chemicals that reduces risks to the health and the environment.[11] In fact DMC (now prepared by oxidative carbonylation of methanol[12]) is an innocuous methylating agent compared to the toxic methyl halides or dimethyl sulfate. Moreover, methylation processes by DMC are intrinsically environmentally benign in that neither organic by-products nor inorganic salts originate; conversely, alkylations by alkyl halides unavoidably lead to stoichiometric amounts of inorganic salts that must be disposed of.

1. Dipartimento di Scienze Ambientali dell'Università di Venezia, Calle Larga S. Marta 2137, I-30123 Venezia, Italy
2. Selva, M.; Marques, C. A.; Tundo, P. *J. Chem. Soc., Perkin Trans. I*, **1994**, 1323.

3. Dictionary of Organic Compounds, Chapman and Hall: New York, 5th ed., Vol. 5, p. 4662, 1982; 6th ed., Vol. 5, p. 5313, 1996.
4. Rieu, J.-P.; Boucherle, A.; Cousse, H.; Mouzin, G. *Tetrahedron* **1986**, *42*, 4095-4131.
5. Parrinello, G.; Stille, J. K. *J. Am. Chem. Soc.* **1987**, *109*, 7122.
6. Castaldi, G.; Giordano, C.; Uggeri, F. *Synthesis* **1985**, 505.
7. Starks, C. M.; Liotta, C. "Phase-Transfer Catalysis: Principles and Techniques"; Academic Press New York, 1978; Chapter 5, pp. 170-196.
8. Mikolajczyk, M.; Grzejszczak, S.; Zatorski, A.; Montanari, F.; Cinquini, M. *Tetrahedron Lett.* **1975**, 3757.
9. Tundo, P.; Selva, M. *CHEMTECH* **1995**, *25,* 31.
10. In particular, the methylation of esters by DMC requires higher temperatures (200-220°C) than those of nitriles (180°C). This behavior has been further confirmed in the methylation of aryloxyacetonitriles and methyl aryloxyacetates by DMC: Bomben, A.; Marques, C. A.; Selva, M.; Tundo, P. *Tetrahedron* **1995**, *51*, 11573.
11. Anastas, P. T.; Williamson, T. In "Green Chemistry: Designing Chemistry for the Environment"; Anastas, P. T.; Williamson, T., Eds.; ACS Symposium Series 626, American Chemical Society: Washington, D.C. 1996, 1-17.
12. Romano, U.; Rivetti, F.; Di Muzio, N. DE Patent 3 045 767, 1981; *Chem. Abstr.* **1981**, *95*, 80141w. See also U.S. Patent 4 318 862, 1979.

Appendix

Chemical Abstracts Nomenclature (Collective Index Number); (Registry Number)

Dimethyl carbonate: Carbonic acid, dimethyl ester (8,9); (616-38-6)

Phenylacetonitrile: Aldrich: Benzyl cyanide: Acetonitrile, phenyl- (8); Benzeneacetonitrile (9); (140-29-4)

2-Phenylpropionitrile: Hydratroponitrile (8); Benzeneacetonitrile, α-methyl- (9); (1823-91-2)

2-Phenylpropionic acid: Hydratropic acid (8); Benzeneacetic acid, α-methyl- (9); (492-37-5)

Table

Monomethylation of Arylacetonitriles and Methyl Arylacetates with Dimethyl Carbonate. [a]
Hydrolysis of 2-Arylpropionitriles and Methyl 2-Arylpropionates to 2-Arylpropionic Acids. [b]

Entry	Substrate	Reaction Time / hr[c]	Temp. (°C) [c]	Conv'n (%) [c]	Product	Yield [d] (%)
1	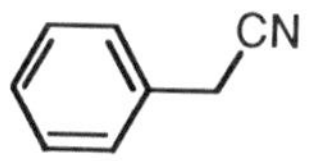 CN	18	180	> 99	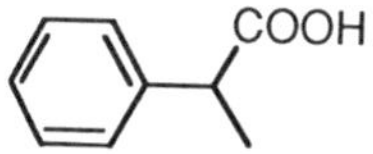 COOH	93
2	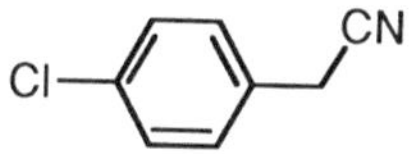 Cl, CN	8.5	180	> 99	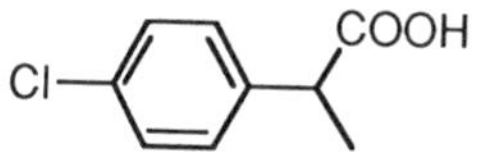 Cl, COOH	94
3	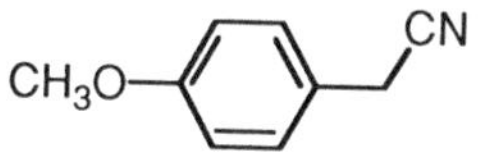 CH_3O, CN	9	180	> 99	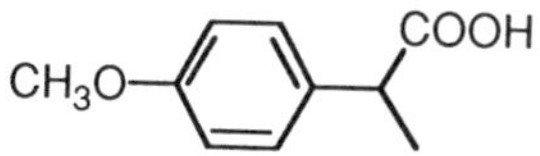 CH_3O, COOH	95
4	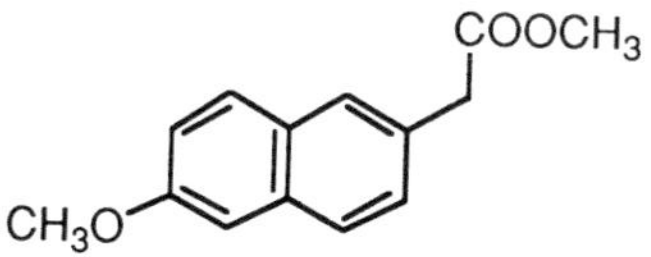 CH_3O, $COOCH_3$	7.5	220	> 98	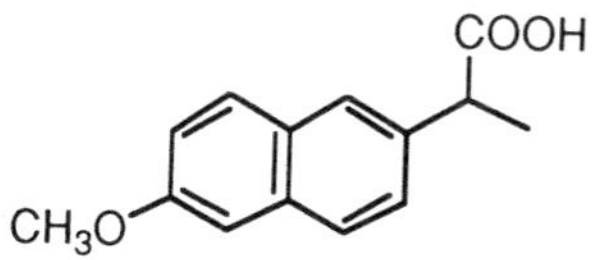 CH_3O, COOH	84

[a]All methylation reactions were carried out in a 500-mL autoclave using a substrate, DMC, and K_2CO_3 in a 1 : 16 : 2 molar ratio, respectively. Entries 1-3 and 4: 12.0 g and 8.0 g of substrate were used, respectively. [b]Hydrolyses of the mono-methyl derivatives were carried out using a 10% aq solution of NaOH (~ 5 mL/g substrate) at reflux temperature. [c]Reaction times, reaction temperature and conversions (determined by GC) refer to the methylation reaction. [d]Yields refer to distilled (bp: 93-94°C/0.9 mm, 116-118°C/0.3 mm, and 130-132°C/0.3 mm, entries 1-3, respectively) or recrystallized [cyclohexane (60 mL/g); mp 150-151°C, entry 4] products.

(tert-BUTYLDIMETHYLSILYL)ALLENE

(Silane, (1,1-dimethylethyl)dimethyl-1,2-propadienyl-)

A. H—C≡C—CH2OH + 3,4-dihydro-2H-pyran —(CSA; CH_2Cl_2, 23°C)→ H—C≡C—CH2OTHP

B. H—C≡C—CH2OTHP —(1) BuLi, THF, −78°C; TBDMS-Cl, 0 → 23°C; 2) *p*-TsOH, MeOH, 23°C)→ TBDMS—C≡C—CH2OH

C. TBDMS—C≡C—CH2OH —($ArSO_2NHNH_2$, PPh_3, DEAD; −15°C, 45 min; 23°C, 5 hr)→ TBDMS(H)C=C=CH2

Ar = *o*-$O_2NC_6H_4$

Submitted by Andrew G. Myers and Bin Zheng.[1]

Checked by Kazuya Matsunaga and Rick L. Danheiser.

1. Procedure

A. 2-Propargyloxytetrahydropyran. An oven-dried, 2-L, three-necked, round-bottomed flask is equipped with a large football-shaped Teflon-coated magnetic stirring bar, a rubber septum, an argon inlet adapter, and an oven-dried, 200-mL, pressure-equalizing addition funnel sealed with a rubber septum. The flask is charged sequentially with 0.116 g (0.500 mmol) of (±)-camphorsulfonic acid (Note 1), 750 mL of dichloromethane (Note 2), and 29.4 mL (0.500 mol) of propargyl alcohol (Note 1) under an argon atmosphere. The flask is cooled to 0°C in an ice-water bath, and a solution of 50.2 mL (0.550 mol) of 3,4-dihydro-2H-pyran (Note 1) in 75 mL of dichloromethane is added dropwise to the reaction mixture over 2 hr. Upon completion of the addition, the ice-water bath is removed, and the reaction mixture is allowed to warm to 23°C. After 2 hr at 23°C, the reaction mixture is transferred to a 2-L separatory funnel and extracted with 100 mL of saturated sodium bicarbonate solution (Note 3). The organic phase is separated and washed with 100 mL of saturated sodium chloride solution, dried over anhydrous sodium sulfate, filtered, and concentrated under reduced pressure (Note 4). The residue is purified by distillation through a 15-cm Vigreux column at reduced pressure (13 mm) to afford 67.0 g (96%) of 2-propargyloxytetrahydropyran as a colorless liquid (Note 5).

B. 3-(tert-Butyldimethylsilyl)-2-propyn-1-ol. An oven-dried, 250-mL, three-necked, round-bottomed flask equipped with a Teflon-coated magnetic stirring bar is sealed under argon with three rubber septa, one of which contains a needle adapter to an argon-filled balloon. The flask is charged with 19.5 g (139 mmol) of 2-propargyloxytetrahydropyran and 140 mL of tetrahydrofuran (THF) (Note 6) and cooled to -78°C in a dry ice-acetone bath. To the well-stirred solution, 13.9 mL (139 mmol) of a 10.0 M solution of butyllithium in hexanes (Note 7) is added slowly via syringe over 15 min. The resulting yellow solution is stirred for 5 min at -78°C, after which time the dry ice-acetone bath is removed and replaced with an ice-water bath, and the reaction mixture is stirred for 15 min at 0°C. One of the septa is removed, 22.0 g (146 mmol) of solid tert-butyldimethylsilyl chloride (Note 8) is added to the reaction mixture in one portion, and the reaction flask is sealed again with a rubber septum. After an additional 2 min at 0°C, the ice-water bath is removed. Within 15 min an exotherm is observed (up to 40°C), followed by a slow return to room temperature (Note 9). The orange-red reaction mixture is poured into a 1-L separatory funnel containing 250 mL of aqueous 10% sodium chloride solution and 250 mL of hexanes. The layers are mixed vigorously and separated. The aqueous layer is extracted with two 125-mL portions of hexanes, and the combined organic extracts are dried over anhydrous sodium sulfate, filtered, and concentrated under reduced pressure. The residue is transferred to a 1-L round-bottomed flask equipped with a Teflon-coated magnetic stirring bar. To this flask, 500 mL of anhydrous methanol (Note 10) is added, followed by 0.530 g (2.79 mmol) of p-toluenesulfonic acid monohydrate (Note 11). The flask is sealed with a rubber septum containing a needle adapter to an argon-filled balloon. The reaction mixture is stirred at 23°C for 2 hr and then quenched by the addition of 100 mL of an aqueous saturated sodium bicarbonate solution. The resulting suspension is stirred for an additional 10 min, after which time the mixture is concentrated under reduced pressure to remove most of the methanol. The residue is transferred to a 1-L separatory funnel containing 250 mL of an aqueous 10% sodium chloride solution, and the mixture is extracted with three 200-mL portions of 1:1 hexanes-ethyl acetate. The combined organic extracts are dried over anhydrous sodium sulfate, filtered, and concentrated under reduced pressure. The residue is purified by distillation at reduced pressure (1.5 mm) using a 15-cm Vigreux column to afford 22.0 g (93%) of 3-(tert-butyldimethylsilyl)-2-propyn-1-ol as a colorless oil that solidifies on standing at 23 °C (Note 12).

C. (tert-Butyldimethylsilyl)allene. An oven-dried, 500-mL, round-bottomed flask equipped with a large football-shaped Teflon-coated magnetic stirring bar is charged with 15.7 g (60.0 mmol) of triphenylphosphine (Note 13) under an argon atmosphere. The flask is sealed with a rubber septum containing a needle adapter to an argon-filled balloon, and 120 mL of THF (Note 6) is added via cannula. The solution is cooled in a -15°C bath (Note 14), and 9.02 mL (57.5 mmol) of diethyl azodicarboxylate (Note 13) is added via syringe over 2 min (Note 15), followed immediately by the addition of a solution of 8.52 g (50.0 mmol) of 3-(tert-butyldimethylsilyl)-2-propyn-1-ol in 18 mL of THF (Note 6) via cannula over 2 min. After an additional 5 min, a solution of 13.0 g (60.0 mmol) of o-nitrobenzenesulfonyl hydrazide (Note 16) in 65 mL of THF (Note 6) is added to the reaction mixture over 5 min via cannula. The resulting orange-red solution is stirred at -15°C for 45 min, after which time the cold mixture is allowed to warm to 23°C and is held at that temperature for 5 hr. During this time, the evolution of dinitrogen is observed. The reaction mixture is poured into a 2-L separatory funnel containing 400 mL of pentane, and the resulting mixture is washed with four 500-mL portions of ice-cold water (Note 17). The organic layer is dried over anhydrous sodium sulfate, filtered, and concentrated by rotary evaporation at 0°C. The residue is purified (Note 18) by flash chromatography using a short column of 230-400 mesh silica gel (60 g, packed dry and eluted with pentane). The fractions containing the product (Note 19) are concentrated by rotary evaporation at 0°C to afford 5.38-5.39 g (70%) of (tert-butyldimethylsilyl)allene as a colorless liquid (Note 20).

2. Notes

1. (±)-10-Camphorsulfonic acid, propargyl alcohol and 3,4-dihydro-2H-pyran were obtained from Aldrich Chemical Company, Inc.; (±)-10-camphorsulfonic acid was used without further purification; propargyl alcohol was dried over potassium carbonate and then was distilled at reduced pressure (18 mm) prior to use; 3,4-dihydro-2H-pyran was dried over sodium carbonate and then distilled under argon (atmospheric pressure).

2. Dichloromethane was obtained from EM Science and was distilled from calcium hydride under an atmosphere of nitrogen.

3. Extraction with bicarbonate removes residual acid and ensures that no hydrolysis of the acetal product takes place during concentration and distillation.

4. Rotary evaporation (at 22 mm) was conducted at or below 23°C to prevent evaporative loss of the product.

5. The product exhibits the following properties: bp 69-70°C/13 mm (lit.[2] 63-65°C/9 mm); TLC R_f = 0.53 (25% ethyl acetate-hexanes); IR (neat, cm^{-1}): 3290, 2943, 1442, 1220, 1120, 1029, 901, 870; ^1H NMR (300 MHz, $CDCl_3$) δ: 1.50-1.88 (m, 6 H), 2.41 (t, 1 H, J = 2.4), 3.51-3.58 (m, 1 H), 3.80-3.88 (m, 1 H), 4.22 (dd, 1 H, J = 15.6, 2.4), 4.28 (dd, 1 H, J = 15.6, 2.4), 4.82 (t, 1 H, J = 3.0); ^{13}C NMR (75 MHz, $CDCl_3$) δ: 18.8, 25.2, 30.0, 53.8, 61.8, 74.0, 79.6, 96.6.

6. Tetrahydrofuran was obtained from EM Science and distilled from sodium benzophenone ketyl under an atmosphere of argon.

7. Butyllithium was purchased from Aldrich Chemical Company, Inc., and titrated with N-benzylidenebenzylamine as indicator according to an established procedure.[3] The use of the highly concentrated organolithium reagent is essential because the use of more dilute solutions of butyllithium in hexanes results in precipitation of the lithium acetylide and therefore less efficient coupling with tert-butyldimethylsilyl chloride.

8. tert-Butyldimethylsilyl chloride was obtained from Lithium Division, FMC Corp., and used without further purification.

9. The checkers observed this exotherm to take place immediately following the addition of tert-butyldimethylsilyl chloride.

10. Anhydrous methanol was obtained from J. T. Baker Inc., and used without further purification.

11. p-Toluenesulfonic acid monohydrate was obtained from Aldrich Chemical Company, Inc., and used without further purification.

12. The product exhibits the following properties: bp 69-72°C/1.5 mm (lit.[4] 68-70°C/0.3 mm); mp 34-37°C (lit.[4] 36-38°C); TLC R_f = 0.43 (25% ethyl acetate-hexanes); IR ($CHCl_3$) cm^{-1}: 3423, 2960, 1506, 1455, 1348, 1273, 1212, 1138, 1012, 958, 830, 715, 679; ^1H NMR (300 MHz, $CDCl_3$) δ: 0.10 (s, 6 H), 0.93 (s, 9 H), 1.91 (bs, 1 H), 4.27 (s, 2 H); ^{13}C NMR (75 MHz, $CDCl_3$) δ: -4.8, 16.5, 26.0, 51.6, 88.8, 104.4.

13. Triphenylphosphine (PPh_3) and diethyl azodicarboxylate (DEAD) were obtained from Aldrich Chemical Company, Inc., and used without further purification.

14. Comparable results are obtained using either a cryobath (-15°C) or a bath containing a mixture of solid sodium chloride and ice (-12 to -18°C) for cooling.

15. Extended stirring of PPh_3 and DEAD can lead to precipitation of the betaine and inferior results. Similarly, more concentrated solutions of PPh_3 and DEAD, or cooling below -15°C, can induce precipitation. For this reason, the indicated protocol and concentrations are recommended. With other substrates, a modified order of addition, involving the addition of DEAD to a solution of PPh_3 and substrate followed by addition of o-nitrobenzenesulfonylhydrazine, may prove beneficial. In the present case this modified order of addition provided nearly identical results (isolated yield 68%).

16. o-Nitrobenzenesulfonyl hydrazide (NBSH) was prepared as follows:[5] Hydrazine monohydrate (12.1 mL, 0.25 mol, 2.5 equiv) is added dropwise to a solution of o-nitrobenzenesulfonyl chloride (22.2 g, 0.10 mol, 1 equiv) in THF (100 mL) at -30°C under an argon atmosphere. During the addition the reaction mixture becomes brown and a white precipitate of hydrazine hydrochloride is deposited. After stirring at -30°C for 30 min, thin-layer chromatographic (TLC) analysis indicates that the sulfonyl chloride has been consumed (2:1 ethyl acetate-hexanes eluent). Ethyl acetate (200 mL, 23°C) is added to the cold reaction solution, and the mixture is washed repeatedly with ice-cold aqueous 10% sodium chloride solution (5 x 150 mL); each wash involved a contact time of ≤ 1 min. The organic layer is dried over sodium sulfate at 0°C, then added slowly to a stirring solution of hexanes (1.2 L) at 23°C over 5 min. o-Nitrobenzenesulfonyl hydrazide precipitates within 10 min as an off-white solid and is collected by vacuum filtration. The filter cake is washed with hexanes (2 x 50 mL, 23°C), and dried under reduced pressure (< 1.5 mm) at 23°C for 14 hr to afford pure NBSH as an off-white powder (17.6 g, 81%); mp 100-101°C; IR (EtOAc) cm^{-1}: 3100-

3400, 1547, 1352, 1165; ^1H NMR (300 MHz, CD_3CN) δ: 3.90 (bs, 2 H), 5.97 (bs, 1 H), 7.78-7.91 (m, 3 H), 8.03-8.17 (m, 1 H); ^{13}C NMR (75 MHz, CD_3CN) δ: 125.8, 130.8, 133.2, 133.4, 135.5, 149.4. Anal. Calcd for $C_6H_7N_3O_4S$: C, 33.18; H, 3.25; N, 19.35. Found: C, 33.41; H, 3.27; N, 19.20; R_f = 0.19 (2:1 ethyl acetate-hexanes). Because solutions of NBSH are unstable at room temperature, the solution of NBSH in tetrahydrofuran should be prepared just prior to addition to the reaction mixture.

17. An orange oil composed of tetrahydrofuran and reaction by-products separates during the work-up and is removed.

18. A short silica gel pad approximately 10 cm in length by 5 cm in diameter is recommended. Isolation of the product by distillation (bp 111-120°C, bath temperature 150°C) resulted in significant material loss presumably due to its thermal decomposition.[6]

19. A suitable stain for detection of the allene (TLC analysis) is basic aqueous potassium permanganate solution.

20. On the basis of ^1H NMR analysis, the product contains 6-10% of pentane that can be removed by further rotary evaporation at 23°C, accompanied by evaporative loss of the product. The submitters obtained the product in 72% yield and report that their material contains 3-5% pentane. The allene product exhibits the following properties: TLC R_f = 0.64 (hexanes); IR (neat, cm^{-1}): 2953, 2858, 1933, 1617, 1471, 1250, 1214, 827, 805, 781; ^1H NMR (300 MHz, C_6D_6) δ: 0.06 (s, 6 H), 0.89 (s, 9 H), 4.26 (d, 2 H, J = 7.2), 4.86 (dd, 1 H, J = 7.2, 7.2); ^{13}C NMR (75 MHz, C_6D_6) δ: -5.8, 17.3, 26.4, 66.9, 78.2, 213.8; high resolution mass spectrum (EI) m/z 154.1176 [(M)+ calcd for $C_9H_{18}Si$: 154.1178].

Waste Disposal Information

All toxic materials were disposed of in accordance with "Prudent Practices in the Laboratory"; National Academy Press; Washington, DC, 1995.

3. Discussion

There are few methods for the stereodefined construction of allenes.[7,8] The procedure described here provides a general and practical method for the synthesis of allenes from propargylic alcohol precursors in a single operation. The transformation involves the Mitsunobu invertive displacement of an alcohol with o-nitrobenzenesulfonyl hydrazide (NBSH), followed by elimination of o-nitrobenzenesulfinic acid to form a propargylic diazene intermediate that undergoes spontaneous sigmatropic elimination of dinitrogen to form an allene. The method is highly efficient and proceeds with complete stereospecificity under mild reaction conditions (neutral pH, reaction temperatures ≤23°C). The method is also well suited for the synthesis of (trialkylsilyl)allenes, including the heretofore difficultly accessible parent (tert-butyldimethylsilyl)allene,[8] as illustrated here. In addition, the method is applicable for the preparation of allenes with a wide variety of sensitive functional groups.[9,10]

1. Division of Chemistry and Chemical Engineering, California Institute of Technology, Pasadena, CA 91125.
2. Jones, R. G.; Mann, M. J. *J. Am. Chem. Soc.* **1953**, *75*, 4048.
3. Duhamel, L.; Plaquevent, J.-C. *J. Org. Chem.* **1979**, *44*, 3404.

4. Danheiser, R. L.; Stoner, E. J.; Koyama, H.; Yamashita, D. S.; Klade, C. A. *J. Am. Chem. Soc.* **1989**, *111*, 4407. In this paper, 3-(tert-butyldimethylsilyl)-2-propyn-1-ol was prepared in one step (76% yield, 95% purity) by treatment of propargyl alcohol with 2.1 equiv of EtMgBr and 1.0 equiv of TBDMSCl in THF at reflux for 8 days.
5. Myers, A. G.; Zheng, B.; Movassaghi, M. *J. Org. Chem.* **1997**, *62,* 7507.
6. See, for example: Levek, T. J.; Kiefer, E. F. *J. Am. Chem. Soc.* **1976**, *98,* 1875.
7. Reviews: (a) Rossi, R.; Diversi, P. *Synthesis* **1973**, 25; (b) "The Chemistry of Ketenes, Allenes, and Related Compounds"; Patai, S., Ed.; Wiley: New York, 1980; (c) "The Chemistry of the Allenes"; Landor, S. R., Ed.; Academic Press: London, 1982; (d) "Allenes in Organic Synthesis"; Schuster, H. F.; Coppola, G. M., Ed.; Wiley: New York, 1984; (e) Pasto, D. J. *Tetrahedron* **1984**, *40*, 2805.
8. For leading references on the preparation and synthetic utility of di- and trisubstituted (trialkylsilyl)allenes, see: (a) Danheiser, R. L.; Tsai, Y.-M.; Fink, D. M. *Org. Synth., Coll. Vol. VIII* **1993**, 471; (b) ref. 4. (c) For a discussion of the preparation of (trimethylsilyl)allene, see: Danheiser, R. L.; Carini, D. J.; Fink, D. M.; Basak, A. *Tetrahedron* **1983**, *39*, 935.
9. Myers, A. G.; Zheng, B. *J. Am. Chem. Soc.* **1996**, *118,* 4492.
10. Generous financial support from the National Science Foundation is gratefully acknowledged.

Appendix

Chemical Abstracts Nomenclature (Collective Index Number); (Registry Number)

(tert-Butyldimethylsilyl)allene: Silane, (1,1-dimethylethyl)dimethyl-1,2-propadienyl- (13); (176545-76-9)

2-Propargyloxytetrahydropyran: 2H-Pyran, tetrahydro-2-(2-propynyloxy)- (8,9); (6089-04-9)

(±)-Camphorsulfonic acid: Bornanesulfonic acid, 2-oxo-, (±)- (8); Bicyclo[2.2.1]heptane-1-methanesulfonic acid, 7,7-dimethyl-2-oxo-, (±)- (9); (5872-08-2)

Propargyl alcohol: 2-Propyn-1-ol (8,9); (107-19-7)

3,4-Dihydro-2H-pyran: 2H-Pyran, 3,4-dihydro- (8,9); (110-87-2)

3-(tert-Butyldimethylsilyl)-2-propyn-1-ol: 2-Propyn-1-ol, 3-[(1,1-dimethylethyl)dimethylsilyl]- (12); (120789-51-7)

Butyllithium: Lithium, butyl- (8,9); (109-72-8)

tert-Butyldimethylsilyl chloride: Silane, chloro(1,1-dimethylethyl)dimethyl- (9); (18162-48-6)

p-Toluenesulfonic acid monohydrate (8); Benzenesulfonic acid, 4-methyl-, monohydrate (9); (6192-52-5)

Triphenylphosphine: Phosphine, triphenyl- (8,9); (603-35-0)

Diethyl azodicarboxylate: Formic acid, azodi-, diethyl ester (8); Diazenedicarboxylic acid, diethyl ester (9); (1972-28-7)

o-Nitrobenzenesulfonyl hydrazide: Benzenesulfonic acid, 2-nitro-, hydrazide (9); (5906-99-0)

N-Benzylidenebenzylamine: Benzylamine, N-benzylidene- (8); Benzenemethanamine, N-(phenylmethylene)- (9); (780-25-6)

Hydrazine HIGHLY TOXIC. CANCER SUSPECT AGENT (8,9); (302-01-2)

o-Nitrobenzenesulfonyl chloride: Benzenesulfonyl chloride, o-nitro- (8); Benzenesulfonyl chloride, 2-nitro- (9); (1694-92-4)

Hydrazine hydrochloride CANCER SUSPECT AGENT (8,9); (14011-37-1)

DIMETHYL SQUARATE AND ITS CONVERSION TO 3-ETHENYL-4-METHOXYCYCLOBUTENE-1,2-DIONE AND 2-BUTYL-6-ETHENYL-5-METHOXY-1,4-BENZOQUINONE

(3-Cyclobutene-1,2-dione, 3-ethenyl-4-methoxy- and 2,5-cyclohexadiene-1,4-dione, 5-butyl-3-ethenyl-2-methoxy- from 3-cyclobutene-1,2-dione, 3,4-dimethoxy-)

A. HO, HO (squaric acid) + $HC(OMe)_3$ $\xrightarrow{MeOH}$ MeO, MeO **1**

B. **1** $\xrightarrow{C_2H_3Li}$ [MeO, MeO, O−] $\xrightarrow{1)\ TFAA;\ 2)\ NH_4Cl}$ MeO **2**

C. **2** $\xrightarrow{1)\ Li—≡—Bu;\ 2)\ 10\%\ NH_4Cl}$ [MeO, OH, ≡—Bu] $\xrightarrow{CH_3CN}$ MeO, Bu **3**

ubmitted by Hui Liu, Craig S. Tomooka, Simon L. Xu, Benjamin R. Yerxa, Robert W. ullivan, Yifeng Xiong, and Harold W. Moore.[1]
hecked by Shino Manabe and Kenji Koga.

1. Procedure

A. Dimethyl squarate. (**1**) An oven-dried, 500-mL, round-bottomed flask equipped with a magnetic stirring bar and an 18-cm reflux condenser topped with a U-shaped drying tube (Note 1) is charged with 20.52 g (180.0 mmol) of squaric acid (Note 2), 180 mL of methanol (Note 3), and 38.78 g (365.4 mmol) of trimethyl orthoformate (Notes 4 and 5). The mixture is heated at reflux (Note 6) with stirring for 4 hr (Note 7). At this time 50 mL of the solvent is slowly removed by distillation (Note 8) over the next 2 hr (Note 9). The resulting solution is heated at reflux for an additional 18 hr. The reaction solution is then concentrated under reduced pressure to remove the volatile components (Note 10). The resulting pale yellow solid is dissolved in 40 mL of methylene chloride (Note 11), and the mixture is purified by flash chromatography on a 40-mm column containing 60 g of silica gel 60 (Note 12) using 600 mL of methylene chloride as the eluant. Removal of the solvent at reduced pressure with the use of a rotary evaporator (Note 13) provides a white solid that is redissolved in 40 mL of methylene chloride. This solution is diluted with 150 mL o anhydrous ether (Note 14). The solvent is removed under reduced pressure (Note 10 to afford 22.75 g (89% yield) of dimethyl squarate as a white solid (Note 15) that is stored under nitrogen, mp 55-56°C (lit.[2] mp 55-57°C) (Notes 16, 17, and 18).

B. 3-Ethenyl-4-methoxycyclobutene-1,2-dione (**2**). A 2-L, flame-dried, round bottomed flask equipped with a magnetic stirring bar, rubber septum, and nitroge bubbler is charged with 10.0 g (70.4 mmol) of dimethyl squarate, **1**, and 1 L of dr tetrahydrofuran (THF) (Note 19). The stirred solution is cooled in a dry ice-aceton bath at -78°C, and 39.0 mL (91.7 mmol) of vinyllithium (2.35 M) (Notes 20 and 21) added via syringe in four portions to the dimethyl squarate solution. After the additio is complete, the yellow solution is stirred for 2 hr (Note 22), whereupon 14.8 mL (1[illegible] mmol) of trifluoroacetic anhydride (Note 23) is added dropwise via syringe. T[illegible] esulting yellow solution is stirred for 30 min and quenched by pouring the cold solution into a 3-L separatory funnel containing 500 mL of aqueous ammonium chloride (10%) (NH_4Cl) and 500 mL of ethyl acetate. The separatory funnel is shaken vigorously and allowed to warm to room temperature. The aqueous layer is removed and washed twice with 200 mL of ethyl acetate. The combined organic layers are washed with 400 mL of brine and dried over solid anhydrous magnesium sulfate. The solution is filtered and concentrated under reduced pressure on a rotary evaporator (Note 24). The resulting oil is dissolved in 30 mL of ethyl acetate and purified by flash chromatography (9.5 x 45.0-cm column) through silica gel (400 g, 32-63 microns) using hexanes/ethyl acetate (7/3) (Note 25) to yield 6.55 g (67%) of the title compound as yellow crystals (Notes 26, 27, and 28) (mp 59-61°C).

C. 2-Butyl-6-ethenyl-5-methoxy-1,4-benzoquinone (**3**). A 100-mL, flame-dried, round-bottomed flask equipped with a magnetic stirring bar, rubber septum and nitrogen bubbler is charged with 1.84 g (22.44 mmol) of 1-hexyne (Note 29) and 50 mL of dry THF. The stirred solution is cooled in a dry ice-acetone bath at -78°C and 20.4 mL (22.44 mmol) of 1.1 M butyllithium (BuLi) (Note 30) is added via syringe. The resulting solution is stirred for 20 min. This solution is transferred via cannula to a solution of 2.58 g (18.70 mmol) of 2-ethenyl-3-methoxycyclobutendione in 350 mL of dry THF in a dry ice-acetone bath. The resulting solution is stirred for 60 min and then quenched with 100 mL of aqueous 10% NH_4Cl. The aqueous layer is separated and extracted with ethyl acetate (2 x 50 mL), and the combined organic layers are washed with brine (150 mL), dried ($MgSO_4$), and concentrated under reduced pressure. The crude cyclobutenone is dissolved in 350 mL of freshly distilled acetonitrile (CH_3CN) (Note 31) and refluxed for 1.5 hr (Note 32). The CH_3CN is then removed under reduced pressure. The resulting red oil is purified by flash chromatography (SiO_2, 110 g, eluting with ethyl acetate:hexanes (1:10) to yield 2.01 g (49% from cyclobutendione) of **3** as a red oil (Note 33).

2. Notes

1. The drying tube contains Drierite as the drying agent.

2. Squaric acid can be obtained from the Aldrich Chemical Company, Inc.

3. Methanol (99.9%, 0.02% H_2O) was used as received from Fisher Scientific Company.

4. Trimethyl orthoformate (98%) was used as received from the Aldrich Chemical Company, Inc. It is important to use the 98% solution rather than the available 99.8% solution in order to avoid unwanted excess of this reagent (Note 5).

5. The use of more trimethyl orthoformate than specified results in the eventual formation of 2,3,4,4-tetramethoxy-2-cyclobuten-1-one as a side product.

6. The temperature of the refluxing solution is approximately 56°C.

7. The initial heterogeneous mixture become homogeneous after approximately 2.5 hr under reflux conditions.

8. The purpose of this operation is to remove the by-product, methyl formate (bp 34°C), by short path distillation.

9. The temperature of the solution after distillation is approximately 63°C.

10. All volatile components are removed under reduced pressure using a rotary evaporator; the crude product is subsequently placed under an oil pump for 2 hr.

11. Methylene chloride (99.9%) was used as received from Fisher Scientific Company.

12. The silica gel (230-400 mesh) was purchased from East Merck Science.

13. The precipitate is very hard and difficult to remove from the flask. By redissolving in methylene chloride/diethyl ether the product is obtained as a fluffy white solid.

14. Anhydrous ether (99.9%, 0.002% H_2O) was used as received from Fisher Scientific Company.

15. The product shows no contaminant at the baseline by TLC analysis and has he following spectral properties: ^{1}H NMR (300 MHz, $CDCl_3$) δ: 4.37 (s); ^{13}C NMR (75 ЛHz, $CDCl_3$) δ: 61.0, 184.5, 189.2.

16. For the first preparation of dimethyl squarate see Cohen.[2]

17. For a recent review on the synthetic utility of squaric acid and related yclobutenones see Moore.[3]

18. The synthesis of dimethyl squarate as outlined here is a general method that an be employed for the synthesis of a variety of other alkyl squarates.[4]

19. Tetrahydrofuran was distilled under nitrogen from benzophenone ketyl.

20. Vinyllithium was purchased from Organometallics Inc. as a 2.35 M solution tetrahydrofuran and used as such. Best results are obtained when a new bottle is sed, and yields are very dependent upon the quality of the reagent.

21. Vinyllithium can also be generated in situ by the dropwise addition of 90 mL a 1.02 N solution of methyllithium in ether (91.48 mmol) to a solution of 4.5 mL 4.63 mmol) of tetravinyltin (Aldrich Chemical Company, Inc.) in tetrahydrofuran (60 L) at 0°C. The solution is stirred at 0°C for 30 min and used directly. The checkers btained a 93% yield of product using vinyllithium thus prepared.

22. A 1-mL aliquot is taken from the reaction flask and 2 drops of trifluoroacetic hydride (Note 23) is added. The solution is then checked against dimethyl squarate TLC (7/3 hex/EtOAc). If dimethyl squarate, **1**, is present an additional 5 mL (11.75 mol) of vinyllithium is added. The solution is allowed to stir for 20 min and the above ocedure is repeated until there is no evidence of dimethyl squarate.

23. Trifluoroacetic anhydride was purchased from the Aldrich Chemical mpany, Inc., and used as such.

24. The water bath should not exceed 35°C because of possible decomposition.

25. The vinylcyclobutenedione **2** will elute from the column as the second llow band.

26. Prolonged exposure to air will result in decomposition of the vinylcyclobutenedione.

27. If a solid is not initially formed the oil will solidify upon standing under vacuum.

28. Spectral properties are as follows: IR ($CHCl_3$) cm^{-1}: 2690, 1790, 1770, 1750, 1620, 1590, 1580, 1460, 1410, 1370, 1350, 1050; ^1H NMR (300 MHz, $CDCl_3$) δ: 4.48 (s, 3 H), 5.87 (dd, 1 H, J = 10.5, 1.9), 6.50 (dd, 1 H, J = 17.6, 1.8), 6.65 (dd, 1 H, J = 17.6, 10.5); ^{13}C NMR (500 MHz, $CDCl_3$) δ: 61.4, 122.1, 128.9, 173.4, 192.6, 192.8, 194.6; MS (EI), m/e (rel intensity) 138 (12), 110 (30), 95 (75), 82 (22), 67 (39), 58 (14), 53 (100). Exact mass calcd for $C_7H_6O_3$: 138.0317, found 138.0335.

29. 1-Hexyne (97% purity) was purchased from Lancaster Chemical Company, Inc., and used as such.

30. Butyllithium was purchased from Aldrich Chemical Company, Inc., as a 1. M solution in hexane and titrated with diphenylacetic acid to indicate a concentration of 1.1 M at the time of use.

31. Acetonitrile was distilled under nitrogen from calcium hydride.

32. The progress of the reaction can be monitored by TLC (ethyl acetate:hexanes = 1:5).

33. The spectral properties are as follows: IR (CCl_4) cm^{-1}: 2959, 2783, 165 1448, 1321, 1215, 1143, 999, 935, 885; ^1H NMR (300 MHz, $CDCl_3$) δ: 0.93 (t, 3 H, J = 7.0), 1.43 (m, 4 H), 2.42 (t, 2 H, J = 6.7), 4.08 (s, 3 H), 5.63 (dd, 1 H, J = 12.0, 2.4), 6.2 (dd, 1 H, J = 17.7, 2.1), 6.41 (s, 1 H), 6.70 (dd, 1 H, J = 17.7, 12.0); ^{13}C NMR (300 MHz, $CDCl_3$) δ: 13.7, 22.3, 28.7, 29.9, 60.8, 124.7, 125.1, 125.4, 130.6, 149.5, 154.4, 183. 187.2; MS (EI) m/e (rel int) 220 (19), 177 (44), 160 (65), 135 (45), 91 (43), 79 (43), (54), 53 (100); exact mass calcd for $C_{13}H_{16}O_3$: 220.1099, found 220.1108.

Waste Disposal Information

All toxic materials were disposed of in accordance with "Prudent Practices in the Laboratory"; National Academy Press; Washington, DC, 1995.

3. Discussion

Dimethyl squarate has recently been used for the versatile and regiospecific synthetic routes to highly substituted benzo- and naphthoquinones and related aromatic compounds.[3]

Two other syntheses of dimethyl squarate have been reported.[2] One such method involves the formation of the disilver salt of squaric acid followed by treatment with methyl iodide. This process, unfortunately, is laborious as well as costly. The other synthesis involves the reaction of squaric acid with diazomethane. This reaction not only gives a lower yield, but is inherently unsuited for large scale preparations. The synthesis presented here provides a convenient and inexpensive route to the preparation of dimethyl squarate, and is also suitable for large scale synthesis.[4] One should take precautions when synthesizing or using dialkyl squarates, particularly dimethyl squarate, since they are known to cause severe contact dermatitis.[5] Thus, all procedures should be carried out in the hood and protective clothing and gloves should be worn.

This procedure describes a synthetic route to vinylcyclobutenediones that are mportant intermediates in the synthesis of other cyclobutenediones (Table) and uinones.[6,7] For example vinylcyclobutenedione **2** readily undergoes 1,6-addition of ucleophiles such as thiols and organocuprates to form more highly substituted roducts as shown in the generalized scheme below.

TABLE

SYNTHESIS OF CYCLOBUTENEDIONES

2 → **4**

	R	% 4
a	Br-	93
b	$HOCH_2CH_2S$-	73
c	C_6H_5S-	86
d	C_6H_5-	43
e	i-C_3H_7-	69
f	$CH_3CO_2(CH_2)_4$-	56
g	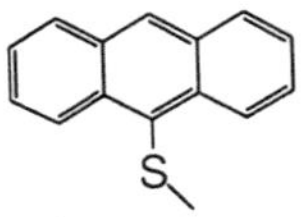	80
h	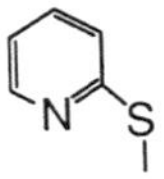	79

The vinylcyclobutenediones **5-8** are representative of other analogs that have been prepared by the general method outlined here.[8]

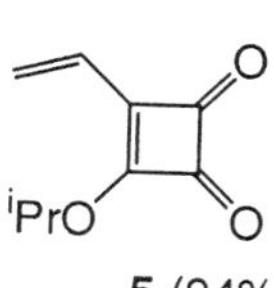

5 (94%) 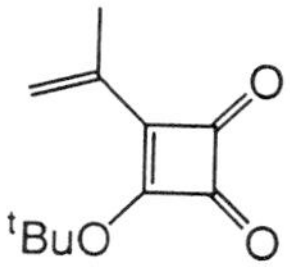 **6** (96%)

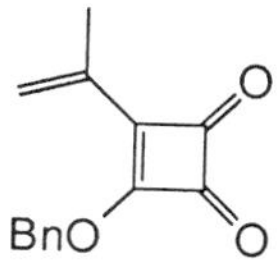

7 (56%) 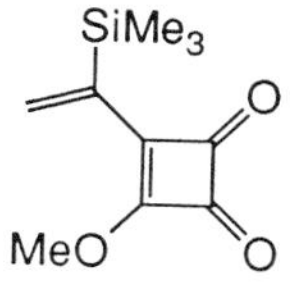 **8** (83%)

1. Department of Chemistry, University of California, Irvine, CA 92717.
2. Cohen, S.; Cohen, S. G. *J. Am. Chem. Soc.* **1966**, *88*, 1533.
3. Moore, H. W.; Yerxa, B. R. *Adv. Strain Org. Chem.* **1995**, *4*, 81-162.
4. Liu, H.; Tomooka, C. S.; Moore, H. W. *Synthetic Comm.* **1997**, *27*, 2177.
5. Avalos, J.; Moore, H. W.; Reed, M. W.; Rodriguez, E. *Journal of Contact Dermatitis* **1989**, *21*, 341.
6. Xu, S.; Yerxa, B. R.; Sullivan, R. W.; Moore H. W. *Tetrahedron Lett.* **1991**, *32*, 1129.
7. For a review on the synthetic utility of cyclobutenone ring expansions see: Moore, H. W.; Yerxa, B. R. *ChemTracts* **1992**, *5*, 273.
8. Moore, H. W. unpublished data.

Appendix

Chemical Abstracts Nomenclature (Collective Index Number); (Registry Number)

Dimethyl squarate: Cyclobutenedione, dimethoxy- (8); 3-Cyclobutene-1,2-dione, 3,4-dimethoxy- (9); (5222-73-1)

3-Ethenyl-4-methoxycyclobutene-1,2-dione: 3-Cyclobutene-1,2-dione, 3-ethenyl-4-methoxy- (12); (124022-02-2)

2-Butyl-6-ethenyl-5-methoxy-1,4-benzoquinone: 2,5-Cyclohexadiene-1,4-dione, 5-butyl-3-ethenyl-2-methoxy- (12); (134863-12-0)

Squaric acid: Cyclobutenedione, dihydroxy- (8); 3-Cyclobutene-1,2-dione, 3,4-dihydroxy- (9); (2892-51-5)

Trimethyl orthoformate: Orthoformic acid, trimethyl ester (8); Methane, trimethoxy- (9); (149-73-5)

Vinyllithium: Lithium, vinyl- (8); Lithium, ethenyl- (9); (917-57-7)

Trifluoroacetic anhydride: Acetic acid, trifluoro-, anhydride (8,9); (407-25-0)

1-Hexyne (8,9); (693-02-7)

Butyllithium: Lithium, butyl- (8,9); (109-72-8)

[3 + 4] ANNULATION USING A [β-(TRIMETHYLSILYL)ACRYLOYL]SILANE AND THE LITHIUM ENOLATE OF AN α,β-UNSATURATED METHYL KETONE: (1R*,6S*,7S*)-4-(tert-BUTYLDIMETHYLSILOXY)-6-(TRIMETHYLSILYL)BICYCLO[5.4.0]UNDEC-4-EN-2-ONE

A. 1. ethyl vinyl ether, p-TsOH; 2. KOBu^t → **1**

B. **1** → 1. BuLi; 2. $^tBuMe_2SiCl$ → **2**

C. **2** → 1. BuLi; 2. Me_3SiCl; 3. p-TsOH, MeOH → **3**

D. **3** + lithium enolate → -80°C to -30°C, THF → **4**

Submitted by Kei Takeda, Akemi Nakajima, Mika Takeda, and Eiichi Yoshii.[1]
Checked by Jing Zhang and Robert K. Boeckman, Jr.

1. Procedure

Caution! Hexamethylphosphoric triamide has been identified as a carcinogen. Glove protection is required during the handling in Part B.

A. 1-(1-Ethoxyethoxy)-1,2-propadiene (**1**) (Note 1). A 300-mL, two-necked round-bottomed flask is equipped with a magnetic stirring bar, a thermometer, and a 100-mL pressure-equalizing addition funnel fitted with a calcium chloride-filled tube. The flask is charged with 100 g (1.39 mol) of ethyl vinyl ether (Note 2) and 50 mg (0.26 mmol) of p-toluenesulfonic acid monohydrate (Note 3). The mixture is then stirred and cooled in an ice-salt bath while 56.1 g (1.00 mol) of propargyl alcohol (Note 4) is added dropwise via the addition funnel over 50 min; the rate of addition adjusted to maintain the temperature between 5°C and 10°C. After the addition is complete, the mixture is stirred at 0°C for 15 min and then quenched by adding of 3 mL of a aqueous saturated solution of potassium carbonate (K_2CO_3). The resulting mixture is dried over K_2CO_3, filtered, and excess ethyl vinyl ether is removed at reduced pressure using a rotary evaporator at temperatures below 25°C. Distillation of the residual liquid under reduced pressure affords 106 - 117 g (88 - 92%) of 3-(1-ethoxyethoxy)-propyne as a colorless clear oil, bp 37-41°C/11 mm (Note 5).

A 200-mL, three-necked, round-bottomed flask, equipped with a magnetic stirring bar, a thermometer, a glass stopper, and a reflux condenser fitted with a nitrogen inlet/outlet adapter, is placed under a nitrogen atmosphere and charged with 98.6 g (0.769 mol) of 3-(1-ethoxyethoxy)-1-propyne and 8.63 g (76.9 mmol) of potassium tert-butoxide (Note 6). The flask is immersed in a preheated oil bath at 70°C with vigorous stirring of the mixture. The internal temperature is maintained at ca. 70°C for 40 min (there is an exothermic reaction; occasional removal of the heating bath may be necessary to moderate the reaction). The reaction mixture is transferred to a 500-mL separatory funnel containing 100 mL of ice-water, and the aqueous nixture is extracted three times with 100-mL portions of diethyl ether (Et_2O). The ;ombined organic phases are washed three times with 100-mL portions of saturated orine, dried over K_2CO_3, filtered, and concentrated under reduced pressure using a otary evaporator at temperatures below 25°C. The residual oil is distilled through a 0-cm Vigreux column under reduced pressure to afford 69.3-74.5 g (70 -75%) of 1-(1-thoxyethoxy)-1,2-propadiene (**1**) as a colorless clear oil, bp 42-43°C/18 mm (Notes 7 nd 8).

B. 1-(tert-Butyldimethylsilyl)-1-(1-ethoxyethoxy)-1,2-propadiene (**2**) (Note 9). A :-L, three-necked, round-bottomed flask, equipped with a magnetic stirring bar, a nermometer, a rubber septum, and a 200-mL pressure-equalizing addition funnel tted with a nitrogen inlet/outlet adapter, is purged with nitrogen and charged with 40.0 (0.312 mol) of 1-(1-ethoxyethoxy)-1,2-propadiene (**1**), 350 mg of 4,4'-thiobis(2-tert-utyl-m-cresol) (Note 10), and 1 L of dry Et_2O (Note 11). The solution is stirred and ooled to -80°C with a hexane-liquid nitrogen bath, and then 220 mL (0.312 mol) of a .42 M hexane solution of butyllithium (BuLi) (Note 12) is added dropwise via a :ainless steel cannula over 20 min. After the addition is complete, the mixture is irred at -80°C for 1 hr, and a solution of 50.8 g (0.337 mol) of tert-butyldimethylsilyl nloride (Note 13) in 180 mL of dry Et_2O is added dropwise via the addition funnel ver 20 min (the funnel is washed with 5 mL of dry Et_2O into the reaction mixture). fter 10 min, a solution of 60 mL of hexamethylphosphoric triamide (Note 14) in 180 L of dry Et_2O is added via a stainless steel cannula over 10 min. After 30 min, the oling bath is removed, and the pale yellow solution is allowed to warm to 0°C over) min whereupon 31 mL of triethylamine (Note 15) is added. The reaction mixture is ansferred to a 3-L separatory funnel containing 1 L of aqueous saturated sodium carbonate ($NaHCO_3$), and the organic phase is separated. The aqueous phase is tracted three times with 300-mL portions of Et_2O, and the combined organic phases e washed successively with 500 mL of water and 500 mL of saturated brine, dried over K_2CO_3, filtered, and concentrated at reduced pressure using a rotary evaporator. The residual liquid is distilled through a 10-cm Vigreux column under reduced pressure to afford 62.8 - 65.4 g (83-86%) of 1-(tert-butyldimethylsilyl)-1-(1-ethoxyethoxy)-1,2-propadiene (**2**) as a colorless clear oil, bp 60-62°C/0.5 mm (Notes 8 and 16).

C. *(E)-1-(tert-Butyldimethylsilyl)-3-trimethylsilyl-2-propen-1-one* (**3**) (Note 17). A 200-mL, two-necked, round-bottomed flask, equipped with a magnetic stirring bar, a thermometer, a rubber septum, and a 10-mL pressure-equalizing addition funnel fitted with a nitrogen inlet/outlet adapter, is placed under a nitrogen atmosphere and charged with 10.0 g (41.2 mmol) of 1-(tert-butyldimethylsilyl)-1-(1-ethoxyethoxy)-1,2-propadiene (**2**), 100 mg of 4,4'-thiobis(2-tert-butyl-m-cresol), and 80 mL of dry tetrahydrofuran (THF) (Note 18). The solution is stirred and cooled to -80°C with a hexane-liquid nitrogen bath, and 30 mL (44.1 mmol) of a 1.47 M hexane solution of BuLi is added dropwise via a stainless steel cannula over 30 min. The mixture is stirred at -80°C for 40 min, and 4.90 g (45.1 mmol) of trimethylsilyl chloride (Note 19) is added dropwise via the addition funnel over 10 min (the funnel is washed with 5 mL of dry THF into the reaction mixture). After 10 min, the cooling bath is removed, and the reaction mixture is allowed to warm to room temperature over 1 hr whereupon 4 mL of triethylamine is added. The reaction mixture is transferred to a 500-mL separatory funnel containing 100 mL of ice-cooled aqueous saturated $NaHCO_3$, and the organic phase is separated. The aqueous phase is extracted three times with 100-mL portions of Et_2O, and the combined organic phases are washed successively with 100 mL of water and 100 mL of saturated brine, dried over K_2CO_3, filtered, and concentrated at reduced pressure using a rotary evaporator to afford the crude 3-trimethylsilyl derivative of **2** as a colorless oil, which is used for the next reaction without further purification.

A 300-mL, one-necked, round-bottomed flask, equipped with a magnetic stirring bar, is charged with the above product, 1.58 g (8.24 mmol) of p-toluenesulfonic acid monohydrate, and 140 mL of methanol (Note 20). The mixture is stirred at room temperature for 15 min and then transferred to a 500-mL separatory funnel containing 100 mL of ice-cooled aqueous saturated $NaHCO_3$. The mixture is extracted three times with 100-mL portions of pentane. The combined organic phases are washed with 100 mL of saturated brine, dried over magnesium sulfate ($MgSO_4$), filtered, and concentrated at reduced pressure using a rotary evaporator. The residual oil is distilled through a 10-cm Vigreux column under reduced pressure to afford 7.4 - 7.6 g (74 - 75% overall) of (E)-1-(tert-butyldimethylsilyl)-3-trimethylsilyl-2-propen-1-one (**3**) as an orange oil, bp 60-63°C/0.4 mm (Notes 21 and 22).

D. (1R,6S*,7S*)-4-(tert-Butyldimethylsiloxy)-6-(trimethylsilyl)bicyclo[5.4.0]-undec-4-en-2-one* (**4**). A 100-mL, three-necked, round-bottomed flask, equipped with a magnetic stirring bar, a thermometer, a rubber septum, and a nitrogen inlet/outlet adapter, is placed under a nitrogen atmosphere and charged with 2.81 mL (2.17 g, 21.4 mmol) of diisopropylamine (Note 23) and 20 mL of dry THF. The solution is stirred and cooled with an ice-water bath while 13.7 mL (20.1 mmol) of a 1.47 M hexane solution of BuLi is added dropwise via a syringe over 5 min. After 10 min, the mixture is cooled to -80°C with a hexane-liquid nitrogen bath, and a solution of 2.58 mL (2.49 g, 20.1 mmol) of 1-acetyl-1-cyclohexene (Note 24) in 20 mL of dry THF is added dropwise via a stainless steel cannula over 15 min. The mixture is then stirred at the same temperature for an additional 30 min, and the solution is maintained at -80°C.

A 500-mL, three-necked, round-bottomed flask, equipped with a stirring bar, a ermometer, a rubber septum, and a nitrogen inlet/outlet adapter, is charged under a trogen atmosphere with 5.34 g (22.0 mmol) of (E)-1-(tert-butyldimethylsilyl)-3-methylsilyl-2-propen-1-one (**3**) and 150 mL of dry THF. The mixture is cooled to

-80°C with a hexane-liquid nitrogen bath, and the THF solution of the lithium enolate of 1-acetylcyclohexene, which was prepared above and kept at -80°C, is added via a stainless steel cannula over 2 min (the enolate-containing flask is washed twice into the reaction mixture with 2-mL portions of dry THF). The reaction mixture is allowed to warm to -30°C over 70 min, and a solution of 1.22 g (20.3 mmol) of acetic acid in 10 mL of dry THF is added rapidly in one portion via a cannula. The resulting mixture is transferred to a 1-L separatory funnel containing 200 mL of aqueous saturated ammonium chloride, and the organic phase is separated. The aqueous phase is extracted twice with 100-mL portions of Et_2O. The combined organic phases are washed with 100 mL of saturated brine, dried over $MgSO_4$, filtered, and concentrated at reduced pressure using a rotary evaporator. The residual crude product is purified by column chromatography (150-325 mesh silica gel, 400 g; elution with hexane-AcOEt = 40:1) to afford 5.8 - 6.0 g (79 - 82%) of **4** as a pale yellow oil crude product, which solidifies on storage in a refrigerator. This material is ~95% pure based upon ^1H NMR spectroscopic analysis. Higher purity material may be obtained by recrystallization of this sample from a minimum amount of Et_2O-hexane affording colorless prisms, mp 74 - 75 °C (Notes 25 and 26).

2. Notes

1. This procedure was reported by Hoff, Brandsma and Arens.[2]

2. Ethyl vinyl ether was purchased from Wako Pure Chemical Co., Ltd. or Aldrich Chemical Company, Inc. and freshly distilled.

3. p-Toluenesulfonic acid monohydrate was purchased from Kanto Chemical Co., Inc. or Aldrich Chemical Company, Inc. and used without purification.

4. Propargyl alcohol was purchased from Nakalai Tesque, Inc. or Aldrich Chemical Company, Inc. and distilled.

5. This material has the following spectral properties: ^{1}H NMR (500 MHz, $CDCl_3$) δ: 1.17 (t, 3 H, J = 7.1, $CH_2\underline{Me}$), 1.29 (d, 3 H, J = 5.3, $CH\underline{Me}$), 2.38 (t, 1 H, J = 2.6, H-1), 3.44-3.51 (m, 1 H, $C\underline{H_2}Me$), 3.58-3.65 (m, 1 H, $C\underline{H_2}Me$), 4.16 (d, 2 H, J = 2.6 H-3), 4.82 (q, 1 H, J = 5.3, $OC\underline{H}Me$); ^{13}C NMR (125 MHz, $CDCl_3$) δ: 15.3 ($CH_2\underline{Me}$), 19.7 ($CH\underline{Me}$), 52.5 (C-3), 60.8 ($C\underline{H_2}Me$), 73.9 (C-1), 80.1 (C-2), 98.7 (OCHO). The checkers obtained material having bp 61-62°C/30 mm.

6. Potassium tert-butoxide was purchased from E. Merck and used as received.

7. This material has the following spectral properties: ^{1}H NMR (500 MHz, $CDCl_3$) δ: 1.20 (t, 3 H, J = 7.1, $CH_2\underline{Me}$), 1.36 (d, 3 H, J = 5.1, $CH\underline{Me}$), 3.46-3.52 (m, 1 H, $C\underline{H_2}Me$), 3.72-3.79 (m, 1 H, $C\underline{H_2}Me$), 4.92 (q, 1 H, J = 5.1, $OC\underline{H}Me$), 5.34 (dd, 1 H, J = 6.0, 8.5, H-3), 5.37 (dd, 1 H, J = 6.0, 8.5, H-3), 6.68 (t, 1 H, J = 6.0, H-1); ^{13}C NMR (125 MHz, $CDCl_3$) δ: 15.3 ($CH_2\underline{Me}$), 20.4 ($CH\underline{Me}$), 62.9 ($C\underline{H_2}Me$), 89.2 (C-3), 99.9 (OCHO), 117.2 (C-1), 201.6 (C-2). The checkers obtained material having bp 55-56°C/35 mm.

8. This compound can be stored in the presence of a trace amount of radical inhibitor such as 4,4'-thiobis(2-tert-butyl-m-cresol) in a freezer for several months without significant decomposition. The checkers recommend that **1** be used as soon as practicable, as they observed significant amounts of polymerization upon storage for 1 month, even in the presence of the radical inhibitor. However, the checkers found that **2** is stable for 3 months in the presence of the inhibitor.

9. This procedure was reported by Reich, Kelly, Olson, and Holtan.[3]

10. 4,4'-Thiobis(2-tert-butyl-m-cresol) was purchased from Tokyo Kasei Kogyo Co., Inc. or Aldrich Chemical Company, Inc. and used as received.

11. Diethyl ether was distilled over calcium hydride and dried over 4 Å molecular sieves.

12. Butyllithium was purchased from Kanto Chemical Co., Inc. or Aldrich Chemical Company, Inc. and standardized by titration using 1,3-diphenylacetone p-tosylhydrazone as an indicator.[4]

13. tert-Butyldimethylsilyl chloride was purchased from Aldrich Chemical Company, Inc. and used as received.

14. Hexamethylphosphoric triamide was purchased from Aldrich Chemical Company, Inc. or Tokyo Kasei Kogyo Co., Inc. and distilled over calcium hydride and then dried over 4 Å molecular sieves.

15. Triethylamine was purchased from Nakalai Tesque, Inc. and distilled over calcium hydride. The checkers used triethylamine purchased from Aldrich Chemical Company, Inc. as received.

16. This material has the following spectral properties: ^{1}H NMR (500 MHz, $CDCl_3$): δ 0.02 (s, 3 H, $SiMe_2$), 0.03 (s, 3 H, $SiMe_2$), 0.91 (s, 9 H, t-Bu), 1.16 (t, 3 H, J = 7.1, $CH_2\underline{Me}$), 1.28 (d, 3 H, J = 5.1, $CH\underline{Me}$), 3.38-3.45 (m, 1 H, $C\underline{H}_2Me$), 3.68-3.74 (m, 1 H, $C\underline{H}_2Me$), 4.92 (q, 1 H, J = 5.1, $C\underline{H}Me$), 5.00 (d, 1 H, J = 8.5, H-3), 5.06 (d, 1 H, J = 8.5, H-3); ^{13}C NMR (125 MHz, $CDCl_3$): δ -6.5 ($SiMe_2$), -6.6 ($SiMe_2$), 15.4 ($CH_2\underline{Me}$) 17.5 ($\underline{C}Me_3$), 20.5 ($CH\underline{Me}$), 26.8 ($C\underline{Me}_3$), 63.1 ($C\underline{H}_2Me$), 84.4 (C-3), 100.1 (OCHO) 125.8 (C-1), 203.0 (C-2). The checkers obtained material having bp 108-109°C/8 mm.

17. This procedure is a modification of Reich's method[3] for the synthesis of 1,3-bis(trimethylsilyl)-2-propen-1-one.

18. Tetrahydrofuran was distilled over calcium hydride and dried over 4 Å molecular sieves.

19. Trimethylsilyl chloride was purchased from Nakalai Tesque, Inc. or Aldrich Chemical Company, Inc. and distilled over calcium hydride.

20. Methanol was purchased from Wako Pure Chemical Co., Ltd. or Aldrich Chemical Company, Inc. and used as received.

21. This material (E)-**3** has the following physical properties: R_f = 0.52 (hexane-Et$_2$O = 10:1). IR (neat) cm^{-1}: 1655, 1250; ^{1}H NMR (500 MHz, CDCl$_3$) δ: 0.09 (s, 9 H, SiMe$_3$), 0.18 (6 H, SiMe$_2$), 0.87 (s, 9 H, t-Bu), 6.64 (d, 1 H, J = 19.2, H-2), 6.80 (d, 1 H, J = 19.2, H-3); ^{13}C NMR (125 MHz, CDCl$_3$) δ: -5.8 (SiMe$_2$), -1.7 (SiMe$_3$), 16.7 (CMe$_3$), 26.7 (CMe$_3$), 145.1 (C-3), 147.1 (C-2), 236.3 (C-1). The checkers obtained material having bp 110-112°C/8 mm.

22. A mixture of **3** and its (Z)-isomer is obtained when trifluoroacetic acid is used for the removal of the 1-ethoxyethyl protective group. A solution of crude 1-(tert-butyldimethylsilyl)-1-(1-ethoxyethoxy)-3-(trimethylsilyl)-1,2-propadiene obtained from 38.7 g (160 mmol) of **2** in 300 mL of THF-water (5:1) is cooled with an ice-water bath, and 91 mL (135 g, 1.18 mol) of trifluoroacetic acid is added in one portion. The mixture is then placed in a refrigerator (ca. 4°C) and, after 12 hr, transferred to a 1-L separatory funnel containing 200 mL of water. The whole is extracted three times with 100-mL portions of pentane. The combined organic phases are washed thoroughly with aqueous saturated NaHCO$_3$ to remove trifluoroacetic acid completely, and then with 100 mL of saturated brine. The pentane solution is dried over MgSO$_4$, filtered, and concentrated at reduced pressure using a rotary evaporator. The residue is subjected to column chromatography (150-325 mesh silica gel, 1 kg; elution with hexane-Et$_2$O, 40:1) to afford 9.89 g (27%) of (Z)-**3** and 17.72 g (46%) of (E)-**3**. (Z)-**3**: bp 60°C/0.45 mm, an orange oil; IR (neat) cm^{-1}: 1655, 1245; R_f = 0.67 (hexane-Et$_2$O, 10:1); ^{1}H NMR (500 MHz, CDCl$_3$) δ: 0.10 (s, 9 H, SiMe$_3$), 0.18 (6 H, SiMe$_2$), 0.91 (s, 9 H, t-Bu), 6.01 (d, 1 H, J = 14.1, H-2), 7.30 (d, 1 H, J = 14.1, H-3); ^{13}C NMR (125 MHz, CDCl$_3$) δ: -7.0 (SiMe$_2$), -0.5 (SiMe$_3$), 17.0 (CMe$_3$), 26.7 (CMe$_3$), 144.6 (C-2), 144.9 (C-3), 238.0 (C-1).

23. Diisopropylamine was purchased from Nakalai Tesque, Inc. or Aldrich Chemical Company, Inc. and distilled over calcium hydride.

24. 1-Acetyl-1-cyclohexene was purchased from Aldrich Chemical Company Inc. and distilled before use.

25. The checkers found that recovery of pure material from the recrystallization is low (~30%) unless care is taken to use the minimum amount of total solvent and the minimum amount of ether. The checkers found use of pure hexanes more satisfactory in terms of recovery; however, the material obtained was pale yellow in color.

26. This material has the following physical properties : R_f = 0.47 (hexane-ethyl acetate (AcOEt) = 15:1); IR (KBr) cm^{-1}: 1705, 1645, 1250; ^1H NMR (500 MHz, C_6D_6) δ: 0.08 (s, 9 H, $SiMe_3$), 0.21 (s, 3 H, $SiMe_2$), 0.26 (s, 3 H, $SiMe_2$), 1.00 (s, 9 H, t-Bu), 1.0 (dddd, 1 H, J = 3.9, 4.3, 13.0, 13.5, H-11), 1.21 (ddddd, 1 H, J = 4.7, 4.7, 13.5, 13.5, 13.5, H-9), 1.42-1.45 (m, 2 H, H-8), 1.52 (br d, 1 H, J = 13.5, H-10), 1.71 (br d, 1 H, J 13.5, H-9), 1.79 (d, 1 H, J = 7.1, H-6), 2.05 (ddddd, 1 H, J = 3.9, 3.9, 13.5, 13.5, 13.5, H-10), 2.13 (br ddd, 1 H, J = 3.9, 3.9, 11.5, H-7), 2.27 (br d, 1 H, J = 13.0, H-11), 2.38 (br dd, 1 H, J = 3.9, 3.9, H-1), 2.85 (dd, 1 H, J = 1.9, 12.2, H-3), 3.67 (dd, 1 H, J = 1.9, 12.2, H-3), 5.00 (ddd, 1 H, J = 1.9, 1.9, 7.1, H-5); ^{13}C NMR (125 MHz, C_6D_6) δ: -4.3, -4.2, -1.8, 18.2, 22.2, 25.8, 26.9, 27.6, 28.8, 31.3, 46.3, 51.3, 57.7, 106.0, 149.9, 203.3. Anal. Calcd. for $C_{20}H_{38}O_2Si_2$: C, 65.51; H, 10.45, Found C, 65.26; H, 10.56. The structure was determined by X-ray crystallographic analysis.[5]

Waste Disposal Information

All toxic materials were disposed of in accordance with "Prudent Practices in the Laboratory"; National Academy Press; Washington, DC, 1995.

3. Discussion

The procedure described here illustrates an efficient preparation of cis-6-alkyl-5-trimethylsilyl-3-siloxy-3-cycloheptenones by reaction of (E)-(β-(trimethylsilyl)acryloylsilane (**3**) with the kinetic lithium enolate of an α,β-unsaturated methyl ketone generated with lithium diisopropylamide in tetrahydrofuran (Table).[5] This new [3 + 4] annulation[6] is also effective with 1-cyclopentenyl and 1-cyclohexenyl methyl ketones, affording all-cis bicyclic cycloheptenones in comparable yields as shown in the scheme. It is remarkable that reaction of **3** with 2'-bromoacetophenone enolate produces benzocycloheptenone, albeit in low yield, showing that a benzenoid unsaturation can also participate in the [3 + 4] annulation. In sharp contrast to the case of (E)-**3**, the reaction of (Z)-**3** under the same conditions is quite slow and affords only 5,6-trans isomer in much lower yield. The stereospecificity in the annulation has been rationalized by intermediacy of cis-1,2-divinylcyclopropanediolate (**6**), which is generated from 1,2-adduct (**5**) by way of a Brook rearrangement/cyclopropanation sequence in a concerted manner. Compound **6** is expected rapidly to undergo a stereospecific oxyanion-accelerated Cope rearrangement to **7**. The low yields in the reaction of (Z)-**3** is attributed to slowness in the initial 1,2-addition step since substantial amount of the starting material is recovered.

The present [3 + 4] annulation methodology for the synthesis of seven-membered carbocycles involves a straightforward procedure that also provides the product functionalities (e.g., masked and unmasked ketone carbonyl and trimethylsilyl groups) that can be transformed to hitherto inaccessible or difficult-to-prepare cycloheptane structures. The prior approach,[7-9] based on Cope rearrangement of cis-1,2-divinylcyclopropane, bears an intrinsic drawback in that there exist a limited number of methods for stereoselective preparation of the substrate.

Me3Si 3 SiMe2Bu^t + OLi R → ^-O ^tBuMe2Si R SiMe3 5 TBDMSO R SiMe3

R3Si Li+ O O R SiMe3 ≡ ^tBuMe2SiO O^- R SiMe3 6 → TBDMSO OLi 7

1. Faculty of Pharmaceutical Sciences, Toyama Medical and Pharmaceutical University, 2630 Sugitani, Toyama 930-0194, Japan.
2. Hoff, S.; Brandsma, L.; Arens, J. F. *Recl. Trav. Chim. Pays-Bas.* **1968**, *87*, 916.
3. Reich, H. J.; Kelly, M. J.; Olson, R. E.; Holtan, R. C. *Tetrahedron* **1983**, *39*, 949.
4. Lipton, M. F.; Sorensen, C. M.; Sadler, A. C.; Shapiro, R. H. *J. Organomet. Chem.* **1980**, *186*, 155.
5. Takeda, K.; Takeda, M.; Nakajima, A.; Yoshii, E. *J. Am. Chem. Soc.* **1995**, *117*, 6400; Takeda, K.; Nakajima, A.;Takeda, M.; Okamoto, Y.; Sato, T.; Yoshii, E.; Koizumi, T.; Shiro, M. *J. Am. Chem. Soc.* **1998**, *120*, 4947.
6. For a review of [4 + 3] cycloadditions, see: Hosomi, A.; Tominaga, Y. In "Comprehensive Organic Synthesis"; Trost, B. M., Fleming, I., Eds.; Pergamon: Oxford, 1991; Vol. 5, pp 593-615.
7. For reviews of the Cope rearrangement of cis-1,2-divinylcyclopropanes, see (a) Davies, Huw M. L. *Tetrahedron* **1993**, *49,* 5203; (b) Piers, E. In "Comprehensive Organic Synthesis"; Trost, B. M., Ed.; Pergamon Press: Oxford U.K., 1991; Vol. 5, pp 971-998.

8. Barluenga, J.; Aznar, F.; Martin, A.; Vázquez, J. T. *J. Am. Chem. Soc.* **1995**, *117*, 9419, and references cited therein.

9. Lee, J.; Kim, H.; Cha, J. K. *J. Am. Chem. Soc.* **1995**, *117*, 9919, and references cited therein.

Appendix
Chemical Abstracts Nomenclature (Collective Index Number); (Registry Number)

Hexamethylphosphoramide HIGHLY TOXIC: Phosphoric triamide, hexamethyl- (8,9); (680-31-9)

1-(1-Ethoxyethoxy)-1,2-propadiene: 1,2-Propadiene, 1-(1-ethoxyethoxy)- (9); (20524-89-4)

Ethyl vinyl ether: Ether, ethyl vinyl (8); Ethene, ethoxy- (9); (109-92-2)

p-Toluenesulfonic acid monohydrate (8); Benzenesulfonic acid, 4-methyl-, monohydrate (9); (6192-2-5)

Propargyl alcohol: 2-Propyn-1-ol (8,9); (107-19-7)

3-(1-Ethoxyethoxy)-1-propyne: 1-Propyne, 3-(1-ethoxyethoxy)- (9); (18669-04-0)

Potassium tert-butoxide: tert-Butyl aclohol, potassium salt (8); 2-Propanol, 2-methyl-, potassium salt (9); (865-47-4)

4,4'-Thiobis(2-tert-butyl-m-cresol): Phenol, 4,4'-thiobis[2-(1,1-dimethylethyl)-3-methyl- (8); m-Cresol, 4,4'-thiobis[2-tert-butyl- (9); (4120-97-2)

Butyllithium: Lithium, butyl- (8,9); (109-72-8)

tert-Butyldimethylsilyl chloride: Silane, chloro(1,1-dimethylethyl)dimethyl- (9); (18162-48-6)

Triethylamine (8); Ethanamine, N,N-diethyl- (9); (121-44-8)

(E)-1-(tert-Butyldimethylsilyl)-3-trimethylsilyl-2-propen-1-one: Silane, (1,1-dimethylethyl)dimethyl[1-oxo-3-(trimethylsilyl)-2-propenyl]- (E)- (11); (83578-66-9)

Trimethylsilyl chloride: Silane, chlorotrimethyl- (8,9); (75-77-4)

Diisopropylamine (8); 2-Propanamine, N-(1-methylethyl)- (9); (108-18-9)

1-Acetyl-1-cyclohexene: Ethanone, 1-(1-cyclohexen-1-yl)- (9); (932-66-1)

Acetic acid (8,9); (64-19-7)

1,3-Diphenylacetone p-tosylhydrazone: p-Toluenesulfonic acid, (α-benzylphenethylidene)hydrazide (8); Benzenesulfonic acid, 4-methyl-, [2-phenyl-1-(phenylmethyl)ethylidene]hydrazide (9); (19816-88-7)

Trifluoroacetic acid: Acetic acid, trifluoro- (8,9); (76-05-1)

TABLE

PREPARATION OF CYCLOHEPTENONES USING [3 + 4] ANNULATION

tone Enolate	(*E*)-**3**		(*Z*)-**3**	
	Product	Yield	Product	Yield
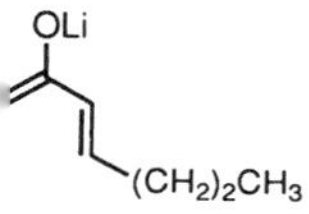 OLi, $(CH_2)_2CH_3$	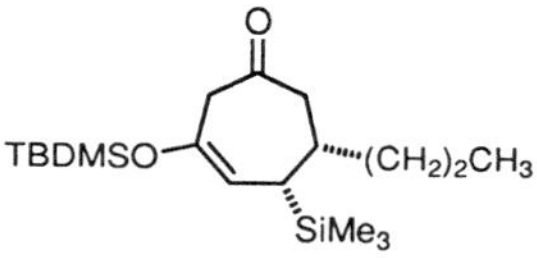 O, TBDMSO, $(CH_2)_2CH_3$, $SiMe_3$	73%	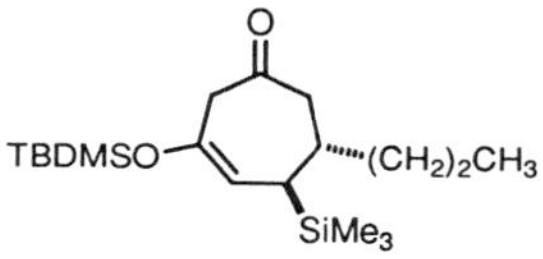 O, TBDMSO, $(CH_2)_2CH_3$, $SiMe_3$	31%
OLi, $CH(CH_3)_2$	O, TBDMSO, $CH(CH_3)_2$, $SiMe_3$	84%	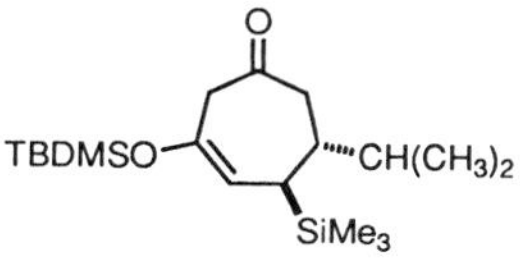 O, TBDMSO, $CH(CH_3)_2$, $SiMe_3$	11%
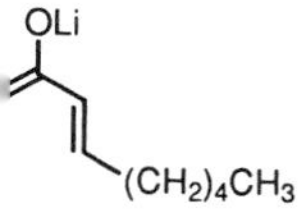 OLi, $(CH_2)_4CH_3$	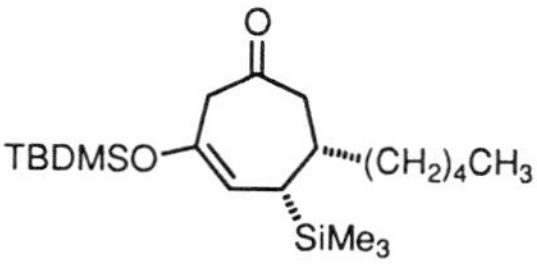 O, TBDMSO, $(CH_2)_4CH_3$, $SiMe_3$	84%	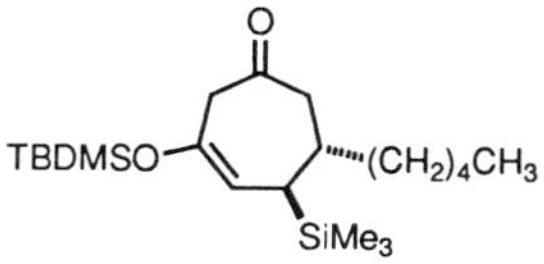 O, TBDMSO, $(CH_2)_4CH_3$, $SiMe_3$	29%
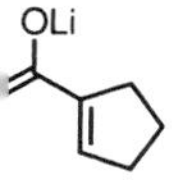 OLi	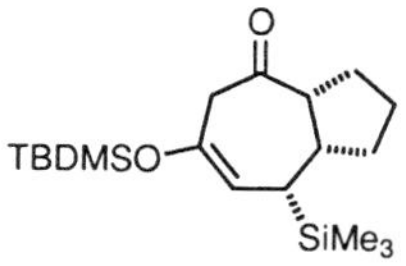 O, TBDMSO, $SiMe_3$	73%		
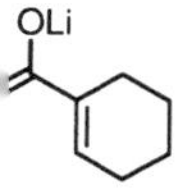 OLi	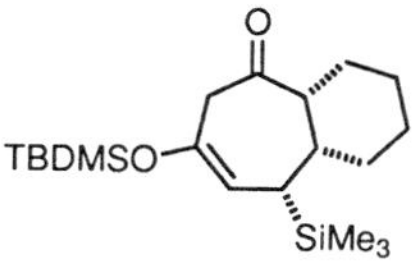 O, TBDMSO, $SiMe_3$	82%		
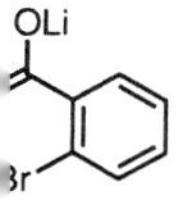 OLi	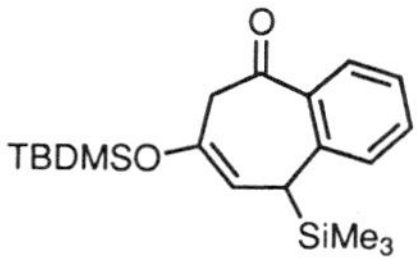 O, TBDMSO, $SiMe_3$	30%		

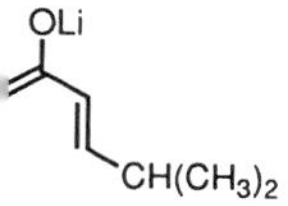

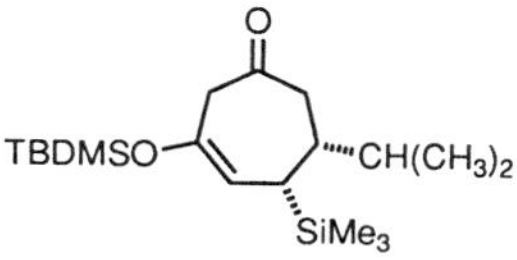

GENERATION OF 1-PROPYNYLLITHIUM FROM (Z/E)-1-BROMO-1-PROPENE: 6-PHENYLHEX-2-YN-5-EN-4-OL

[1-Hexen-4-yn-3-ol, 1-phenyl-, (E)-])

A. CH_3CH=CHBr —(BuLi, THF, -78°C, 2 hr)→ [CH_3C≡CLi]

B. [CH_3C≡CLi] + PhCH=CHCHO —(THF, -78°C, 30 min)→ PhCH=CHCH(OH)C≡CCH_3

Submitted by Dominique Toussaint and Jean Suffert.[1]

Checked by Robin R. Frey and Stephen F. Martin.

1. Procedure

A. and B. In a dry, 500-mL, two-necked flask flushed with argon, fitted with a magnetic stirring bar and a 250-mL pressure-equalizing addition funnel is placed 18.65 g of (Z/E)-1-bromo-1-propene (0.15 mol) (Note 1) in 100 mL of tetrahydrofuran (THF, Note 2). The flask is cooled to -78°C with a dry ice-acetone bath, and 140 mL of butyllithium (BuLi, 1.57 M in hexane, 0.22 mol) (Note 3) is added dropwise over 30 min. The funnel is rinsed with an additional 10-mL portion of THF. The milky white suspension (Note 4) is stirred at -78°C for another 2 hr. Freshly distilled trans-cinnamaldehyde (13.21 g, 0.1 mol) in 50 mL of THF is added dropwise over 10 min, and the funnel is rinsed with 10 mL of THF. The solution is stirred for 30 min at -78°C (Note 5), quenched by the addition of 50 mL of aqueous saturated ammonium chloride (NH_4Cl), allowed to warm to room temperature and poured into a 1-L separatory funnel containing 100 mL of water and 100 mL of ether. The layers are separated, and the aqueous phase is extracted with three 100-mL portions of ether. The combined organic layers are washed with two 100-mL portions of brine, dried over anhydrous sodium sulfate and filtered. The solvent is removed by rotary evaporation leaving 17.17 g of a yellow oil that is almost pure based upon ^1H NMR and GC (crude yield: >99%) (Notes 6 and 7). Extensive purification can be achieved by flash chromatography on silica gel eluting with 20% ether in hexane (Note 8) to leave a yellow oil that solidifies in the freezer to yield 15.88 g of a pale yellow solid (mp 40-42°C, yield, 92%) (Note 9).

2. Notes

1. 1-Bromo-1-propene was purchased from Lancaster Synthesis Inc. (mixture of isomers, technical grade) and distilled prior to use (bp 58-62°C, 760 mm).

2. THF was distilled from sodium/benzophenone ketyl under nitrogen prior to use.

3. Butyllithium was purchased from Aldrich Chemical Company, Inc., and titrated with N-pivaloyl-o-toluidine.[2] The checkers observed that use of less concentrated solutions of BuLi resulted in longer reaction times.

4. In some cases no suspension was observed (only a yellowish solution was obtained), but the reactions worked equally well.

5. The progress of the reaction can be monitored by TLC by quenching an aliquot with a mixture of aqueous saturated NH_4Cl/ether and eluting with Et_2O/hexane 20/80 (R_f trans-cinnamaldehyde = 0.40, R_f product = 0.25). Visualization can be achieved with vanillin (25 g/L of ethanol containing 1 mL of concd sulfuric acid) and heating on a hot plate.

6. ^1H NMR (300 MHz) and ^{13}C NMR (75 MHz) spectra were recorded in $CDCl_3$ solution on a Varian Unity Plus 300 MHz spectrometer.

7. The purity of the crude solid was determined to be 97% by GC (column conditions: SE-30 column, 25 m x 0.32 mm, He 0.8 kg/cm^2 carrier gas). The recovered oil solidified upon standing in the freezer. Attempts to recrystallize the resulting off-white material (in cyclohexane or pentane/ethyl acetate) only met with oiling.

8. An 8-cm diameter column packed with 15 cm of silica was used. Some decomposition is observed during purification.

9. This material was >99% pure as determined by GC (column conditions: SE-30 column, 25 m x 0.32 mm, He 0.8 kg/cm^2 carrier gas) and showed the following spectroscopic characteristics: IR ($CHCl_3$) cm^{-1}: 3597, 2242, 1632; MS (CI) m/z 173.0958 [$C_{12}H_{12}O$+H (M+1) requires 173.0966], 172, 155, 133. ^1H NMR (300 MHz, $CDCl_3$) δ: 1.89 (d, 3 H, J = 2.1), 2.24 (d, 1 H, J = 5.9), 5.00-5.03 (br m, 1 H), 6.28 (dd, 1 H, J = 15.7, 5.9), 6.73 (d, 1 H, J = 15.7), 7.21-7.42 (comp, 5 H); ^{13}C NMR (75 MHz) δ: 3.6, 63.1, 65.2, 82.8, 126.7, 127.9, 128.5, 128.7, 131.4, 136.1. Anal. Calcd for $C_{12}H_{12}O$: C, 83.64; H, 7.02. Found: C, 83.56; H, 6.96.

Waste Disposal Information

All toxic materials were disposed of in accordance with "Prudent Practices in the Laboratory"; National Academy Press; Washington, DC, 1995.

3. Discussion

The use of 1-propynyllithium in the synthesis of natural and unnatural compounds has been extensive, and a number of procedures have been reported for its generation. The most common method uses propyne gas, which may be metallated with lithium in liquid ammonia and other solvents,[3] or butyllithium[4] or lithium hydride in dimethyl sulfoxide (DMSO).[5] However, propyne is expensive, and it is important to have a more economical source of 1-propynyllithium. In some cases, propyne has been replaced by the inexpensive welding gas mixture MAPP (Methyl Acetylene, Propadiene, Propene), which contains up to 13.5% of propyne.[6] The anion can also be prepared by direct metallation of allene with BuLi at -78°C.[7] 1-Propynyllithium has also been generated by the reaction of 1-chloro-1-propene with BuLi or sec-BuLi, but the subsequent reaction with methyl iodide gave at best a 50% yield of product.[8] Moreover, 1-chloro-1-propene is expensive, and, because of its low boiling point (37°C), it is somewhat inconvenient to use. For example, the conditions required for generating propynyllithium from 1-chloro-1-propene involve use of a liquid nitrogen - ethanol cooling bath (-110°C). This technical difficulty somewhat limits the scale and utility of this procedure. The method of Gribble, et al. for generating 1-propynyllithium uses 1,2-dibromopropane and 3 equiv of lithium diisopropylamide (LDA).[9] The presence of such a large excess of base does not allow the addition of 1-propynyllithium to highly functionalized electrophiles. Recently a new procedure for generating 1-propynyllithium was reported that involved the reaction of an allenic telluride with BuLi at -70°C, followed by heating the mixture at 66°C and quenching the anion with an electrophile such as benzaldehyde or cyclohexenone.[10]

The present procedure provides a method for the easy generation of 1-propynyllithium from an inexpensive, commercially available starting material. The anion is prepared in anhydrous THF in high yield by reaction of the commercially available mixture of (Z/E)-1-bromopropene with BuLi at -78°C. Its reaction with various electrophiles such as aldehydes or Weinreb amides[11] is clean and efficient to afford secondary alcohols and ketones respectively (Method a, Table). The 1-propynyllithium generated in this way can be transmetallated with $CeCl_3$[12] (Method b, Table) or $ZnCl_2$ [in the presence of $Pd(PPh_3)_4$]), Method c, Table][13] to add to enolizable ketones and acid chlorides, respectively. In all cases yields were high (see Table).[14]

TABLE I

ADDITION OF 1-PROPYNYLLITHIUM TO VARIOUS ELECTROPHILES

Entry	Starting Compound	Method	Product		Yield (%)
1	O, H	a	3a	HO, CH_3	94
2	O, H	a	3b	OH, CH_3	94
3	O, H	a	3c	HO, CH_3	95
4	O	b	4a	HO, CH_3	92
5	O, O	b	4b	HO, CH_3, O	89
6	O	b	4c	HO, CH_3	90
7	O	b	4d	HO, CH_3	95
8	O	b	4e	HO, CH_3	88
9	THPO, O	b	4f	HO, CH_3, THPO	86
10	OMe, N, N, O, Ph, O, O	a	5a	CH_3, N, O, Ph, O, O	86
11	OMe, OMe, N, O	a	5b	OMe, O, CH_3	89
12	O, Cl	c	5c	O, CH_3	90
13	O, Cl	c	5d	O, CH_3	78

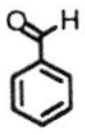

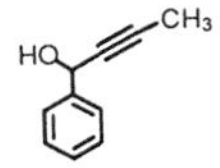

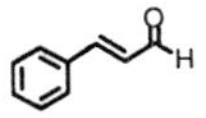

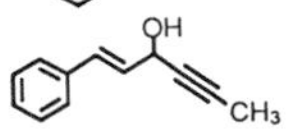

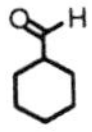

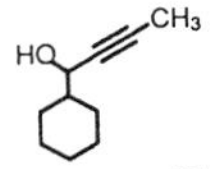

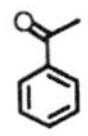

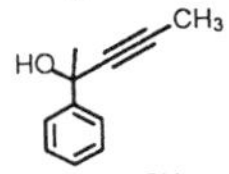

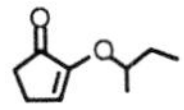

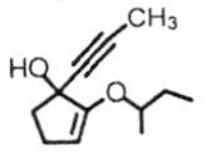

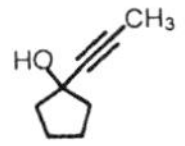

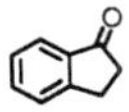

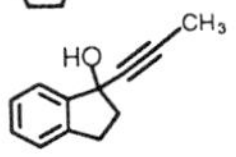

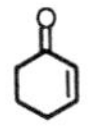

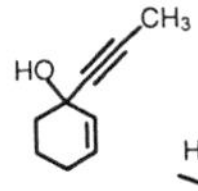

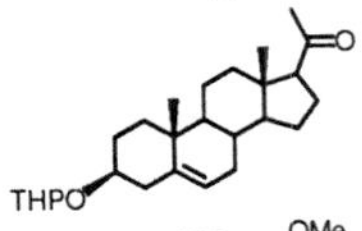

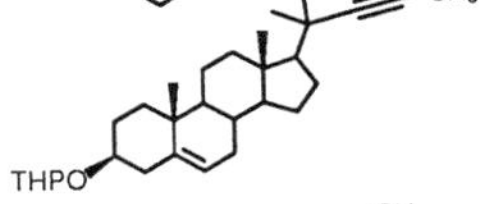

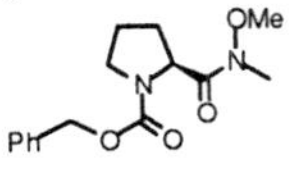

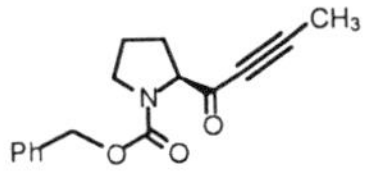

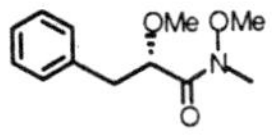

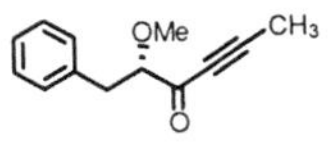

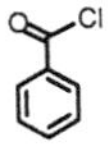

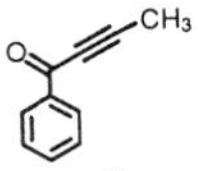

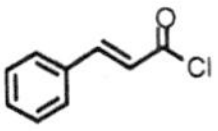

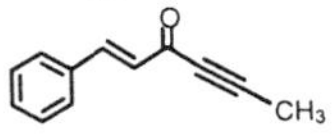

1. Laboratoire de Pharmacochimie Moléculaire - UPR 421 du CNRS, Centre de Neurochimie, 5, rue Blaise Pascal 67084 Strasbourg Cedex, France. Present address: Laboratoire de Pharmacochimie de la Communication Cellulaire (ERS 655 du CNRS) 74, Route du Rhin B.P. 2467401 Illkirch Cedex, France.
2. Suffert, J. *J. Org. Chem.* **1989**, *54*, 509.
3. Starowieyski, K. B.; Chwojnowski, A.; Kusmierek, Z. *J. Organomet. Chem.* **1980**, *192*, 147; Robinson, J. A.; Flohr, H.; Kempe, U. M.; Pannhorst, W.; Retey, J. *Liebigs Ann. Chem.* **1983**, 181; Verkruijsse, H. D.; Brandsma, L. *Synth. Commun.* **1991**, *21*, 235.
4. Berger, H. O.; Noeth, H.; Rub, G.; Wrackmeyer, B. *Chem. Ber.* **1980**, *113*, 1235; Bender, S. L.; Detty, M. R.; Haley, N. F. *Tetrahedron Lett.* **1982**, *23*, 1531; Blunt, J. W.; Hartshorn, M. P.; Soong, L. T.; Munro, M. H. G.; Vannoort, R. W.; Vaughan, J. *Aust. J. Chem.* **1983**, *36*, 1387; Hooz, J.; Calzada, J. G.; McMaster, D. *Tetrahedron Lett.* **1985**, *26*, 271; Maignan, C.; Guessous, A.; Rouessac, F. *Bull. Soc. Chim. Fr.* **1986**, *78*, 645; Stang, P. J.; Boehshar, M.; Wingert, H.; Kitamura, T. *J. Am. Chem. Soc.* **1988**, *110*, 3272; Perri, S. T.; Dyke, H. J.; Moore, H. W. *J. Org. Chem.* **1989**, *54*, 2032; Marshall, J. A.; Wang, X.-j. *J. Org. Chem.* **1991**, *56*, 960.
5. Braude, E. A.; Coles, J. A. *J. Chem. Soc.* **1951**, 2078; Tarrant, D.; Savory, J.; Iglehart, E. S. *J. Org. Chem.* **1964**, *29*, 2009; Kriz, J.; Benes, M. J.; Peska, J. *Collect. Czech. Chem. Commun.* **1967**, *32*, 398; Corey, E. J.; Kirst, H. A. *Tetrahedron Lett.* **1968**, 5041; Pant, B. C.; Davidsohn, W. E.; Henry, M. C. *J. Organomet. Chem.* **1969**, *16*, 413.
6. Jauch, J.; Schmalzing, D.; Schurig, V.; Emberger, R.; Hopp, R.; Köpsel, M.; Silberzahn, W.; Werkhoff, P. *Angew. Chem., Int. Edit. Engl.* **1989**, *28*, 1022.
7. Keinan, E.; Bosch, E. *J. Org. Chem.* **1986**, *51*, 4006.
8. Nelson, D. J. *J. Org. Chem.* **1984**, *49*, 2059.

9. Gribble, G. W; Joyner, H. H.; Switzer, F. L. *Synth. Commun.* **1992**, *22*, 2997.
10. Kanda, T.; Ando, Y.; Kato, S.; Kambe, N.; Sonoda, N. *Synlett* **1995**, 745.
11. Nahm, S.; Weinreb, S. M. *Tetrahedron Lett.* **1981**, *22* , 3815.
12. Imamoto, T.; Sugiura, Y.; Takiyama, N. *Tetrahedron Lett.* **1984**, *25*, 4233.
13. Negishi, E.-i.; Bagheri, V.; Chatterjee, S; Luo, F.-T.; Miller, J. A.; Stoll, A. T. *Tetrahedron Lett.* **1983**, *24*, 5181; Verkruijsse, H. D.; Heus-Kloos, Y. A.; Brandsma, L. *J. Organomet. Chem.* **1988**, *338*, 289.
14. Suffert, J.; Toussaint, D. *J. Org. Chem.* **1995**, *60*, 3550.

Appendix

Chemical Abstracts Nomenclature (Collective Index Number); (Registry Number)

1-Propynyllithium: Lithium, 1-propynyl- (8,9); 4529-04-8)

(Z/E)-1-Bromo-1-propene: 1-Propene, 1-bromo- (8,9); (590-14-7)

6-Phenylhex-2-yn-5-en-4-ol: 1-Hexen-4-yn-3-ol, 1-phenyl-, (E)- (10); (63124-68-5)

Butyllithium: Lithium, butyl- (8,9); (109-72-8)

trans-Cinnamaldehyde: Cinnamaldehyde, (E)- (8); 2-Propenal, 3-phenyl-, (E)- (9); (14371-10-9)

N-Pivaloyl-o-toluidine: Propanamide, 2,2-dimethyl-N-(2-methylphenyl)- (10); (61495-04-3)

6-CHLORO-1-HEXENE AND 8-CHLORO-1-OCTENE

(1-Hexene, 6-chloro- and 1-Octene, 8-chloro-)

A. $CH_2{=}CHCH_2Br \xrightarrow[Et_2O]{Mg} CH_2{=}CHCH_2MgBr{\cdot}2\ Et_2O$

B. $CH_2{=}CHCH_2MgBr{\cdot}2\ Et_2O \xrightarrow[65°C]{THF} CH_2{=}CHCH_2MgBr{\cdot}2\ THF$

C. $CH_2{=}CHCH_2MgBr{\cdot}2\ THF + Cl(CH_2)_nBr \xrightarrow[65°C]{THF} CH_2{=}CHCH_2(CH_2)_nCl$

$n = 3,5$

Submitted by Pierre Mazerolles, Paul Boussaguet, and Vincent Huc.[1]
Checked by Frédéric Denonne and Léon Ghosez.

1. Procedure

A. Allylmagnesium bromide (ethereal complex solution). A dry, 3-L, three-necked, round-bottomed flask is equipped with a sealed mechanical stirrer (Note 1), 500-mL, pressure-equalizing, dropping funnel and a reflux condenser, the top of which is connected to a calcium chloride drying tube. The flask is charged with a large excess of magnesium turnings (90.00 g, 3.75 g-atom) (Note 2) and 150 mL of dry diethyl ether. To the stirred mixture is added dropwise a solution of allyl bromide (Note 3) (181.60 g, 1.50 mol) in 1.5 L of dry diethyl ether. At the end of the addition (8 hr) the mixture is stirred for 1 hr, whereupon the Grignard reagent solution is transferred under nitrogen, with a cannula, into a 2-L, two-necked, round-bottomed flask equipped with a 500-mL dropping funnel and a reflux condenser fitted with a swan neck (distillation) adapter connected with a descending condenser (Note 4).

B. Allylmagnesium bromide (THF complex solution). After removal of the water in the vertical reflux condenser, the solution of the Grignard reagent is heated with a water-bath (45°C, then progressively up to 65°C) with stirring to remove the uncomplexed diethyl ether (2 hr). Dry tetrahydrofuran (THF, 500 mL) is added rapidly (10 min) to the gray pasty residue. A vigorous reaction occurs, and the decomplexed diethyl ether is easily removed by distillation. The reaction is completed by heating the mixture with stirring for 1 hr to give a fluid gray THF solution of allylmagnesium bromide (Note 5).

C. 6-Chloro-1-hexene. The swan neck and the descending condenser are removed, the vertical reflux condenser is fitted with a drying tube, and water is again circulated through the condenser. To the stirred mixture heated at 50-60°C, pure 1-bromo-3-chloropropane (Note 6) (150.00 g, 0.95 mol) is added at a rate sufficient to maintain a good reflux in the condenser. At the end of the addition (0.5 hr), the mixture is boiled for 1 hr and then cooled. Excess Grignard reagent is destroyed by *slow* addition of water, while cooling the mixture, to produce two clear layers. The aqueous phase is extracted with pentane (3 x 300 mL each), and the organic phase is dried over calcium chloride. Removal of the solvent (water bath at 45°C, then progressively to 95°C, Vigreux column 2") gives 183.40 g of oily residue. The remainder of the solvent is removed under vacuum (up to 45°C/147 mm). Distillation of the residue in a Claisen flask gives the following fractions: Head fraction, from 50°C to 74°C/130 mm (21.16 g) containing 60% of chlorohexene (Note 7); central fraction, 74-75°C/130 mm (93.00 g, 0.78 mol, 82% yield) consisting of almost pure chlorohexene (Note 8); residue, clear yellow, about 1 mL.

8-Chloro-1-octene. Similarly, the reaction of 1-bromo-5-chloropentane (Note 9) (88.37 g, 0.47 mol) with a solution of allylmagnesium bromide prepared as above from allyl bromide (90.75 g, 0.75 mol) and magnesium turnings (45.00 g, 1.87 mol), gives, after the usual treatment, 60.14 g of 8-chloro-1-octene (86% yield), bp 75-76°C/20 mm (Note 10).

2. Notes

1. Stirring with a large magnetic bar and an efficient magnetic stirrer (IKA mod. MAG. RET) is possible, but the magnetic bar turns with some difficulty at the beginning of the reaction.

2. A large excess of magnesium turnings is necessary to prevent the formation of biallyl (1,5-hexadiene) by a Würtz-coupling reaction. Small magnesium turnings (3-4-mm size, 0.5-mm thickness) were obtained with a lathe from a pure magnesium bar (A. Weber Métaux). The checkers used Aldrich magnesium turnings (98%).

3. The reaction starts easily; the addition of a crystal of iodine is not necessary. To minimize the formation of hexadiene, the addition of allyl bromide must be slow enough to maintain the temperature of the flask below the boiling point of ether. Allyl bromide (Fluka "purum") even when stored in the dark in the presence of silver wool contains some high-boiling material and must be purified by distillation. Pure product (bp 71°C) is used just after distillation.

4. Unreacted magnesium, washed with water, alcohol and ether and dried in an oven, weighs 56.16 g (94% of the theoretical yield).

5. Such a large amount of THF is not necessary to remove the complexed ether, but if a smaller quantity is used, the mixture solidifies on cooling.

6. 1-Bromo-3-chloropropane (Fluka "purum") was dried and distilled before use (bp 56°C/ 34 mm).

7. The amount was estimated by VPC. As the principal impurity is THF, the yield of chlorohexene may be increased to 90% by careful distillation.

8. The product obtained contains traces of THF and a small amount (3-5%) of 6-bromo-1-hexene (identified by mass spectroscopy). Usually this by-product is not an

impurity, especially when 6-chloro-1-hexene is used to prepare the corresponding Grignard reagent. The spectral data are as follows: ^{1}H NMR (80 MHz, $CDCl_3$) δ: 1.79 (m, 6 H), 3.51 (t, 2 H, J = 8.0), 4.98 (m, 2 H), 5.71 (m, 1 H); ^{13}C NMR (20 MHz, $CDCl_3$) δ: 26.1, 32.0, 32.9, 44.8, 114.8, 136.1.

9. 1-Bromo-5-chloropentane is easily prepared from tetrahydropyran according to the process described by Newman and Wotiz,[20] bp 89°C/17 mm; n_D^{20} 1.4842. The checkers purchased 1-bromo-5-chloropentane from Aldrich Chemical Company, Inc.

10. Like 6-chloro-1-hexene, 8-chloro-1-octene contains a small amount (3-5%) of the corresponding bromide (8-bromo-1-octene), which is not a problem in some reactions.

Waste Disposal Information

All toxic materials were disposed of in accordance with "Prudent Practices in the Laboratory"; National Academy Press; Washington, DC, 1995.

3. Discussion

Haloalkenes, $CH_2=CH-(CH_2)_nX$ (n = 3-7; X = Cl, Br), are interesting synthons for use: in the preparation of long chain alkenols and alkenoic acids;[2] in cyclization reactions,[3-9] as intermediates in the synthesis of pheromones;[10-11] in the preparation of silica gel having a functional surface;[12] and for the synthesis of ω-iodochloroalkanes,[13] and organogermanium dendrimers.[14] They are usually prepared from the corresponding alkenols.

An interesting synthesis of this type of compound is the reaction of allyl bromide with a mono Grignard reagent of a dihalide, especially a chlorobromide:[15] However this synthesis is only suitable for long chains, n ≥ 4: n = 4, 45% yield; n = 6, 60% yield.

$$CH_2=CHCH_2Br \quad + \quad BrMg(CH_2)_nCl \xrightarrow{\text{Ether}} CH_2=CH-(CH_2)_{n+1}Cl$$

On the other hand, bromoalkanes react with saturated[16] or ethylenic[17] Grignard reagents in tetrahydrofuran in the presence of a suitable catalyst (dilithium tetrachlorocuprate[18-19]):

$$RMgBr + Br(CH_2)_nBr \xrightarrow[THF]{Li_2CuCl_4} R(CH_2)_nBr + R(CH_2)_nR$$

n = 3, 4, 5, 6, 10 (54-96% yield)

It was shown that alkyl bromides react *without a catalyst* with allylmagnesium bromide *in THF*. However, because allyl bromide reacts quantitatively with magnesium in THF to give 1,5-hexadiene by a Würtz-coupling reaction, allylmagnesium bromide can be obtained in this solvent only by solvent exchange:

$$CH_2{=}CHCH_2Br + Mg \xrightarrow{THF} CH_2{=}CHCH_2\text{-}CH_2CH{=}CH_2$$

$$CH_2{=}CHCH_2Br + Mg \xrightarrow{Et_2O} CH_2{=}CHCH_2MgBr\cdot 2\ Et_2O \xrightarrow[65°C]{THF} CH_2{=}CHCH_2MgBr\cdot 2\ THF$$

The reaction with the readily available 1-bromo-3-chloropropane is a direct, inexpensive method for preparing of 6-chlorohexene in high yields (80-90%); this reaction may be readily extended to 1-bromo-5-chloropentane[20] and to other accessible ω-bromochloroalkanes.[21]

With dibromoalkanes the corresponding dienes are formed in good yield:

$$Br(CH_2)_5Br + CH_2{=}CHCH_2MgBr \xrightarrow{THF} H_2C{=}CHCH_2\text{-}(CH_2)_5\text{-}CH_2CH{=}CH_2$$

(85.2%)

With aralkyl dibromides, only alkyl bromides react:

$$BrCH_2-C_6H_4-Br + CH_2{=}CHCH_2MgBr \xrightarrow{THF} CH_2{=}CHCH_2CH_2-C_6H_4-$$

(excess) (87.7%)

1. Laboratoire d'Hétérochimie Fondamentale et Appliquée, Université Paul Sabatier, 31062 TOULOUSE Cedex, France.
2. Gaubert, P.; Linstead, R. P.; Rydon, H. N. *J. Chem. Soc.* **1937**, 1971.
3. Liao, Z.-K.; Kohn, H. *J. Org. Chem.* **1985**, *50*, 1884.
4. Bailey, W. F.; Gagnier, R. P.; Patricia, J. J. *J. Org. Chem.* **1984**, *49*, 2098.
5. Ley, S. V.; Lygo, B. *Tetrahedron Lett.* **1982**, *23*, 4625.
6. McIntosh, J. M. *J. Org. Chem.* **1982**, *47*, 3777.
7. Toi, H.; Yamamoto, Y.; Sonoda, A.; Murahashi, S.-I. *Tetrahedron* **1981**, *37*, 2261.
8. Gössinger, E.; Imhof, R.; Wehrli, H. *Helv. Chim. Acta* **1975**, *58*, 96.
9. Oude-Alink, B. A. M.; Chan, A. W. K.; Gutsche, C. D. *J. Org. Chem.* **1973**, *38*, 1993.
10. Doherty, A. M.; Ley, S. V.; Lygo, B.; Williams, D. J. *J. Chem. Soc., Perkin Trans. 1* **1984**, 1371.
11. Smith, L. M.; Smith, R. G.; Loehr, T. M.; Daves, G. D., Jr.; Daterman, G. E.; Wohleb, R. H. *J. Org. Chem.* **1978**, *43*, 2361.
12. Matlin, S. A.; Gandham, P. S. *J. Chem. Soc., Chem. Commun.* **1984**, 798.
13. Kabalka, G. W.; Gooch, E. E., III *J. Org. Chem.* **1980**, *45*, 3578.
14. Huc, V.; Boussaguet, P.; Mazerolles, P. *J. Organometal. Chem.* **1996**, *521*, 253.
15. Noël, M.; Combret, J. C.; Leroux, Y.; Normant, H. *C.R. Acad. Sci., Paris, Ser. C.* **1969**, *268*, 1152.
16. Friedman, L.; Shani, A. *J. Am. Chem. Soc.* **1974**, *96*, 7101.

17. Subramaniam, C. S.; Thomas, P. J.; Mamdapur, V. R.; Chadha, M. S. *Tetrahedron Lett.* **1978**, 495.

18. Tamura, M.; Kochi, J. *Synthesis* **1971**, 303.

19. Tamura, M.; Kochi, J. *J. Am. Chem. Soc.* **1971**, *93,* 1485.

20. Newman, M. S.; Wotiz, J. H. *J. Am. Chem. Soc.* **1949**, *71*, 1292.

21. Hahn, R. C. *J. Org. Chem.* **1988**, *53*, 1331.

Appendix

Chemical Abstracts Nomenclature (Collective Index Number); (Registry Number)

6-Chloro-1-hexene: 1-Hexene, 6-chloro- (8,9); (928-89-2)

8-Chloro-1-octene: 1-Octene, 8-chloro- (8,9); (871-90-9)

Allylmagnesium bromide: Magnesium, bromo-2-propenyl- (8,9); (1730-25-2)

Magnesium (8,9); (7439-95-4)

Allyl bromide: 1-Propene, 3-bromo- (8,9); (106-95-6)

1-Bromo-3-chloropropane: Propane, 1-bromo-3-chloro- (8,9); (109-70-6)

1-Bromo-5-chloropentane: Pentane, 1-bromo-5-chloro- (9); (54512-75-3)

6-Bromo-1-hexene: 1-Hexene, 6-bromo- (8,9); (2695-47-8)

USE OF CERIUM(III) CHLORIDE IN THE REACTIONS OF CARBONYL COMPOUNDS WITH ORGANOLITHIUMS OR GRIGNARD REAGENTS FOR THE SUPPRESSION OF ABNORMAL REACTIONS: 1-BUTYL-1,2,3,4-TETRAHYDRO-1-NAPHTHOL

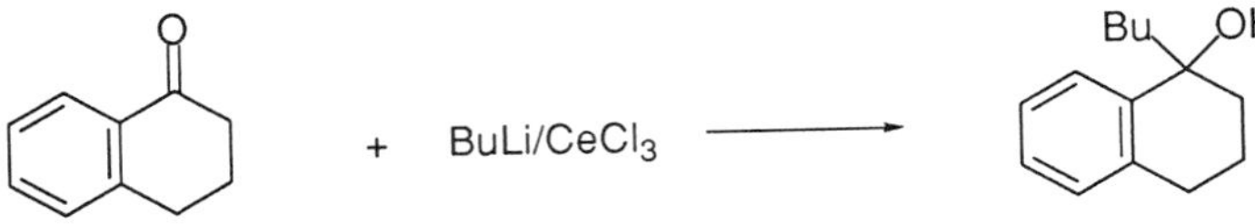

Submitted by Nobuhiro Takeda and Tsuneo Imamoto.[1]

Checked by Scott Ballentine and David J. Hart.

1. Procedure

1-Butyl-1,2,3,4-tetrahydro-1-naphthol. A 500-mL, three-necked, round-bottomed flask is equipped with two glass stoppers and a three-way stopcock (Notes 1 and 2). The flask is connected to a trap (Note 3) that is cooled at -78°C in a dry ice-ethanol bath and attached to a vacuum pump (Figure 1). The flask is charged with powdered cerium(III) chloride heptahydrate ($CeCl_3 \cdot 7H_2O$, 44.7 g, 0.12 mol) (Note 4) and evacuated to 0.1-0.2 mm. After gradual warming to 90°C over 30 min with an oil bath, the flask is heated at 90-100°C for 2 hr with intermittent shaking (Note 5). The system is filled with dry argon and cooled to room temperature. The solid is transferred to a mortar and quickly pulverized with a pestle. The resulting white powder and a magnetic stirring bar (Note 6) are placed in the original flask. Gradual warming to 90°C at 0.1-0.2 mm over 30 min, followed by further evacuating at 90-100°C for 1.5 hr with intermittent shaking, gives cerium(III) chloride monohydrate ($CeCl_3 \cdot H_2O$) (Note 7). The cerium(III) chloride monohydrate is gradually warmed to

140°C over 30 min under reduced pressure (0.1-0.2 mm) without stirring (Note 8). Heating at 140-150°C/0.1-0.2 mm for 2 hr with gentle stirring (Note 9) affords a fine, white powder of anhydrous cerium(III) chloride (Note 10). While the flask is still hot, the area that is not immersed in the oil bath is heated by the use of a heat gun in order to remove traces of water. After introduction of argon into the flask, the resulting anhydrous cerium(III) chloride is cooled to room temperature. One of the glass stoppers is replaced by a rubber septum in a stream of dry argon.

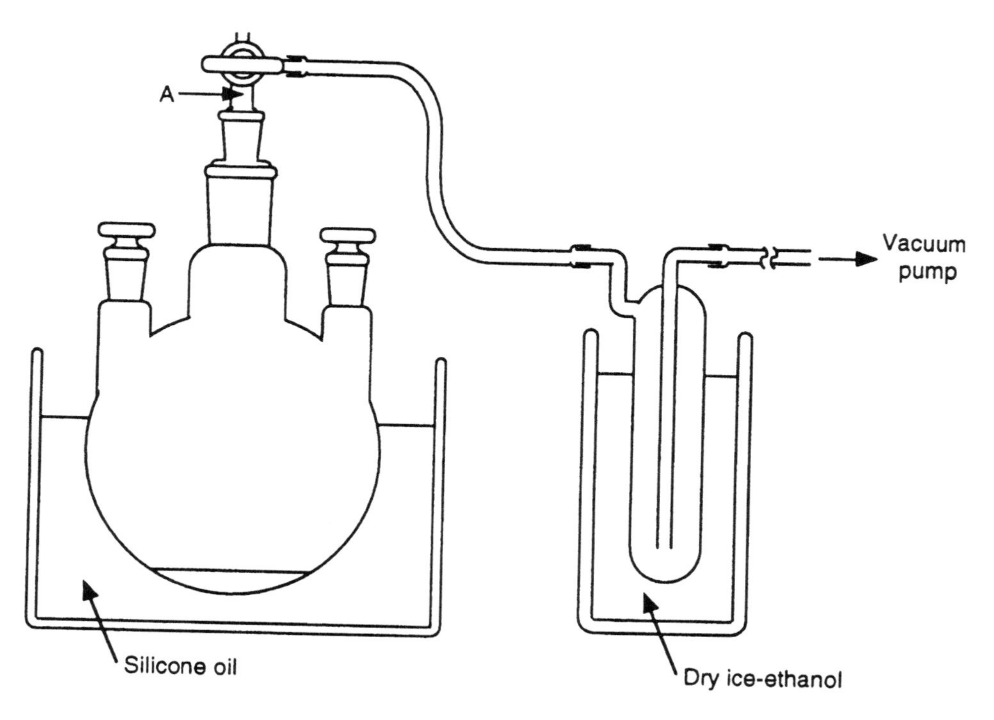

Figure 1. Apparatus for the preparation of anhydrous cerium(III) chloride.

Tetrahydrofuran (200 mL) (Note 11) is added to the powder of anhydrous cerium(III) chloride in one portion at 0°C in an ice-water bath with vigorous stirring (Note 12). After the mixture is stirred under argon atmosphere at room temperature overnight (Note 13), the resulting milky suspension is cooled to -78°C with a dry ice-ethanol bath. A 1.54 M hexane solution of butyllithium (78 mL, 0.12 mmol) is added dropwise over 15 min to the suspension, so that the temperature of the reaction mixture remains at -78°C. The resulting yellow suspension is stirred at the same temperature for 30 min, and α-tetralone (13.5 mL, 14.8 g, 0.101 mol) (Note 14) is added to the mixture at -78°C over 15-30 min. After the mixture is stirred at the same temperature for 30 min, the dry ice-ethanol bath is removed, and the pale yellow suspension is warmed to room temperature. An aqueous 5% solution of acetic acid (200 mL) is added to the suspension with vigorous stirring. The reaction mixture is transferred to a 1-L separatory funnel, and ethyl acetate (100 mL) is added. The organic layer is separated, and the water layer is extracted with ethyl acetate (2 x 50 mL). The combined organic layers are washed sequentially with brine (50 mL), a saturated aqueous solution of sodium bicarbonate (50 mL), and brine (50 mL) and dried over anhydrous magnesium sulfate. After evaporation of the solvent by a rotary evaporator under reduced pressure, distillation of the residual oil gives 19.1-19.9 g of 1-butyl-1,2,3,4-tetrahydro-1-naphthol (92-97%) (Notes 15 and 16) as a colorless oil, 103-104°C/0.5 mm.

2. Notes

1. The joints of the apparatus are lubricated with silicone grease to avoid permeation of moisture from outside. All temperatures refer to external temperatures.

2. Neither cotton nor glass wool should be attached to part "A" of the three-way stopcock (Figure 1), because adhesion of cerium(III) chloride to them, caused by bumping of the powder, prevents reducing the pressure.

3. The trap is attached in a reverse manner to prevent clogging of the trap by ice.

4. Cerium(III) chloride heptahydrate was purchased from Japan Yttrium Co., Ltd. or Aldrich Chemical Company, Inc., and ground in a mortar with a pestle just before use.

5. The temperature of the oil bath must be kept below 100°C to minimize the hydrolysis of cerium(III) chloride heptahydrate to CeOCl.

6. A powerful magnetic stirrer and a large rugby ball-shaped stirring bar are used.

7. When the resulting cerium(III) chloride (730-779 mg) is dissolved in 10 mL of dry methanol, Karl Fischer analysis of the solution shows that the cerium(III) chloride contains 6.5-7.6% of water which is equal to 95-113% of the water expected from the formula $CeCl_3 \cdot H_2O$.

8. If the cerium(III) chloride powder is stirred at this stage, it sprays around the flask. This causes not only the loss of a large amount of cerium(III) chloride, but also clogs of the three-way stopcock with the powder.

9. A small amount of cerium(III) chloride is lost by bumping; however, it does not affect the yield of the product.

10. Karl Fischer titration of a dry methanol solution (15 mL) of the resulting cerium(III) chloride (731-817 mg) shows that it contains 0.71-0.94% of water and can be represented by a formula $CeCl_3(H_2O)_{0.10-0.13}$. This result indicates that this procedure affords practically anhydrous cerium(III) chloride, although Evans, et al. reported that gradual heating to 150°C at 0.03 mm over 3 hr, followed by further

evacuating at 150°C/0.03 mm for 12 hr, resulted in the production of cerium(III) chloride monohydrate.[2]

Anhydrous cerium(III) chloride can be stored in a sealed vessel. It is dried under reduced pressure (0.1-0.2 mm) at 140-150°C for 1 hr just before use.

11. Tetrahydrofuran was freshly distilled under argon from sodium and benzophenone.

12. Addition of tetrahydrofuran to anhydrous cerium(III) chloride without either cooling or vigorous stirring results in the formation of a hard cake, which prevents the formation of the milky suspension.

13. Instead of stirring overnight, the suspension may be prepared by sonication for more than 1 hr.[3]

14. α-Tetralone was purchased from Tokyo Kasei Kogyo Co., Ltd. or Aldrich Chemical Company, Inc., and distilled under reduced pressure before use.

15. The preparation of anhydrous cerium(III) chloride[4] is necessary to obtain a satisfactory result. For example, the use of cerium(III) chloride monohydrate in the same reaction causes a significant lowering of the yield (34%) of the desired alcohol (the recovery of the starting ketone: 54%). The use of the reverse addition procedure (see Method B in the Table) also resulted in the formation of the alcohol in poor yield (36%; the recovery of the ketone: 58%).

16. The spectral and physical data of this material are as follows: IR (neat) cm^{-1}: 3390; ^{1}H NMR (400 MHz, $CDCl_3$) δ: 0.88 (t, 3 H, J = 7.3), 1.10-1.40 (m, 4 H), 1.70-2.05 (s, 7 H), 2.65-2.85 (m, 2 H), 7.00-7.07 (m, 1 H), 7.10-7.22 (m, 2 H), 7.45-7.52 (m, 1 H); ^{13}C NMR (75.4 MHz, $CDCl_3$) δ: 13.9 (q), 19.7 (t), 23.1 (t), 26.3 (t), 29.9 (t), 35.9 (t), 42.1 (t), 72.4 (s), 126.09 (d), 126.12 (d), 126.9 (d), 128.8 (d), 136.7 (s), 142.4 (s); Anal. Calcd for $C_{14}H_{20}O$: C, 82.30; H, 9.87. Found: C, 82.13; H, 9.97.

Waste Disposal Information

All toxic materials were disposed of in accordance with "Prudent Practices in the Laboratory"; National Academy Press; Washington, DC, 1995.

3. Discussion

Reactions of carbonyl compounds with organolithiums or Grignard reagents normally give the corresponding alcohols via nucleophilic addition to the carbonyl group.[5] Nevertheless, the reactions are often accompanied by so-called abnormal reactions such as enolization, reduction, condensation, conjugate addition, and pinacol coupling. Several attempts to suppress these undesired reactions have been made by changing solvents[6] or using additives.[7] These methods are not always efficient, although they are effective in some cases. On the other hand, the use of cerium(III) chloride as an additive results in the efficient suppression of these abnormal reactions,[8,9] and many applications of this method to practical organic syntheses have been reported.[10]

A typical example of an effective synthesis of a tertiary alcohol from an easily enolizable ketone by taking advantage of the organolithium/cerium(III) chloride system is illustrated here, along with experimental details for the preparation of anhydrous cerium(III) chloride. The reaction of α-tetralone with butyllithium alone affords 1-butyl-1,2,3,4-tetrahydro-1-naphthol in poor yield (26%), because of enolization, which gives the starting ketone (55%) (Table, Entry 1). The efficient production of the desired alcohol (92-97%) in the reaction with butyllithium/cerium(III) chloride (Table, Entry 2) suggests the lower basicity of this system. The use of butylmagnesium bromide instead of butyllithium in this reaction also results in the efficient transformation of the starting ketone to the corresponding alcohol (Table, Entry 5). The reaction with butylmagnesium bromide/cerium(III) chloride can be performed at 0°C, in contrast to the reaction with butyllithium/cerium(III) chloride at 0°C, which gives a complex mixture. The reverse addition procedure (Table, Method B) affords the desired alcohol in moderate or satisfactory yield (Table, Entries 3 and 6).

This method is also effective in enhancing 1,2-addition to α,β-enones (Table, Entries 7 and 8) as well as suppressing metal-halogen exchange (Table, Entries 9 and 10) and aldol-type condensation (Table, Entries 11 and 12). Furthermore, this is applicable to the reaction of sterically congested ketones (Table, Entries 13 and 14). Various organolithiums and Grignard reagents such as alkyl, alkenyl, alkynyl, and aryl-lithiums and magnesium halides can be employed for this method.[8-10] The reactions with these organolithiums and Grignard reagents are carried out at -78°C and 0°C, respectively, although vinylic Grignard reagents are treated with cerium(III) chloride at -78°C because of the rapid decomposition of the Grignard reagents on contact with cerium(III) chloride at 0°C.[9]

1. Department of Chemistry, Faculty of Science, Chiba University, Inage, Chiba 263, Japan.
2. Evans, W. J.; Feldman, J. D.; Ziller, J. W. *J. Am. Chem. Soc.* **1996**, *118*, 4581.
3. (a) Utimoto, K.; Nakamura, A.; Matsubara, S. *J. Am. Chem. Soc.* **1990**, *112*, 8189; (b) Greeves, N.; Lyford, L. *Tetrahedron Lett.* **1992**, *33*, 4759.
4. Some methods for the preparation of anhydrous cerium(III) chloride have been reported. See: (a) Bunnelle, W. H.; Narayanan, B. A. *Org. Synth., Coll. Vol. VIII* **1993**, 602; (b) Dimitrov, V.; Kostova, K.; Genov, M. *Tetrahedron Lett.* **1996**, *37*, 6787, and ref. 8-10.
5. For reviews, see: (a) Lai, Y.-H. *Synthesis* **1981**, 585; (b) Wakefield, B. J. In "Comprehensive Organometallic Chemistry"; Wilkinson, G.; Stone, F. G. A.; Abel, E. W., Eds.; Pergamon: Oxford, 1982, Vol. 7, pp. 1-110.

6. (a) Gocmen, M.; Soussan, G. *J. Organomet. Chem.* **1974**, *80*, 303; (b) Canonne, P.; Foscolos, G. B.; Lemay, G. *Tetrahedron Lett.* **1979**, 4383; (c) Canonne, P.; Foscolos, G.; Caron, H.; Lemay, G. *Tetrahedron* **1982**, *38*, 3563.
7. (a) Swain, C. G.; Boyles, H. B. *J. Am. Chem. Soc.* **1951**, *73*, 870; (b) McBee, E. T.; Pierce, O. R.; Higgins, J. F. *J. Am. Chem. Soc.* **1952**, *74*, 1736; (c) Chastrette, M.; Amouroux, R. *Bull. Soc. Chim. Fr.* **1970**, 4348; (d) Chastrette, M.; Amouroux, R. *J. Chem. Soc., Chem. Commun.* **1970**, 470; (e) Weidmann, B.; Seebach, D. *Angew. Chem., Int. Ed. Engl.* **1983**, *22*, 31.
8. Imamoto, T.; Kusumoto, T.; Tawarayama, Y.; Sugiura, Y.; Mita, T.; Hatanaka, Y.; Yokoyama, M. *J. Org. Chem.* **1984**, *49*, 3904.
9. Imamoto, T.; Takiyama, N.; Nakamura, K.; Hatajima, T.; Kamiya, Y. *J. Am. Chem. Soc.* **1989**, *111*, 4392.
10. For recent reviews, see: (a) Imamoto, T. "Lanthanides in Organic Synthesis"; Academic Press: London, 1994; pp. 80-97; (b) Molander, G. A. *Chem. Rev.* **1992**, *92*, 29; (c) Imamoto, T. In "Comprehensive Organic Synthesis"; Trost, B. M.; Fleming, I., Eds.; Pergamon: Oxford, 1991; Vol. 1, Chapter 1.8, pp. 231-250.

Appendix

Chemical Abstracts Nomenclature (Collective Index Number); (Registry Number)

Cerium(III) chloride heptahydrate: Cerium chloride heptahydrate (8); Cerium chloride ($CeCl_3$), heptahydrate (9); (18618-55-8)

Cerium(III) chloride monohydrate: Cerium chloride ($CeCl_3$), monohydrate (10); (64332-99-6)

Cerium(III) chloride, anhydrous: Cerium chloride (8,9); (7790-86-5)

Butyllithium: Lithium, butyl- (8,9); (109-72-8)

α-Tetralone: 1(2H)-Naphthalenone, 3,4-dihydro- (8,9); (529-34-0)

TABLE

CERIUM(III) CHLORIDE-PROMOTED CARBONYL ADDITIONS[a]

Entry	Ketones	Reagents	Method[b]	Products	Yield (%)[c,d]
1	α-Tetralone	BuLi		Bu OH	26[e] (55[e])
2[f]	α-Tetralone	BuLi/$CeCl_3$	A	same as above	92-97
3	α-Tetralone	BuLi/$CeCl_3$	B	same as above	80[e] (11[e])
4	α-Tetralone	BuMgBr		same as above	28[e] (23[e])[g]
5	α-Tetralone	BuMgBr/$CeCl_3$	A	same as above	96
6	α-Tetralone	BuMgBr/$CeCl_3$	B	same as above	92[e] (2[e])
7[h]	PhCH=CHCOPh	MeLi		PhCH=CHC(OH)(Me)Ph Ph(Me)$CHCH_2COPh$	54 22
8[h]	PhCH=CHCOPh	MeLi/$CeCl_3$	A	PhCH=CHC(OH)(Me)Ph Ph(Me)$CHCH_2COPh$	97 1
9[i]	p-IC_6H_4COMe	BuLi		p-IC_6H_4C(OH)(Me)Bu	trace
10[i]	p-IC_6H_4COMe	BuLi/$CeCl_3$	A	p-IC_6H_4C(OH)(Me)Bu	93
11[h]	$PhCH_2CO_2Me$	i-PrMgBr		$PhCH_2COCH(Ph)CO_2Me$ $PhCH_2COPr$-i $PhCH_2C(OH)(Pr\text{-}i)_2$	71 (10) 19 0
12[h]	$PhCH_2CO_2Me$	i-PrMgBr/$CeCl_3$	A	$PhCH_2COCH(Ph)CO_2Me$ $PhCH_2COPr$-i $PhCH_2C(OH)(Pr\text{-}i)_2$	0 0 97
13[h]	Et_3CCOMe	MeMgBr		$Et_3CC(OH)Me_2$	0
14[h]	Et_3CCOMe	MeMgBr/$CeCl_3$	A	$Et_3CC(OH)Me_2$	95

[a]The reactions with organolithiums and Grignard reagents are carried out at -78°C and (
respectively. [b]Method A: the carbonyl compound is added to the mixture of $CeCl_3$ and
organolithium or Grignard reagents; Method B: the organolithium or Girgnard reagent is adde
the mixture of $CeCl_3$ and the carbonyl compound. [c]Isolated yield unless otherwise stated. [d]
figures in parentheses indicate the yields of the recovered starting material. [e]Determined by
NMR. [f]The reaction procedure is described in the text. [g]The reduction product, α-tetralol, is for
in 39% yield. [h]See Ref. 9. [i]See Ref. 8.

REGIOSELECTIVE MONOALKYLATION OF KETONES VIA THEIR MANGANESE ENOLATES: 2-BENZYL-6-METHYLCYCLOHEXANONE FROM 2-METHYLCYCLOHEXANONE

(Cyclohexanone, 2-methyl-6-(phenylmethyl)-)

1) LDA, THF, -78°C, 2 hr
2) $MnCl_2 \cdot 2LiCl$, -78°C, 30 min

OMnCl

OMnCl

$PhCH_2Br$, THF-NMP
rt, 2 hr

Ph

Submitted by Gérard Cahiez,[1] François Chau, and Bernard Blanchot.
Checked by Jari Yli-Kauhaluoma and Rick L. Danheiser.

1. Procedure

A 500-mL, three-necked, round-bottomed flask is equipped with a mechanical tirrer, 100-mL pressure-equalizing dropping funnel, and a Claisen head fitted with a w-temperature thermometer and a nitrogen inlet (Note 1). The flask is charged with 0 mL of tetrahydrofuran (THF) (Note 2) and 5.55 g (55 mmol) of diisopropylamine Note 3). The resulting solution is cooled to -15°C, and 34.4 mL (55 mmol) of a 1.6 M olution of butyllithium in hexane (Note 4) is added dropwise over a 10-min period. he reaction mixture is stirred at -15°C for 30 min and then cooled to -78°C in a dry ice-acetone bath. A solution of 5.6 g (50 mmol) of 2-methylcyclohexanone (Note 5) in 20 mL of THF is added over 5 min, and the reaction mixture is stirred for 2 hr at -78°C. A solution of 55 mmol of $MnCl_2 \cdot 2LiCl$ (dilithium tetrachloromanganate) in 80 mL of THF (Note 6) is added dropwise over a 15-min period, and after 30 min the resulting clear brown solution is allowed to warm to room temperature.

To this solution of the manganese enolate are added successively (each over 5 min) 80 mL of 1-methyl-2-pyrrolidinone (NMP) (Note 7) and 12.2 g (71.3 mmol) of benzyl bromide (Note 8). The resulting mixture is stirred for 2 hr and hydrolyzed by the dropwise addition of 100 mL of a 1 M aqueous hydrochloric acid solution (HCl) over 15 min. Petroleum ether (35-60°C, 100 mL) is added, and the aqueous layer is separated and extracted three times with 50-mL portions of diethyl ether. The combined organic layers are washed with 100 mL of an aqueous saturated sodium carbonate solution, dried over anhydrous magnesium sulfate, filtered, and concentrated under reduced pressure with a rotary evaporator to afford 13.7-22.3 g of a brown oil (Note 9). Short path distillation under reduced pressure affords 8.8-8.9 g (87-88%) of 2-benzyl-6-methylcyclohexanone as a pale yellow oil, bp 95-100°C (0.3 mm) (Note 10).

2. Notes

1. The apparatus is flame-dried under a stream of dry nitrogen or argon. A slight positive pressure of nitrogen or argon is maintained with an oil bubbler throughout the reaction.

2. THF was freshly distilled from sodium benzophenone ketyl under a nitrogen atmosphere.

3. Diisopropylamine (99%, Aldrich Chemical Company, Inc.) was distilled from calcium hydride prior to use.

4. Butyllithium (1.6 M solution in hexane) was purchased from Acros Organics and titrated immediately before use according to the procedure of Watson and Eastham.[2]

5. 2-Methylcyclohexanone (99%) was purchased from Aldrich Chemical Company, Inc., and distilled prior to use.

6. A solution of 55 mmol of the ate complex $MnCl_2 \cdot 2LiCl$ is prepared by stirring a suspension of 6.93 g of anhydrous manganese chloride ($MnCl_2$) (Note 11) and 4.65 g of anhydrous lithium chloride (LiCl) (Note 12) in 80 mL of THF at room temperature until an amber solution is obtained. It should be noted that the rate of dissolution (formation of the ate-complex Li_2MnCl_4) is very dependent on both the grain size of the two salts ($MnCl_2$ and LiCl) and their purity. When unpulverized Aldrich Chemical Company, Inc., or Acros Organics material is used it is necessary to stir for 4 to 24 hr to obtain complete dissolution; on the other hand, with finely pulverized anhydrous $MnCl_2$ obtained by drying analytical grade manganese chloride tetrahydrate (e.g., manganese chloride tetrahydrate purum p.a. Fluka, Inc.), it is possible to obtain complete dissolution after only 5 to 10 min. The formation of the ate-complex in this case is exothermic.

7. The submitters purchased 1-methyl-2-pyrrolidinone (NMP, 99%) from Aldrich Chemical Company, Inc. and distilled it prior to use. The checkers used 99.5% NMP (Aldrich Chemical Company, Inc.) without further purification.

8. Benzyl bromide (99%) was purchased by the submitters from Aldrich Chemical Company, Inc., and distilled prior to use (*caution: lachrymator!*). The checkers purified benzyl bromide by filtration through activated neutral alumina (EM Science, ca. 2 g of alumina/12 g of benzyl bromide).

9. The weight of crude product varies depending on how much of the NMP is removed during the work up. The residual NMP is easily separated from 2-benzyl-6-methylcyclohexanone during the subsequent purification by distillation. Alternatively,

residual NMP can be removed during the workup by extracting the combined organi phases with four 50-mL portions of 1 M HCl prior to the aqueous Na_2CO_3 wash.

10. The checkers determined this material to consist of a mixture of 2-benzyl-6- methylcyclohexanone (94-97%) and 2-benzyl-2-methylcyclohexanone (3-6%). In 1 runs, the submitters found the yield to range from 85 to 90% and the regioselectivit (ratio of 2-benzyl-6-methylcyclohexanone to 2-benzyl-2-methylcyclohexanone) t range from 93:7 to 97:3. The regioisomeric ratio was determined by ^1H NM according to House[3] and by GC (capillary column SGE CYDEX B 25 m x 0.22 mm i.d. 0.25 μm film thickness, 165°C), retention time of 2-benzyl-6-methylcyclohexanone 14.53 min, retention time of 2-benzyl-2-methylcyclohexanone: 14.99 min. The spectr properties of the product were as follows: ^1H NMR (C_6D_6, 500 MHz) δ: 0.92-1.16 (m, H), 1.03 (d, 3 H, J = 6.4), 1.30-1.34 (m, 1 H), 1.59-1.63 (m, 1 H), 1.72-1.77 (m, 1 H 1.83-1.90 (1 H, app. sept, J = 6.2), 2.17-2.21 (m, 1 H), 2.42 (dd, 1 H, J = 8.3, 13.7), 3.3 (dd, 1 H, J = 5.1, 13.9), 7.07-7.19 (m, 5 H); ^{13}C NMR ($CDCl_3$, 125 MHz) δ: 14.5, 25. 34.6, 35.4, 37.3, 45.6, 52.5, 125.7, 128.1 (2C), 129.0 (2C), 140.6, 213.5; IR (thin film cm^{-1}: 3082(m), 3058(m), 3023(m), 1710(s), 1604(m), 1495(m), 1451(m), 1375(m 920(w), 745(m), 705(m). Anal. Calcd for $C_{14}H_{18}O$: C, 83.12; H, 8.97. Found: C, 83.0 H,8.90.

11. Manganese(II) chloride tetrahydrate, purum p.a. (Fluka, Inc.) was fine ground using a mortar and pestle and then dried by heating at 180-200°C at 0.01-0. mm in a vacuum oven for 10 hr prior to use. The checkers dried 13.8 g manganese(II) chloride tetrahydrate by heating in a 100-mL flask with magnetic stirrir at 205°C/0.1 mm for 15 hr. The submitters found that it was sometimes necessary grind the dried material again under a dry atmosphere before use. The anhydrous s is very hygroscopic and must be protected against moisture (a well-closed bottle adequate); it can, however, be handled *very quickly* in air without special precaution

12. Anhydrous lithium chloride (99%), purchased from Aldrich Chemical Company, Inc., was finely pulverized with a mortar and pestle and then dried by heating at 200°C under reduced pressure (0.1-0.01 mm) for 8 hr before use. The salt is hygroscopic and must be handled *very quickly*.

Waste Disposal Information

All toxic materials were disposed of in accordance with "Prudent Practices in the Laboratory"; National Academy Press; Washington, DC, 1995.

3. Discussion

The procedure described here illustrates a general and very convenient method[4] to carry out the regioselective monoalkylation of ketones via their Mn-enolates. A comparison with the classical procedure previously reported by House in *Organic Syntheses* to prepare the 2-benzyl-6-methylcyclohexanone via the corresponding Li-enolates[3] clearly shows that the Mn-enolate gives a higher yield of desired product since the regioselectivity is better and the formation of polyalkylated products is not observed.

OMetal

$PhCH_2Br$

THF^a, 20°C

$PhCH_2$ — A + B — CH_2Ph

Reaction Conditions	Yield (%)	Regioselectivity A/B	Polyalkylated Product (%)
Li-Enolate[a]	42-45	76/24	18-19
Mn-Enolate	85	95/5	< 1

[a]The reaction described by House[3] was performed in dimethoxyethane. The same results have been obtained by using THF.

The other regioisomer, the 2-benzyl-2-methylcyclohexanone, can also be selectively obtained in good yield from the more substituted Mn-enolate.[4d] In fact,

OSiMe$_3$ —— 1) MeLi, 2) $MnCl_2$ ——> OMnCl —— $PhCH_2Br$, THF-NMP ——> 2-CH_2Ph-2-methylcyclohexanone

89%

Regioisomeric purity: 95%[a]

(a) Regioisomeric purity of the starting silyl enol ether was 95%.

these results prove that the deprotonation equilibrium that is responsible for the formation of side-products from Li-enolates does not exist under the reaction conditions in the case of Mn-enolates.

As a rule, this procedure also compares favorably with the other methods previously reported to achieve regioselective monoalkylation of ketones.[5] It gives similar or higher yields and selectivities, and it is clearly easier to carry out since no toxic, expensive or hazardous material such as Et_3Al (Al-enoates), KH then Et_3B (B-enolates), Bu_3SnCl (Sn-enolates) or Et_2Zn (Zn-enolates) is required. Moreover, large excess of alkylating reagents (Al- and Sn-enolates) is not required.

As shown, Mn-enolates are easily and quantitatively obtained from Li-enolates by transmetalation.[4b,d,e] They can also be prepared by deprotonation of ketones with Mn-amides.[4a,c]

1) PhMeNMnCl, THF, 20°C, 1 hr
2) $(EtCO)_2O$; 20°C, 3 hr

EtCOO — 90%*

*Only a 40% yield is obtained by using iso-Pr_2NMnCl instead of PhMeNMnCl.

Note that only Mn-amides prepared from aromatic amines, ArRNH or Ar_2NH, give quantitative yields of enolization products. A procedure using only a catalytic amount of aromatic amine has also been described.[6]

The deprotonation reactions occur regio- and stereoselectively to give the less-substituted Z enolates that can be readily silylated or acylated to afford mainly the less-substituted Z silyl enol ethers or Z enol esters in high yields.[4a, 7]

Pent

1) PhMeNMnPh, THF, -10°C, 1 hr
2) Me_3SiCl or $(EtCO)_2O$, -10°C to rt, 30 min

OZ, Pent

Z = Me_3Si: 90% (Z/E = 97:3; regioselectivity ≥99%)
Z = EtCO: 95% (Z/E = 93:7; regioselectivity ≥99%)

The regioselective monoalkylation of ketones described above has wide applicability. The monoalkylated products are regioselectively obtained in high yields by reacting Mn-enolates prepared in THF from a wide range of ketones with various reactive organic halides in the presence of a polar cosolvent such as DMSO, NMP, sulfolane, DMF, MeCN[4a-d] (Tables I and II) as well as DMPU.[4e] With less reactive alkylating reagents (e.g., BuBr) the reaction rate is slower, and the reaction generally leads to lower yields (Table I). Note that alkyl sulfonates do not undergo reaction. (Table I).

Mn-enolates can also be hydroxyalkylated (Table III). They react easily with a vast array of aldehydes (even enolizable or α,β-unsaturated aldehydes), to give syn-aldol products in good yields.[8] The stereoselectivity obtained from Mn- and Li-enolates are very similar.

$R^1CH_2COCH_2R^2$ → 1) PhMeNMnMe, -10°C, 1 hr; 2) -78°C, RCHO, 5 min → $R^1CH_2COCH(R^2)CH(OH)R$

α-Halogeno ketones lead to β-keto epoxides.[8]

Pr(OMnCl)C=CHEt + RCOCH$_2$X → THF, -78°C to rt → PrCOCH(Et)C(R)(epoxide)

R=Ph; X=Br; 92%
R=Me; X=Cl; 73%

Finally, Mn-enolates are useful synthetic reagents for Michael additions.

PrCH$_2$C(OMnCl)=CHEt + BuCH=C(COOEt)$_2$ → THF; -10°C, 1 hr → PrCH$_2$COCH(Et)CH(Bu)CH(COOEt)$_2$ 83%

TABLE I

MONOALKYLATION OF Mn-ENOLATES OBTAINED BY DEPROTONATION OF KETONES WITH Mn-AMIDES

Ketone	Solvent (Alkylation Step)[a]	Reaction Conditions	Alkylating Agent[b]	Yield[c] (%)
BuCOBu	THF/DMSO	20°C, 2 hr	MeI	93
PrCOPr	THF	20°C, 3 hr	$PhCH_2Br$	43
"	THF/NMP	20°C, 2 hr	"	86
"	THF/DMSO	"	"	91
"	THF/NMP	"	$CH_2{=}CHCH_2Br$	81
"	THF/DMSO	"	$PhCH{=}CHCH_2Br$	86
"	"	"	—Br[d]	87
"	THF/NMP	-30°C, 2 hr	$BrCH_2COOEt$	88
"	THF/DMSO	50°C, 2 hr	BuI	67
"	"	50°C, 24 hr	BuBr	48
"	"	"	$BuOSO_2Ph$	0

[a]Deprotonation step: PhMeNMnZ (Z = Cl, Ph), 0°C, 1 hr. [b]1.25 equiv. [c]Yield of isolated product. [d]Only the S_N2 product is obtained. The geometry of the allylic double bond is retained (Z > 99%).

TABLE II

REGIOSELECTIVE MONOALKYLATION OF Mn-ENOLATES OBTAINED BY DEPROTONATION OF UNSYMMETRICAL KETONES WITH Mn-AMIDES

Ketone	Alkylating Agent[a]	Monoalkylated Ketone(%)[b]	Regioselectivity[c]
iso-PrCOHex	CH_2=$CHCH_2Br$	80	> 99:1
iso-PrCOHex	$PhCH_2Br$	85	> 97:3
$PhCH_2$(Et)CHCOPr	CH_2=$CHCH_2Br$	89	>99:1
2-Me cyclohexanone	$PhCH_2Br$	90	93:7

[a]Alkylation step: THF-NMP or THF-DMSO, 20°C, 1 hr. [b]Yield of isolated product. [c]Ratio of $\alpha\alpha'/\alpha\alpha$-disubstituted ketones.

TABLE III

REACTION OF Mn-ENOLATES WITH ALDEHYDES

Ketone	Aldehyde	Yield (%)[a]	Syn/Anti
EtCOEt	PrCHO	83	66/33
"	Et_2CHCHO	86	77/23
"	MeCH=CHCHO	86	64/36
"	Furfural	79	71/29
"	PhCHO	94	71/29
PrCOPr	PhCHO	88	88/12
tert-BuCOPr	PhCHO	81	99/1
PrCOPr	PrCHO	86	85/15
BuCOBu	PrCHO	83	51/49
PhCOPr	PrCHO	87	72/28

[a]Yield of isolated product.

1. Ecole Supérieure de Chimie Organique et Minerale (E.S.C.O.M.) Département de Chimie, 13 Boulevard de l'Hautil, F-95092 Cergy Pontoise, France.
2. Watson, S. C.; Eastham, J. F. *J. Organomet. Chem.* **1967**, *9*, 165-168. See also Ref. 3.
3. Gall, M.; House, H. O. *Org. Synth., Coll. Vol. VI* **1988**, 121-130.
4. (a) Cahiez, G.; Figadère, B.; Tozzolino, P.; Cléry, P. Fr. Pat. Appl. 1988, 88/15,806; Eur. Pat. Appl. 1990, EP 373, 993; *Chem. Abstr.* **1991**, *114*, 61550y; (b) Cahiez, G.; Cléry, P.; Laffitte, J. A. Fr. Pat. Appl. 1991, 91/11,814; PCT Int. Appl. 1993, WO 93 06,071; *Chem. Abstr.* **1993**, *119*, 116519f; (c) Cahiez, G.; Figadère, B.; Cléry, P. *Tetrahedron Lett.* **1994**, *35*, 3065-3068; (d) Cahiez, G.; Chau, K.; Cléry, P. *Tetrahedron Lett.* **1994**, *35*, 3069-3072; (e) Reetz, M. T.; Haning, H. *Tetrahedron Lett.* **1993**, *34*, 7395-7398.
5. **Sn and Al-enolates**: Tardella, P. A. *Tetrahedron Lett.* **1969**, 1117-1120; **B-enolates:** Negishi, E.-i.; Chatterjee, S. *Tetrahedron Lett.* **1983**, *24*, 1341-1344; **Zn-enolates**: Morita, Y.; Suzuki, M.; Noyori, R. *J. Org. Chem.* **1989**, *54*, 1785-1787.
6. Cahiez, G.; Kanaan, M.; Cléry, P. *Synlett* **1995**, 191-192.
7. Cahiez, G.; Figadère, B.; Cléry, P. *Tetrahedron Lett.* **1994**, *35*, 6295-6298.
8. Cahiez, G.; Cléry, P.; Laffitte, J. A. Fr. Demande FR 2/671,085, Fr. Pat. Appl. 1990, 90/16,413; *Chem. Abstr.* **1993**, *118*, 69340b.

Appendix

Chemical Abstracts Nomenclature (Collective Index Number); (Registry Number)

2-Benzyl-6-methylcyclohexanone: Cyclohexanone, 2-benzyl-6-methyl- (8); Cyclohexanone, 2-methyl-6-(phenylmethyl)- (9); (24785-76-0)

2-Methylcyclohexanone: Cyclohexanone, 2-methyl- (8,9); (583-60-8)

Diisopropylamine (8); 2-Propanamine, N-(1-methylethyl)- (9); (108-18-9)

Butyllithium: Lithium, butyl- (8,9); (109-72-8)

Dilithium tetrachloromanganate ($MnCl_2 \cdot 2\ LiCl$; $Cl_4Mn \cdot 2\ Li$): Manganate (2-), tetrachloro-, dilithium, (I-4)- (9); (57384-24-4)

1-Methyl-2-pyrrolidinone: 2-Pyrrolidinone, 1-methyl- (8,9); (872-50-4)

Benzyl bromide: Toluene, α-bromo- (8); Benzene, (bromomethyl)- (9); (100-39-0)

Manganese(II) chloride: Manganese chloride (8,9); (7773-01-5)

Lithium chloride (8,9); (7447-41-8)

2-Benzyl-2-methylcyclohexanone: Cyclohexanone, 2-benzyl-2-methyl- (8); Cyclohexanone, 2-methyl-2-(phenylmethyl)- (9); (1206-21-9)

COPPER-CATALYZED CONJUGATE ADDITION OF FUNCTIONALIZED ORGANOZINC REAGENTS TO α,β-UNSATURATED KETONES: ETHYL 5-(3-OXOCYCLOHEXYL)PENTANOATE

(Pentanoic acid, 5-(3-oxocyclohexyl)-, ethyl ester)

$I\text{-}(CH_2)_4CO_2Et$ → 1. Zn(0), THF; 2. MeLi, −78°C; 3. 10 mol% $Me_2Cu(CN)Li_2$; 4. 3 TMS–Cl; 5. 2-cyclohexenone, −78°C, 4.5 hr; 6. Bu_4NF → 3-oxocyclohexyl-$(CH_2)_4CO_2Et$

Submitted by B. H. Lipshutz,[1] M. R. Wood, and R. Tirado.

Checked by Ying Huang and David J. Hart.

1. Procedure

Ethyl 5-(3-oxocyclohexyl)pentanoate. A 50-mL, round-bottomed flask equipped with a magnetic stir bar and a septum, is cooled under a stream of argon (Note 1), and charged with 3.0 g of zinc powder (46 mmol, Note 2), followed by 16 mL of tetrahydrofuran (THF) (Note 3). To this mixture is added 155 μL of 1,2-dibromoethane (338 mg, 1.8 mmol, Note 4), and the stirred slurry is heated to ca. 65°C (Note 5) for 1 min using a heat gun. The flask is then placed in a water bath (ca. 23°C) with continued magnetic stirring. After one minute, 155 μL of chlorotrimethylsilane (TMS Cl) (133 mg, 1.2 mmol, Note 6) is added, and the slurry is stirred for 15 min (Note 7). The flask is wrapped in aluminum foil and 6.7 mL of neat ethyl 5-iodovalerate (10.24 g, 40.0 mmol, Note 8) is slowly added dropwise over 10 min. The flask is placed in 35°C oil bath and stirred overnight. Initially, the reaction is very exothermic, and TH can be seen slowly refluxing around the neck of the round-bottomed flask. Dry Ice used to cool the neck of the flask to avoid leaching the septum with THF. Twelve hours later (overnight, Note 9), when the reaction is judged complete by TLC (Note 10), the aluminum foil and the oil bath are removed. The presence of a white "dust" that never quite settles is noted. The reaction is allowed to cool to room temperature and 20 mL of THF is added to the stirred slurry. Stirring is stopped, the solids are allowed to settle as much as possible (*vida supra*), and the THF solution of the alkylzinc iodide is transferred via canula to a clean, dry 250-mL, round-bottomed flask equipped with a stir bar and a septum. Two additional washes of the remaining excess zinc with THF (29 mL each) are carried out to ensure complete transfer and proper dilution of the alkylzinc iodide (0.38 M). The solution is cooled to -78°C, and methyllithium (MeLi) in ether (26 mL, 36 mmol, Note 11) is added dropwise over 15 min.

While the alkylzinc is being formed, a 25-mL, round-bottomed flask, equipped with a stir bar and septum is cooled under argon, and charged with 178.9 mg of copper(I) cyanide (CuCN) (2.0 mmol, Note 12), followed by 10 mL of THF. The mixture is cooled to -78°C and MeLi in ether (2.86 mL, 4.0 mmol, Note 11) is added slowly over 10 min. When the addition of MeLi is complete, the slurry is gently warmed until the reaction mixture becomes homogeneous at which point it is recooled to -78°C. With both solutions at -78°C, the higher-order dimethylcyanocuprate is transferred via canula to the alkylzinc reagent. After 5 min at -78°C, 7.77 mL of neat TMS-Cl (6.65 g, 1.3 mmol, Note 6) is added dropwise over 10 min, followed by 10 min of stirring. inally, 2.02 mL of neat 2-cyclohexen-1-one (2.01 g, 20.9 mmol, Note 13) is added ropwise over 15 min via syringe. After 4.5 hr, the reaction is complete as determined y TLC (Note 14) and it is quenched by pouring the reaction mixture into 200 mL of pH buffer (Note 15) and 200 mL of diethyl ether (Et_2O) in a 1-L separatory funnel. An dditional 25 mL of ether is used to complete the transfer of the reaction mixture to the eparatory funnel. The aqueous layer is further extracted with an additional 50 mL of ther. The combined organic layers are shaken for 5 min with 50 mL of 1 M tetrabutylammonium fluoride (TBAF) in THF (Note 16), followed by three washes with 100 mL of brine. The combined organic phases are dried over anhydrous magnesium sulfate ($MgSO_4$). Gravity filtration, concentration under reduced pressure, and flash chromatography on silica gel (Note 17) using 10:1 (petroleum ether : ethyl acetate) affords 4.02 g of slightly contaminated product (Note 18). Warming this material at 50°C on a Kugelrohr apparatus for 2 hr at 0.5 mm affords 3.46 g (73%) of pure 5-(3-oxocyclohexyl)pentanoate as a pale-yellow liquid (Note 19).

2. Notes

1. All reactions are carried out under an inert atmosphere of argon (Linde prepurified grade) using oven-dried glassware, syringes, needles and canulas (at least 8 hr, in an oven at 120°C), employing standard syringe/septa techniques. After dry solids are added to reaction flasks by briefly removing the septum, argon is passed through the flask to purge it of atmospheric gases for ca. 15 min.

2. Zinc, powder, -100 mesh, 99.998%, was purchased from the Aldrich Chemical Company, Inc., and quickly weighed out on a bench-top balance.

3. Tetrahydrofuran is freshly distilled from sodium benzophenone ketyl under a nitrogen atmosphere.

4. 1,2-Dibromoethane, 99+%, purchased from the Aldrich Chemical Company, Inc., was used as received.

5. Gentle refluxing is seen on the sides of the round-bottomed flask, at which time the hot air of the heat gun is directed away from the flask, until the refluxing stops.

6. Chlorotrimethylsilane, redistilled, 99+%, was purchased from the Aldrich Chemical Company, Inc., distilled from calcium hydride (CaH_2) under argon, and stored under a Teflon-taped polyethylene cap.

7. Gas evolution is occasionally observed at this point.

8. Ethyl 5-iodovalerate is prepared from commercially available ethyl 5-bromovalerate (Aldrich Chemical Company, Inc.) through a Finkelstein reaction in acetone with sodium iodide. To 12.5 g (59.5 mmol) of ethyl 5-bromovalerate in 150 mL of dry acetone is added a total of 44.8 g (299 mmol) of solid sodium iodide in three equal portions. The solution is warmed under reflux (oil bath temperature of 67°C) for 40 hr and then cooled to room temperature. The mixture is partitioned between 200 mL of diethyl ether and 200 mL of water. The aqueous phase is extracted with three 100-mL portions of ether. The combined organic phases are washed with 50 mL of 10% aqueous sodium bisulfite, 50 mL of brine, dried ($MgSO_4$), and concentrated under reduced pressure. The residual oil is distilled (170-180°C/0.5 mm) to give 14.3 g (94%) of the iodide: ^1H NMR ($CDCl_3$) δ: 1.24 (t, 3 H, J = 7), 1.72 (m, 2 H), 1.85 (m, 2 H), 2.31 (t, 2 H, J = 7), 3.17 (t, 2 H, J = 7), 4.12 (q, 2 H, J = 7). The product is stored over freshly cut pieces of copper wire to ensure dryness and long-term purity.

9. The reaction time for complete zinc insertion is significantly less than 12 hr (usually 1-4 hr), but additional time at 35°C did not diminish the quality of the zinc reagent (unless the zinc reagent contained an enolizable ketone group). An overnight reaction time, for the zinc oxidative insertion, was used solely for convenience.

10. The starting iodide is no longer seen by TLC under UV-light, $R_f = 0.70$ in 10:1 petroleum ether : ethyl acetate on pre-coated silica gel 60 F_{254} plates (EMx Science), 0.25-mm layer thickness. Further evidence of complete zinc insertion is seen in the clumping of the excess zinc into small, shiny metallic balls.

11. MeLi in ether (1.4 M, low halide) was purchased from the Aldrich Chemical Company, Inc. and titered against distilled 2-pentanol, with 1,10-phenanthroline as indicator.

12. Copper(I) cyanide, 99%, purchased from the Aldrich Chemical Company, Inc., was used as received and stored in an Abderhalden desiccator over potassium hydroxide. It is quickly weighed out on a bench-top balance.

13. 2-Cyclohexen-1-one, 95+%, was purchased from the Aldrich Chemical Company, Inc. and distilled prior to use.

14. Only a trace of 2-cyclohexen-1-one, (R_f = 0.16 in 10:1 petroleum ether : ethyl acetate) could be seen (see Note 10 for type of TLC plates used). A new spot at R_f = 0.50 appeared corresponding to the TMS enol ether product if the TLC plate was eluted immediately after spotting. If, however, the TLC plate was not eluted for minutes after spotting, significant cleavage of the TMS enol ether occurred on the silica gel, allowing observation of the final ketone product at R_f = 0.11.

15. Aqueous pH 7 buffer was purchased from Fisher Scientific Company. An acidic workup should be avoided so as to prevent the formation of HCN.

16. 1 M TBAF in THF (5% water) was purchased from the Aldrich Chemical Company, Inc., and used as received.

17. Flash chromatography was performed on ICN BioMedical's, ICN Silica, 32-63, 60 Å, using ca. 150 g of silica in a 2-in diameter column.

18. Impurities consisted of a small amount of 2-cyclohexen-1-one and another impurity that displayed a triplet at δ 0.9 in its NMR spectrum. These volatile impurities could be removed by warming the crude product under reduced pressure for several hours.

19. The spectral data are as follows: ^{1}H NMR δ: 1.22 (t, 3 H, J = 7, CH_3), 1.3[illegible] (m, 5 H), 1.54-2.04 (m, 6 H), 2.19-2.40 (m, 4 H), 2.24 (t, 2 H, J = 7, CH_2CO_2), 4.07 (q, [illegible] H, J = 7, $O\underline{CH_2}CH_3$); ^{13}C NMR δ: 14.1 (q), 24.8 (t), 26.0 (t), 26.1 (t), 31.1 (t), 34.1 (t), 36.1 (t), 38.3 (d), 41.3 (t), 47.9 (t), 60.1 (t), 173.4 (s), 211.6 (s); IR (neat) cm^{-1}: 293[illegible], 2861, 1734, 1188; MS (EI), m/e (rel. intensity) 226 (M^+, 2), 181 (7), 135 (5), 101 (7), 9[illegible] (8), 97 (100), 82 (6), 81 (7), 67 (7), 55 (16); HRMS (EI) calcd for [M^+, $C_{13}H_{22}O$[illegible] 226.1563; found: 226.1569.

Waste Disposal Information

All toxic materials were disposed of in accordance with "Prudent Practices in the Laboratory"; National Academy Press; Washington, DC, 1995.

3. Discussion

Organocopper chemistry[2] has steadily evolved so that it now includes the preparation and coupling reactions of cuprates bearing ligands that contain electrophilic centers.[3,4,5] Functional groups (FG) such as esters, ketones, nitriles, and halides can be incorporated into lower order cyanocuprates, **2**, via metathesis between precursor organozinc halides **1** and CuCN.[3] Use of zinc in this scheme allows for generation of organometallics such as **1** that would otherwise be difficult to prepare in the corresponding lithiated form (**3**) en route to lithiocyanocuprates **4**.[6] The reactivity patterns of cuprates **2** reflect the importance of the gegenion in Michael-type additions, for while their lithio counterparts **4** react with most enones quickly at low temperatures,[2] zinc halide cuprates are relatively sluggish.[7] Most significantly, greater than stoichiometric amounts of copper are normally required. The procedure described here[8] addresses these shortcomings, as well as those of related procedures which, e.g., rely on excesses of hexamethyl phosphoramide (HMPA).[9]

$$\underset{\mathbf{1}}{FG{\sim\sim}CH_2\text{-}ZnI} \quad + \quad CuCN \cdot 2LiCl \longrightarrow \underset{\mathbf{2}}{FG{\sim\sim}CH_2\text{-}Cu(CN)\mathbf{ZnI}}$$

$$\underset{\mathbf{3}}{FG{\sim\sim}CH_2\text{-}Li} \quad + \quad CuCN \cdot 2LiCl \dashrightarrow \underset{\mathbf{4}}{FG{\sim\sim}CH_2\text{-}Cu(CN)\mathbf{Li}}$$

The key feature that allows the cuprate-catalyzed procedure described is the facility with which ligands on zinc and copper undergo exchange.[8] Thus, when zinc halide **1** is converted to the mixed zinc species **5** and then exposed to 5 mol% $Me_2Cu(CN)Li_2$ at low temperatures, catalytic amounts of FG ~$CH_2Cu(CN)Li$ (**4**) are produced via transmetallation (along with Me_3ZnLi).[10] In the presence of TMS-Cl as

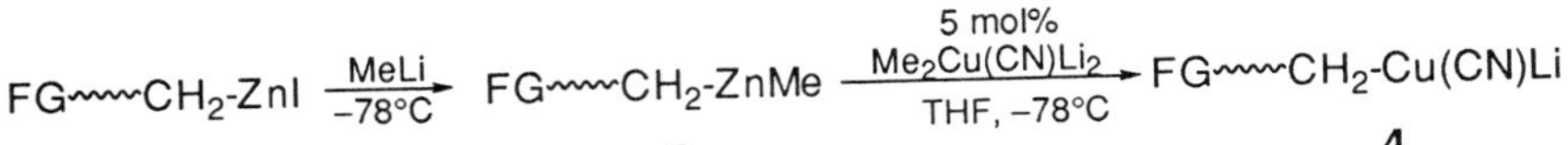

$$\underset{\mathbf{1}}{FG\sim CH_2\text{-}ZnI} \xrightarrow[-78°C]{MeLi} \underset{\mathbf{5}}{FG\sim CH_2\text{-}ZnMe} \xrightarrow[THF,\ -78°C]{5\ mol\%\ Me_2Cu(CN)Li_2} \underset{\mathbf{4}}{FG\sim CH_2\text{-}Cu(CN)Li}$$

an activating additive,[11] **4** delivers the functionalized ligand to the β-position of an enone (Scheme 1). The intermediate enolate is trapped by the Me_3SiCl present, the Si-O bond being cleaved on workup with fluoride ion to arrive at the product ketone. As anticipated from earlier studies,[12] **4** is the most reactive among several species in solution.[10] The overall process is made catalytic in Cu(I) by virtue of formation of a silyl enol ether, thereby releasing the metal for recycling.

Scheme 1

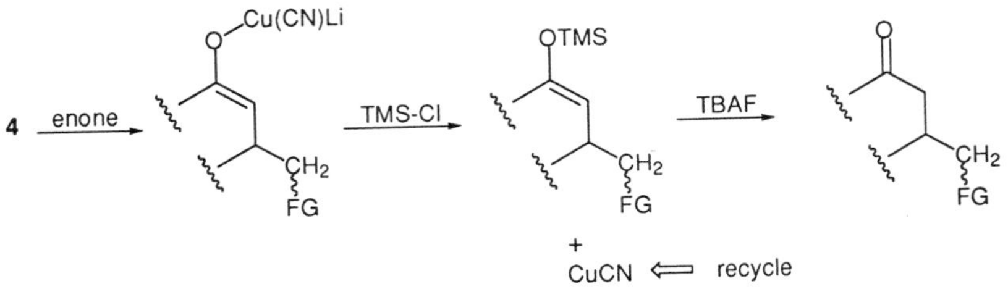

4 —enone→ [O-Cu(CN)Li enolate, CH₂~FG] —TMS-Cl→ [OTMS enol ether, CH₂~FG] + CuCN ⇐ recycle —TBAF→ [ketone, CH₂~FG]

Table I[8] highlights several other examples that demonstrate the scope of the coupling process. Cases of simple ethyl esters (entries 1-3), pivaloates (entries 4,5), chlorides (entries 6,7), a nitrile (entry 8), and a ketone (entry 9), all participate readily. Note the transfer of an acyl silane[13] (entry 10), and the fact that these couplings are effected under very mild conditions and afford good isolated yields.

1. Department of Chemistry, University of California, Santa Barbara, Santa Barbara, CA 93106-9510.
2. Lipshutz, B.H.; Sengupta, S. *Org. React.* **1992**, *41*, 135-631.
3. Klement, I.; Knochel, P.; Chau, K.; Cahiez, G. *Tetrahedron Lett.* **1994**, *35*, 1177; Rozema, M.J.; Eisenberg, C.; Lütjens, H.; Ostwald, R.; Belyk, K.; Knochel, P. *Tetrahedron Lett.* **1993**, *34*, 3115; Rozema, M. J.; Sidduri, A.; Knochel, P. *J. Org. Chem.* **1992**, *57*, 1956; Knochel, P.; Yeh, M. C. P.; Berk, S. C.; Talbert, J. *J. Org. Chem.* **1988**, *53*, 2390.
4. Hanson, M. V.; Brown, J. D.; Rieke, R. D.; Niu, Q. J. *Tetrahedron Lett.* **1994**, *35*, 7205; Rieke, R. D.; Stack, D. R.; Dawson, B. T.; Wu, T.-C. *J. Org. Chem.* **1993**, *58*, 2483; Stack, D. E.; Klein, W. R.; Rieke, R. D. *Tetrahedron Lett.* **1993**, *34*, 3063; Stack, D. E.; Dawson, B. T.; Rieke, R. D. *J. Am. Chem. Soc.* **1992**, *114*, 5110; **1991**, *113*, 4672; Ebert, G. W.; Klein, W. R. *J. Org. Chem.* **1991**, *56*, 4744; Rieke, R. D.; Wu, T. C.; Stinn, D. E.; Wehmeyer, R. M. *Synth. Commun.* **1989**, *19*, 1833; Rieke, R. D.; Wehmeyer, R. M.; Wu, T.-C.; Ebert, G. W. *Tetrahedron* **1989**, *45*, 443.
5. For a general review on the preparation of highly reactive metal powders, see Rieke, R. D. *Science* **1989**, *246*, 1260.
6. For the preparation of alkenyl- and aryllithium reagents containing electrophilic centers, see Klement, I.; Rottlaender, M.; Tucker, C. E.; Majid, T. N.; Knochel, P.; Venegas, P.; Cahiez, G., *Tetrahedron*, **1996**, *52*, 7201.

7. For a recent review on organozinc chemistry, see Knochel, P.; Singer, R. D. *Chem. Rev.* **1993**, *93*, 2117.

8. Lipshutz, B. H.; Wood, M. R.; Tirado, R. *J. Am. Chem. Soc.* **1995**, *117*, 6126.

9. Nakamura, E. *Synlett* **1991**, 539; Kuwajima, I.; Nakamura, E. *Top Curr. Chem.* **1990**, *155*, 1; Tamaru, Y.; Tanigawa, H.; Yamamoto, T.; Yoshida, Z.-i. *Angew. Chem., Int. Ed. Engl.* **1989**, *28*, 351; Nakamura, E.; Kuwajima, I. *Org. Synth., Coll. Vol. VIII* **1993**, 277; Nakamura, E.; Aoki, S.; Sekiya, K.; Oshino, H.; Kuwajima, I. *J. Am. Chem. Soc.* **1987**, *109*, 8056.

10. Lipshutz, B. H.; Wood, M. R. *Tetrahedron Lett.* **1994**, *35*, 6433.

11. Corey, E. J.; Boaz, N. W. *Tetrahedron Lett.* **1985**, *26*, 6015, 6019; Alexakis, A.; Berlan, J.; Besace, Y. *Tetrahedron Lett.* **1986**, *27*, 1047; Nakamura, E.; Matsuzawa, S.; Horiguchi, Y.; Kuwajima, I. *Tetrahedron Lett.* **1986**, *27*, 4029.

12. Lipshutz, B. H.; Wilhelm, R. S.; Kozlowski, J. A. *J. Org. Chem.* **1984**, *49*, 3938.

13. Lipshutz, B. H.; Lindsley, C.; Susfalk, R.; Gross, T. *Tetrahedron Lett.* **1994**, *35*, 8999, and references therein. In this case, TBAF could not be used in the workup to cleave the silyl enol ether. Instead, the keto product was isolated by allowing the crude material to sit at the top of the silica gel column for 30 min prior to elution.

Appendix

Chemical Abstracts Nomenclature (Collective Index Number); (Registry Number)

Zinc (8,9); (7440-66-6)

1,2-Dibromoethane: Ethane, 1,2-dibromo- (8,9); (106-93-4)

Chlorotrimethylsilane: Silane, chlorotrimethyl- (8,9); (75-77-4)

Ethyl 5-iodovalerate: Pentanoic acid, 5-iodo-, ethyl ester (9); (41302-32-3)

Methyllithium: Lithium, methyl- (8,9); (917-54-4)

Copper(I) cyanide: Copper cyanide (8,9); (544-92-3)

2-Cyclohexen-1-one HIGHLY TOXIC: (8,9); (930-68-7)

Tetrabutylammonium fluoride: Ammonium, tetrabutyl-, fluoride (8); 1-Butanaminium, N,N,N-tributyl-, fluoride (9); (429-41-4)

Ethyl 5-bromovalerate: Valeric acid, 5-bromo-, ethyl ester (8); Pentanoic acid, 5-bromo-, ethyl ester (9); (14660-52-7)

Sodium iodide (8,9); (7681-82-5)

2-Pentanol (8,9); (6032-29-7)

1,10-Phenanthroline (8,9); (66-71-7)

Table I. Cuprate-Catalyzed 1,4-Additions of Organozinc Reagents[8]

Entry	Iodide	Enone	Functionalized Product	Yield (%)
1	I-$(CH_2)_4Cl$	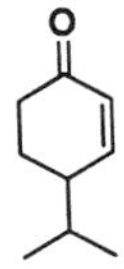 O	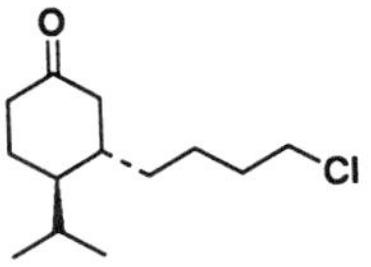 O, Cl	89
2	I-$(CH_2)_4Cl$	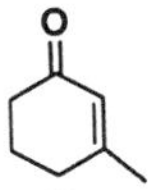 O	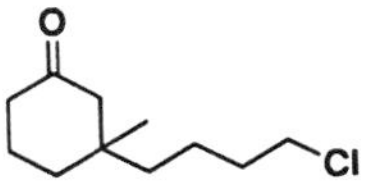 O, Cl	83
3	I-$(CH_2)_5CN$	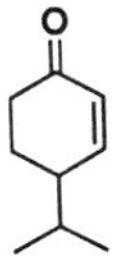 O	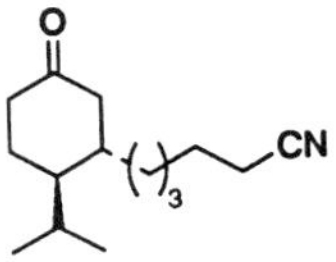 O, ( )3, CN	85
4	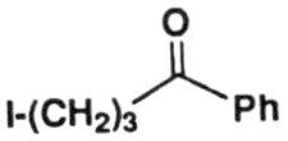 I-$(CH_2)_3$C(=O)Ph	 O	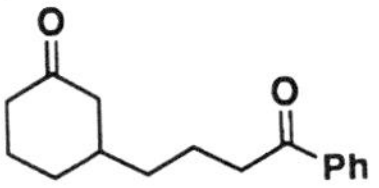 O, O, Ph	85
5	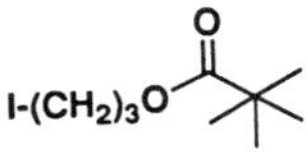 I-$(CH_2)_3$OC(=O)C$(CH_3)_3$	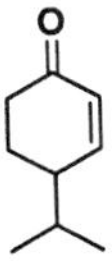 O	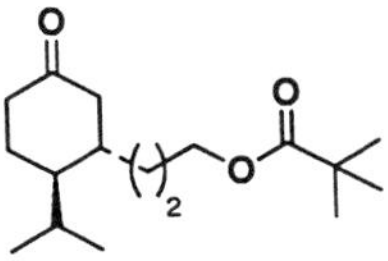 O, ( )2, O, O	74
6	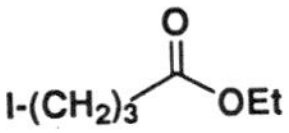 I-$(CH_2)_3$C(=O)OEt	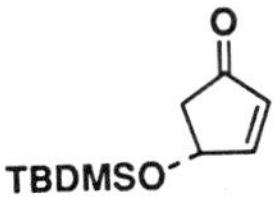 O, TBDMSO	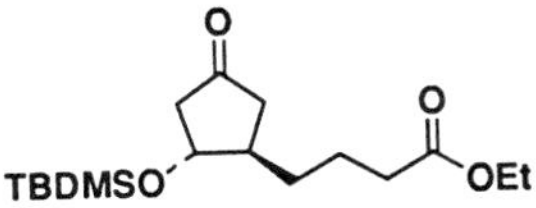 O, TBDMSO, O, OEt	81
7	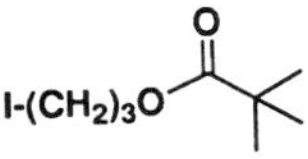 I-$(CH_2)_3$OC(=O)C$(CH_3)_3$	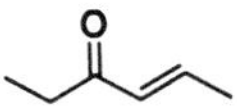 O	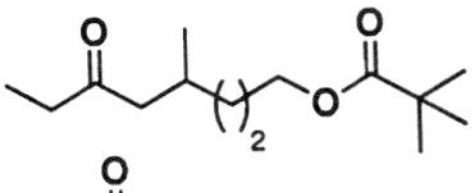 O, ( )2, O, O	85
8	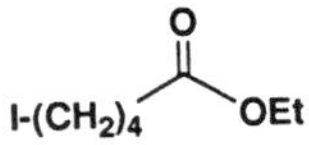 I-$(CH_2)_4$C(=O)OEt	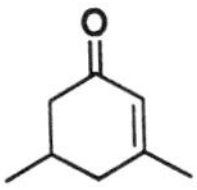 O	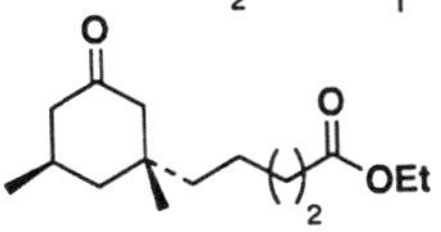 O, ( )2, O, OEt	72
9	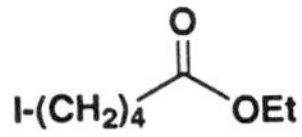 I-$(CH_2)_4$C(=O)OEt	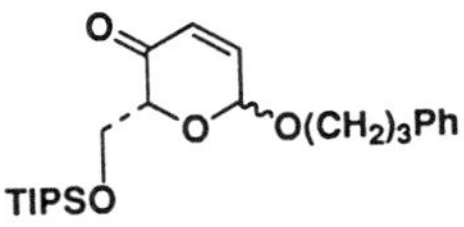 O, O, $O(CH_2)_3Ph$, TIPSO	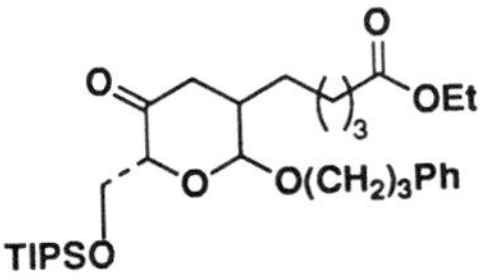 O, ( )3, O, OEt, O, $O(CH_2)_3Ph$, TIPSO	83
10	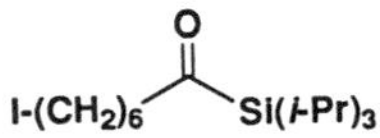 I-$(CH_2)_6$C(=O)Si(i-Pr)$_3$	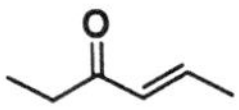 O	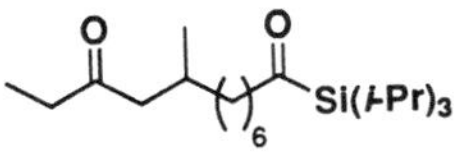 O, ( )6, O, Si(i-Pr)$_3$	74

ISOMERIZATION OF β-ALKYNYL ALLYLIC ALCOHOLS TO FURANS CATALYZED BY SILVER NITRATE ON SILICA GEL: 2-PENTYL-3-METHYL-5-HEPTYLFURAN

[Furan, 5-heptyl-3-methyl-2-pentyl-]

A. $Et_4N^+ Br^-$, HBr / CH_2Cl_2

B. $(Ph_3P)_2PdCl_2$ / CuI, Et_2NH / 1-nonyne

C. 10% $AgNO_3$ on silica / hexane

Submitted by James A. Marshall and Clark A. Sehon.[1]

Checked by Scott A. Frank and William R. Roush.

1. Procedure

Caution! Because of the corrosive and toxic nature of the reagents, steps A and should be conducted in an efficient fume hood. Eye protection and protective othing should be worn while performing these experiments.

A. 2-Bromo-1-octen-3-ol.[2] A one-necked, 250-mL, round-bottomed flask uipped with a magnetic stirring bar and a rubber septum is charged with 70 mL of methylene chloride (CH_2Cl_2) and 14.71 g (69.7 mmol) of anhydrous tetraethylammonium bromide (Et_4NBr) (Note 1). The flask and contents are cooled to 0°C, weighed, and 7.05 g (87.1 mmol) of gaseous hydrogen bromide (HBr) is introduced by needle through the septum into the stirred suspension (Note 2). The solid dissolves during the addition of HBr. To this solution is added 10.2 mL (69.7 mmol) of freshly distilled 1-octyn-3-ol (Note 3), and the reaction is allowed to warm to room temperature. The reaction is closely monitored by thin-layer chromatography (Note 4) until starting material is consumed. The reaction mixture is cooled to 0°C, and 26 mL of triethylamine (Et_3N) is carefully added to the solution. The mixture is diluted with 200 mL of water and 200 mL of ether, and the layers are separated. The organic layer is washed with saturated sodium bicarbonate and brine, dried over magnesium sulfate, and concentrated under reduced pressure. Purification of the crude product by flash chromatography (Note 5) followed by bulb to bulb distillation (95°C, 0.5 mm) affords 8.91 g (62%) of 2-bromo-1-octen-3-ol as a clear colorless oil (Note 6).

B. 7-Methylene-8-hexadecyn-6-ol.[3] A one-necked, 500-mL, round-bottomed flask equipped with a magnetic stirring bar and a rubber septum is charged with 8.65 g (41.8 mmol) of 2-bromo-1-octen-3-ol in 210 mL of diethylamine (Note 7). A stream of argon is bubbled through the stirred solution for 10 min at room temperature. To this solution is added 1.47 g (2.09 mmol) of dichlorobis(triphenylphosphine)palladium(II) [$(Ph_3P)_2PdCl_2$] (Note 8), 0.80 g (4.18 mmol) of copper(I) iodide (CuI) (Note 9), and 6.91 mL (42.1 mmol) of 1-nonyne (Note 10). The reaction is closely monitored by thin-layer chromatography until all the vinyl bromide is consumed (Notes 11, 12). The reaction mixture is poured into a 2-L Erlenmeyer flask and diluted with 500 mL of ether. Saturated ammonium chloride is added to the stirred solution until the evolution of gas ceases. The organic layer is washed with two portions of aqueous 10% hydrochloric acid, and the combined aqueous layers are extracted with 200 mL of ether. The organic extracts are dried over magnesium sulfate and concentrated unde reduced pressure. Purification of the crude product by flash chromatography (Note 13) on silica gel (5% ethyl acetate-hexane) yields 9.94 g (95%) of 7-methylene-8-hexadecyn-6-ol as a light yellow oil (Notes 14, 15).

C. 2-Pentyl-3-methyl-5-heptylfuran.[4] A one-necked, 500-mL, round-bottomed flask equipped with a magnetic stirring bar and a rubber septum is charged with 9.64 g (38.5 mmol) of 7-methylene-8-hexadecyn-6-ol in 200 mL of hexane. To this solution is added 6.53 g (3.85 mmol) of 10% silver nitrate on silica gel (Note 16); the reaction is protected from light (Note 17) and stirred for 40 min at room temperature. Then 50 mL of ether is added, the mixture is filtered, and the filtrate is concentrated under reduced pressure. The recovered silver catalyst can be used for another reaction if desired, in which case a more extended reaction time (2-2.5 hr) may be required. The product is purified by flash chromatography (Note 18) on deactivated silica gel (Note 19) to afford 9.24 g (96%) of 2-pentyl-3-methyl-5-heptylfuran as a clear light yellow oil (Note 20). In the case at hand, when the above reaction was repeated with the recycled catalyst, a 91% yield of the furan product was obtained after chromatography.

2. Notes

1. The salt is crushed and then azeotropically dried with benzene three times, filtered, dried under vacuum for 2 days, and stored under argon over phosphorus pentoxide.

2. The flask is removed from the ice bath, wiped dry, and weighed every 3-5 min until the desired weight is achieved. Also, to relieve pressure, the flask is vented with an exit needle connected to tygon tubing placed in the back of the hood.

3. 1-Octyn-3-ol (96%) was purchased from Aldrich Chemical Company, Inc. It can be fractionally distilled to remove minor impurities.

4. TLC analysis is performed on E. Merck silica gel 60F-254 glass plates of 0.25-mm thickness purchased from EM Reagents. The eluting solvent was 15% ethyl acetate-hexane.

5. The checkers found that the addition of HBr to the starting propargyl alcohol provided an ca. 20 : 1 mixture of 2-bromo-1-octen-3-ol and the isomeric (Z)-1-bromo-1-octen-3-ol in 86-90% yield (isolated by distillation of the crude product). It proved necessary to purify 2-bromo-1-octen-3-ol by chromatography since products deriving from (Z)-1-bromo-1-octen-3-ol were completely inseparable at all subsequent stages of this procedure. The purification was performed by using a 90 x 500-mm column packed with 350 g of silica gel wetted with 1 L of hexane. The crude compound was charged as a solution in 75 mL of CH_2Cl_2. The column was eluted with 500 mL of hexane, followed by 1 L of 5% ethyl acetate/hexane, and then 2 L of 10% ethyl acetate/hexane. 2-Bromo-1-octen-3-ol eluted first [R_f = 0.64 (25% ethyl acetate-hexane)] followed by (Z)-1-bromo-1-octen-3-ol [R_f = 0.53 (25% ethyl acetate-hexane)].

6. Spectral analysis for 2-bromo-1-octen-3-ol is as follows: IR (film) cm^{-1}: 3358, 2930, 1626, 1465, 896; ^1H NMR (400 MHz, $CDCl_3$) δ: 0.87 (t, 3 H, J = 6.8), 1.24-1.37 (m, 6 H), 1.57-1.69 (m, 2 H), 1.86 (d, 1 H, J = 6.0), 4.06 (q, 1 H, J = 6.3), 5.54 (d, 1 H, J = 1.9), 5.85 (dd, 1 H, J = 0.74, J = 1.9); ^{13}C NMR (100 MHz, $CDCl_3$) δ: 14.0, 22.5, 24.8, 31.5, 35.5, 76.1, 116.9, 137.6.

7. Diethylamine was purified by distillation from calcium hydride (CaH_2).

8. $(Ph_3P)_2PdCl_2$ was purchased from the Aldrich Chemical Company, Inc..

9. CuI purchased from the Aldrich Chemical Company, Inc. was washed with tetrahydrofuran in a Soxhlet extractor overnight, dried under vacuum overnight, and stored under argon over calcium sulfate.

10. The reaction mixture turned bright yellow after addition of the palladium catalyst. The color changed to greenish brown after the CuI was added and became yellow upon addition of the 1-nonyne.

11. Extended reaction times cause colored by-products to form, that are difficult to separate. TLC analysis was performed on E. Merck silica gel 60F-254 glass plates of 0.25-mm thickness purchased from EM Reagents. The eluting solvent was 20% ethyl acetate-hexane; in this solvent system, the R_f's of 7-methylene-8-hexadecyn-6-ol and 2-bromo-1-octen-3-ol are 0.68 and 0.55, respectively.

12. If the reaction was not complete within 2 hr, the checkers added an additional 0.25 equiv of 1-nonyne. The reaction was typically complete within 30-45 min following this addition.

13. A 70 x 370-mm column packed with 500 g of silica gel was used.

14. Spectral analysis for 7-methylene-8-hexadecyn-6-ol is as follows: IR (film) cm^{-1}: 3363, 2929, 2858, 2225, 1614, 1465, 902; ^{1}H NMR (400 MHz, $CDCl_3$) δ: 0.75-0.98 (m, 6 H), 1.09-1.40 (m, 14 H), 1.41-1.66 (m, 5 H), 2.31 (t, 2 H, J = 6.9), 4.03 (q, 1 H, J = 6.3 Hz), 5.33 (d, 2 H, J = 7.5); ^{13}C NMR (100 MHz, $CDCl_3$) δ: 14.0, 14.0, 19.2, 22.5, 22.6, 25.0, 28.6, 28.7, 28.8, 31.6, 31.7, 35.9, 74.8, 77.7, 92.7, 119.2, 135.2.

15. The checkers obtained a 92% yield for this reaction.

16. The 10% silver nitrate on silica gel (200 mesh) was purchased from the Aldrich Chemical Company, Inc.

17. The flask was wrapped with aluminum foil.

18. A 70 x 370-mm column packed with 300 g of silica gel was eluted with hexane; in this solvent system, the product has R_f = 0.65.

19. The silica gel was deactivated by flushing with 1 L of 5% triethylamine-hexane solution followed by 2 L of hexane to remove excess triethylamine.

20. Spectral analysis for 2-pentyl-3-methyl-5-heptylfuran is as follows: IR (film) cm^{-1}: 2927, 2856, 1577, 1467, 792; ^{1}H NMR (300 MHz, $CDCl_3$) δ: 0.93 (t, 3 H, J = 6.8), 0.94 (t, 3 H, J = 6.8), 1.23-1.45 (m, 12 H), 1.57-1.67 (m, 4 H), 1.94 (s, 3 H), 2.54 (t, 2 H, J = 7.3), 2.57 (t, 2 H, J = 7.3), 5.77 (s, 1 H); ^{13}C NMR (100 MHz, $CDCl_3$) δ: 9.8,

14.0, 14.0, 22.4, 22.6, 25.9, 28.0, 28.2, 28.4, 29.1, 29.6, 31.4, 31.8, 107.6, 113.7, 149.3, 153.4. Anal. Calcd for $C_{17}H_{30}O$: C, 81.53; H, 12.07. Found: C, 81.42; H, 11.99.

Waste Disposal Information

All toxic materials were disposed of in accordance with "Prudent Practices in the Laboratory"; National Academy Press; Washington, DC, 1995.

3. Discussion

The present procedure evolved from our previous work on the conversion of allenals, allenones, and allenylcarbinols to furans and 2,5-dihydrofurans with catalytic silver nitrate ($AgNO_3$) in acetone.[5-10] It has also been shown that allenylcarbinols can be converted to 2,5-dihydrofuran under these conditions.[11] β- and γ-Alkynyl allylic alcohols can also be isomerized to furans under strongly basic conditions with potassium tert-butoxide in tetrahydrofuran-tert-butyl alcohol-18-crown-6 or hexamethylphosphoramide (KO-t-Bu in THF-t-BuOH-18-crown-6 or HMPA).[12] The $AgNO_3$/silica gel method is milder, faster, and more efficient than the previously reported procedures.[13] Moreover, it offers the potential advantage of catalyst recovery and possible applicability to a flow system in which a packed column, protected from light, could serve as the reactor.[4]

1. University of Virginia, Department of Chemistry, McCormick Road, Charlottesville, VA 22901.
2. Cousseau, J. *Synthesis* **1980**, 805.
3. Sonogashira, K.; Tohda, Y.; Hagihara, N. *Tetrahedron Lett.* **1975**, 4467.
4. Marshall, J. A.; Sehon, C. A. *J. Org Chem.* **1995**, *60*, 5966.

5. Marshall, J. A.; Robinson, E. D. *J. Org. Chem.* **1990**, *55*, 3450.
6. Marshall, J. A.; Wang, X-j. *J. Org. Chem.* **1991**, *56*, 960.
7. Marshall, J. A.; Wang, X-j. *J. Org. Chem.* **1992**, *57*, 3387.
8. Marshall, J. A.; Pinney, K. G. *J. Org. Chem.* **1993**, *58*, 7180.
9. Marshall, J. A.; Yu, B.-C. *J. Org. Chem.* **1994**, *59*, 324.
10. Marshall, J. A.; Bartley, G. S. *J. Org. Chem.* **1994**, *59*, 7169.
11. Olsson, L.-I.; Claesson, A. *Synthesis* **1979**, 743.
12. Marshall, J. A.; DuBay, W. J. *J. Org. Chem.* **1993**, *58*, 3435.
13. For a general overview with leading references, see: Danheiser, R. L.; Stoner, E. J.; Koyama, H.; Yamashita, D. S.; Klade, C. A. *J. Am. Chem. Soc.* **1989**, *111*, 4407.

Appendix

Chemical Abstracts Nomenclature (Collective Index Number); (Registry Number)

2-Pentyl-3-methyl-5-heptylfuran: Furan, 5-heptyl-3-methyl-2-pentyl- (13); (170233-67-7)

2-Bromo-1-octen-3-ol: 1-Octen-3-ol, 1-bromo-. (E)- (9); (52418-90-3)

Tetraethylammonium bromide: Ammonium, tetraethyl-, bromide (8); Ethaniminium, N,N,N-triethyl-, bromide (9); (71-91-0)

Hydrogen bromide: Hydrobromic acid (8,9); (10035-10-6)

1-Octyn-3-ol (8,9); (818-72-4)

7-Methylene-8-hexadecyn-6-ol: 8-Hexadecyn-6-ol, 7-methylene- (13); (170233-66-6)

Dichlorobis(triphenylphosphine)palladium(II); Palladium, dichlorobis(triphenylphosphine)- (8,9); (13965-03-2)

Copper(I) iodide: Copper iodide (8,9); (7681-65-4)

1-Nonyne (8,9); (3452-09-3)

Silver nitrate ~ 10 wt, % on silica gel: Nitric acid silver(+) salt (9); (7761-88-8)

Phosphorus pentoxide: Phosphorus oxide (8,9); (1314-56-3)

(Z)-1-Bromo-1-octen-3-ol: 1-Octen-3-ol, 1-bromo-, (Z)- (11); (87937-09-5)

2-CHLOROPHENYL PHOSPHORODICHLORIDOTHIOATE

(Phosphorodichloroidothioic acid, O-(2-chlorophenyl)ester)

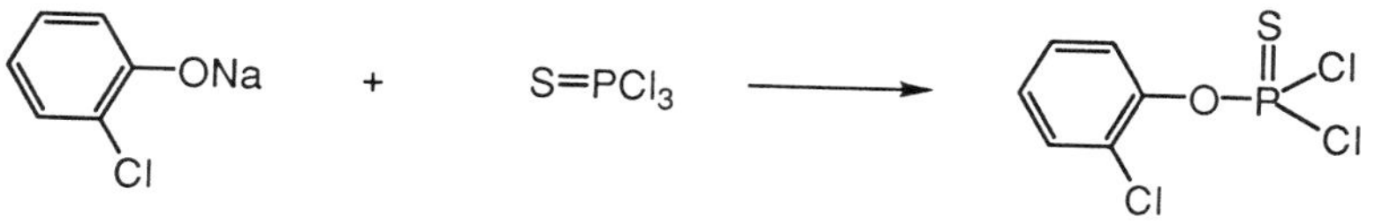

Submitted by Vasulinga T. Ravikumar and Bruce Ross.[1]

Checked by Adam R. Renslo and Rick L. Danheiser.

1. Procedure

A 2-L, three-necked, round-bottomed flask equipped with a 250-mL addition funnel, glass stoppers, and a magnetic stirring bar is charged with thiophosphoryl chloride (271 g, 1.60 mol) (Note 1), tetrabutylammonium bromide (3.2 g; 0.01 mol) (Note 1), and 400 mL of dichloromethane (Note 2), and the resulting solution is cooled in an ice bath at 0-5°C. 2-Chlorophenol (51.4 g, 0.400 mol) (Note 1) is added to a magnetically stirred solution of sodium hydroxide (25%, 350 mL, Note 1) in a 500-mL Erlenmeyer flask cooled in an ice bath at 0-5°C, and the resulting solution is transferred to the addition funnel and added slowly over a period of 30 min to the reaction mixture. The resulting two-phase mixture is stirred for 8 hr while the bath is allowed to warm to room temperature, and the aqueous layer is then separated and extracted with 100 mL of dichloromethane. The combined organic layers are washed with 100 mL of brine, dried over magnesium sulfate, and filtered. The dichloromethane is removed by rotary evaporation, and the residue is distilled under reduced pressure through a short-path still head (Note 3) to give 93.2-98.4 g (89-94%) of 2-chlorophenyl phosphorodichloridothioate as a colorless oil (Note 4).

2. Notes

1. Reagent grade 2-chlorophenol, sodium hydroxide, thiophosphoryl chloride, and tetrabutylammonium bromide were purchased from Aldrich Chemical Company, Inc., and used without purification.

2. HPLC-grade dichloromethane was purchased from Mallinckrodt Inc. and used without further purification.

3. Two traps cooled in liquid nitrogen are connected between the distillation apparatus and the vacuum pump to trap excess thiophosphoryl chloride.

4. 2-Chlorophenyl phosphorodichloridothioate has the following properties: bp 90-93°C (0.2 mm); IR (neat) cm^{-1}: 3160, 1580, 1475, 1450, 1260, 1210, 1060, 1040, 940, 780, 760, 720; ^{1}H NMR ($CDCl_3$, 300 MHz) δ: 7.22-7.34 (m, 2 H), 7.43-7.51 (m, 2 H); ^{13}C NMR ($CDCl_3$, 75 MHz) δ: 122.5, 122.6, 126.7, 126.8, 127.7, 127.8, 127.91, 127.95, 131.09, 131.12, 146.7, 146.9; ^{31}P NMR ($CDCl_3$, 202 MHz) δ: 54.4. Anal. Calcd for $C_6H_4Cl_3OPS$: C, 27.56; H, 1.54. Found: C, 27.57; H, 1.42.

Waste Disposal Information

All toxic materials were disposed of in accordance with "Prudent Practices in the Laboratory"; National Academy Press; Washington, DC, 1995.

3. Discussion

Deoxyribonucleotides, deoxyribonucleotide phosphorothioates, modified DNA and analogs have wide applications in molecular biology, antisense applications antigene therapy, etc. Three methods are available for the synthesis o oligonucleoside phosphorothioates: phosphoramidite, H-phosphonate, and phospho triester approaches. Different protecting groups, most of which are base labile, have been used for the phosphoramidite approach. In the H-phosphonate approach, no protecting group is involved. In the phosphotriester approach, aryl groups are used as O-protecting groups, and deprotection occurs via a nucleophilic attack on the phosphorus center with the aryloxy group being the leaving group. Because of their base labile nature, most of the groups used in the phosphoramidite approach are not suitable for the synthesis of aryl phosphorodichloridothioates, which are used as the starting material in the phospho triester approach. The previously reported[2,3] routes for the synthesis of aryl phosphorodichloridothioates involve drastic conditions such as refluxing or the use of liquid sulfur dioxide, and are not amenable to very large scale synthesis. A much simpler alternative method involving phase transfer reaction is described here.

Phase transfer catalysis[4] is a valuable tool in organic synthesis. The process is exemplified by the convenient synthesis of 2-chlorophenyl phosphorodichloridothioate. Using this phase transfer reaction, a number of dichloridothioates of substituted phenyl, benzyl, thiophenyl, and thiobenzyl alcohols are accessible. The phosphorodichloridothioate reacts with various coupling reagents to form activated species that are useful in the synthesis of oligonucleotide phosphorothioates via the phosphotriester approach as illustrated below.[5,6]

DMTO Base OH — S, Cl, Cl, O–P, Cl; 1-HOBt; HO Base OH → DMTO Base, O, O–P=S, Cl, O Base OH

The procedure shown here describes the preparation of a fully protected phosphorothioate triester dimer. The dimer can then be coupled subsequently in a similar way to form elongated oligomers.

1. Department of Chemistry, Isis Pharmaceuticals, Carlsbad, CA 92008.
2. Tolkmith, H. *J. Org. Chem.* **1958**, *23*, 1682.
3. Sindona, A. P. G.; Uccella, N. *Nucleosides Nucleotides* **1991**, *10*, 615.
4. Starks, C. M.; Liotta, C. L.; Halpern, M. "Phase-Transfer Catalysis: Fundamentals, Applications, and Industrial Perspectives"; Chapman & Hall: New York, 1994.
5. Marugg, J. E.; Van den Bergh, C.; Tromp, M.; Van der Marel, G. A.; Van Zoest, W. J.; Van Boom, J. H. *Nucleic Acids Res.* **1984**, *12*, 9095.
6. Kemal, O.; Reese, C. B.; Serafinowska, H. T. *J. Chem. Soc., Chem. Commun.* **1983**, 591.

Appendix

Chemical Abstracts Nomenclature (Collective Index Number); (Registry Number)

2-Chlorophenyl phosphorodichloridothioate: Phosphorodichlorodiothioic acid, O-(2-chlorophenyl)ester (10); (68591-34-4)

Thiophosphoryl chloride: HIGHLY TOXIC (8,9); (3982-91-0)

Tetrabutylammonium bromide (8); 1-Butanaminium, N,N,N-tributyl-, bromide (9); (1643-19-2)

2-Chlorophenol: Phenol, o-chloro- (8); Phenol, 2-chloro- (9): (95-57-8)

Sodium hydroxide (8,9); (1310-73-2)

VITAMIN D_2 FROM ERGOSTEROL

(9,10-Secoergosta-5,7,10(19),22-tetraen-3-ol,(3β)- from Ergosta-5,7,22-trien-3-ol,(3β)-)

A.

1, **Ergosterol** (20 mM)

1) hν, t-BuOMe, p-Me_2N-C_6H_4-CO_2Et (4 mM)

2) hν (uranium filter) 9-acetylanthracene (0.2 mM)

2

3) Δ, MeOH

4) pyridine, 3,5-$(NO_2)_2$-C_6H_4-COCl

3

B.

3

NaOH / EtOH

4, Vitamin D_2

Submitted by Masami Okabe.[1]

Checked by Gilles Chambournier and David J. Hart.

1. Procedure

Caution! Light from a mercury lamp is damaging to the eyes and skin. Suitable precautions, such as wearing an appropriate face shield, UV radiation protective eyewear and gloves, and surrounding the reaction vessel with aluminum foil, should be taken. It is recommended that one work in a hood and that the hood sash be covered with aluminum foil.

A. 3,5-Dinitrobenzoate of vitamin D_2 (**3**). A 2-L photo-reaction vessel equipped with a quartz immersion well, a thermometer, an argon-inlet tube, a mineral oil outlet-bubbler, a mechanical stirrer, and supported in an adequately sized Dewar (Note 1), is charged with 13.5 g (34 mmol) of ergosterol (**1**) (Note 2), 1.31 g (6.8 mmol) of ethyl p-dimethylaminobenzoate, and 1.7 L of tert-butyl methyl ether (tert-BuOMe) (Note 3). The mixture is stirred at room temperature overnight with gentle bubbling of argon (Note 4). A 450-watt Hanovia medium-pressure mercury lamp is inserted into the well, through which a fast stream of water is continuously passed (Note 5). The solution is cooled in a dry ice-ethanol bath and stirred vigorously (Note 6). When the temperature of the solution reaches 0°C, the lamp is turned on, and irradiation is continued at 0°C to -20°C for 4 hr (Note 7). The lamp is turned off, a solution of 75 mg (0.34 mmol) of 9-acetylanthracene (Note 8) in 3 mL of tert-BuOMe is added to the solution, and a uranium filter (Note 9) is inserted into the arc housing (Note 10). After 10 min, the lamp is started again, and the mixture is irradiated at 0°C to -20°C for 1 hr through the uranium filter (Note 11). The cold, pale yellow solution of pre-vitamin **2** thus obtained (Note 12) is transferred to a 3-L, round-bottomed flask and concentrated at 20-25°C under reduced pressure. The residue is transferred again to a 500-mL, round-bottomed flask using tert-BuOMe to allow quantitative transfer. The solution is concentrated and dried at room temperature under high vacuum (1 mm) for 30 min to give approximately 15.4 g of a yellow resin. The flask is filled with argon and equipped with a magnetic stirrer. Methanol (100 mL) is added, and the mixture is shaken to give a stirrable suspension. The suspension is then stirred for 45 min at room temperature and stored in a freezer overnight. After the mixture is stirred at -30°C (Note 13) for 30 min, it is quickly filtered through a 60-mL sintered-glass funnel of coarse porosity. The collected solid is washed with 20 mL of cold methanol (Note 14). Ergosterol (1.83 g, 14% recovery, mp 145-151°C, 99.4% pure) is recovered by washing this solid with absolute ethanol (30 mL at room temperature). The filtrate and cold-methanol wash are transferred to a 500-mL, round-bottomed flask equipped with a magnetic stirrer, an argon-inlet tube, and a reflux condenser. The flask is flushed with argon, and the orange solution is heated under reflux for 6 hr (Note 15) and then stirred at 35-40°C overnight (Note 16). The mixture is concentrated at 30°C under reduced pressure, and the residual methanol is removed by coevaporation with 50 mL of toluene at 30°C to give approximately 14.6 g of an orange-tan oil. The flask is filled with argon and then equipped with a magnetic stirrer. The residue is dissolved in 40 mL of pyridine, and the solution is cooled in an ice-water bath. Solid 3,5-dinitrobenzoyl chloride (9.0 g, 39 mmol) (Note 17) is added in small portions over 5 min followed by 20 mL of pyridine to rinse the walls of the flask, and the mixture is stirred at 0°C for 20 min. The very thick suspension obtained is shaken and then allowed to stand at 0°C for a further 20 min, whereupon methanol (30 mL) is added to the cold mixture. The mixture is allowed to stand at 0°C for 5 min, and then it is shaken for about 5 min to give a stirrable suspension. After the orange suspension is stirred at 0°C for 1.5 hr, it is diluted by the dropwise addition of 150 mL of methanol over 15 min and stirred at 0°C for another hour. The yellow solid is collected by filtration, washed with 50 mL of ice-cold methanol, and dried at room temperature under high vacuum for 2 hr. The yellow-orange solid is transferred to a 250-mL, round-bottomed flask equipped with a magnetic stirrer and an argon-inlet tube. The flask is flushed with argon, and the solid is suspended in 50 mL of absolute ethanol. The suspension is stirred at room temperature for 15 min and at 0°C for 45 min. The solid is collected by filtration, washed with 20 mL of cold methanol, and dried at room temperature under high vacuum overnight to give 8.3-10.1 g (41-50%) of **3** as a yellow solid, mp 139-141°C (lit.[2] 147-149°C) (Notes 18 and 19).

B. Vitamin D_2 (**4**). A 500-mL, round-bottomed flask equipped with a magnetic stirrer and an argon-inlet tube is charged with 10.1 g (17.1 mmol) of **3** and 171 mL of absolute ethanol. The flask is flushed with argon, and 1.88 mL (18.8 mmol) of 10 N sodium hydroxide is added. After the purple suspension is stirred at room temperature for 45 min, it is cooled with an ice-water bath. Then, 75 mL of ethanol-water (EtOH-H_2O)(2:5) is added over 5 min, and the mixture is stirred for 1 hr with ice-water cooling. To the resulting rose-colored suspension is added 100 mL of EtOH-H_2O (2:5) dropwise over 30 min. The mixture is stirred at 0°C for 30 min and then stored in a refrigerator overnight. The precipitate (Note 12) is collected by filtration using a 60-mL sintered-glass funnel of coarse porosity, washed quickly with 50 mL of cold EtOH-H_2O (3:2) (Note 20), and dried at room temperature under high vacuum to give 6.55 g of an off-white solid. This material is transferred to a 500-mL, round-bottomed flask equipped with a magnetic stirrer and an argon-inlet tube. The flask is flushed with argon, and the solid is dissolved in 200 mL of methanol (MeOH) at room temperature. The solution is cooled with an ice-water bath, and 15 mL of methanol-water (MeOH-H_2O) (4:6) is added to give a cloudy solution. After stirring at 0°C for 1 hr, the resulting white suspension is diluted by the dropwise addition of 85 mL of MeOH-H_2O (4:6) over 45 min. After 2 hr at 0°C, the solid is collected by filtration using a 60-mL sintered-glass funnel of coarse porosity, washed with 50 mL of cold MeOH-H_2O (4:1) (Note 20), and dried at room temperature under high vacuum overnight to give 5.9-6.2 g (87-91%) of **4** as a white solid, mp 112-114°C (lit.[2] 114.5-117°C) (Notes 21 and 22). The overall yield of vitamin D_2 from ergosterol is 44% (51% based on the recovered ergosterol) (Note 2).

2. Notes

1. The reactor is available from Ace Glass Inc. [reaction vessel (#7851-17), immersion well (#7854-28, 290 mm), Teflon bearing (#8066-24), and stirring shaft (#8068-303)]. It is similar to that shown in Figure 1 (*Org. Synth., Coll. Vol. V* **1973**, p.529) with an additional stirring chamber. The submitter used a 4-mm I.D. tube for an argon-inlet in order to avoid clogging; the tube should reach near to the bottom of the vessel. The checkers used a 20 x 45-cm (ID x height) Dewar, available from Cole-Parmer Instrument Co. #H-03774-54).

2. Ergosterol (**1**), obtained from Aldrich Chemical Company, Inc. (mp 134-142°C, ε 8,030 at 282 nm in EtOH), was purified before use as follows: 24.4 g of **1** was suspended in 200 mL of ethanol (EtOH), and the mixture was stirred at room temperature for 3 hr prior to filtration. The collected solid was washed with 40 mL of EtOH and dried under high vacuum (1.0 mm) to give 19.3 g (79% recovery) of **1** as a white solid (mp 147-153°C, ε 11,900 at 282 nm in EtOH). The submitter observed that when **1** purchased from Kaneka Co. (mp 147-153°C, ε 11,560 at 282 nm in EtOH) was used as received, a better quality of vitamin D_2 (**4**) (mp 114-115°C, 99.8% pure) was obtained in a better overall yield of 48% (55% based on the recovered ergosterol). The checkers used ergosterol obtained from Acros Organics (mp 156-158°C), which was purified as described above (mp 154-158°C).

3. Ethyl p-dimethylaminobenzoate and tert-butyl methyl ether (HPLC grade) were obtained from Aldrich Chemical Company, Inc., and used as received.

4. The overnight stirring with bubbling of argon to remove oxygen is probably too long, but was done simply for convenience, thereby allowing the irradiation to be carried out the next day. The checkers found that 4 hr was sufficient.

5. Water flow must be very fast to avoid freezing, which could result in breakage of the photochemical reactor and generation of a hazardous situation. Water flow should be monitored continuously during the course of the reaction.

6. Efficient stirring is necessary to achieve relatively homogeneous temperature distribution throughout the reaction mixture. The submitter recommends that the mixture be kept below 0°C to prevent thermal isomerization of **1** to vitamin D_2, which produces a variety of photoproducts upon irradiation.

7. The checkers monitored the reaction temperature at 5-min intervals and added dry ice to the bath each time the temperature approached 0°C. Approximately 30 pounds of dry ice are required over the 4-hr irradiation period. Using ^1H NMR analyses, the submitter judged the conversion to be 80-85% after 4 hr of irradiation with the apparatus described. After 3 hr, a 1:2:1 mixture of **1**:**2**:tachy-isomer (see Discussion) was obtained. The diagnostic peaks in the ^1H NMR spectra are listed in Table. The R_f values on silica gel TLC using 1:4 EtOAc-hexane are as follows: **1** (0.29), **4** (0.34), tachy-isomer (0.34), **2** (0.42), ethyl p-dimethylaminobenzoate (0.47), and 9-acetylanthracene (0.53), using short wave UV detection.

8. 9-Acetylanthracene was purchased from Aldrich Chemical Company, Inc. and used as received.

9. A cylindrical uranium filter (31-mm O.D., 2.5-mm thickness) was obtained from Houde Glass Co., Inc.

10. *Caution! The lamp is very hot.* The lamp should be allowed to cool for 10 min before restarting to prevent damage.

11. Based on ^1H NMR analyses, the photosensitized isomerization of the tachy-isomer into **2** was complete after 40 min of irradiation (see Discussion).

12. The submitter recommends that pre-isomer (**2**) and vitamin D_2 (**4**) not be exposed to air at room temperature for more than 30 min, since these compounds are relatively easily oxidized by air.

13. The checkers used a dry ice-ethylene glycol bath at -25°C.

14. The checkers cooled the methanol to -70°C in a dry ice-acetone bath.

15. After 3 hr of reflux, the ratio of **2** to **4** was ca. 1:3, whereas after 6 hr, it reached about 1:5.

16. At this point, the ratio of **2** to **4** was greater than 1:6.

17. 3,5-Dinitrobenzoyl chloride was purchased from Aldrich Chemical Company, Inc. and pyridine (A.C.S. certified) was obtained from Fisher Scientific Co. They are used as received.

18. The elemental analysis and spectral properties of **3** are as follows: Anal. Calcd for $C_{35}H_{46}N_2O_6$: C, 71.16; H, 7.85; N, 4.74. Found: C, 71.05; H, 7.89; N, 4.58; IR (KBr) cm^{-1}: 1733, 1546, 1342; ^1H NMR ($CDCl_3$) δ: 0.56 (s, 3 H), 0.82 (d, 3 H, J = 6.5), 0.84 (d, 3 H, J = 6.5), 0.92 (d, 3 H, J = 6.8), 1.02 (d, 3 H, J = 6.6), 1.25-2.17 (m, 16 H), 2.32 (m, 1 H), 2.50 (m, 1 H), 2.59 (dd, 1 H, J = 12.2 and 6.8), 2.73 (dd, 1 H, J = 12.2 and 4.5), 2.80 (dd, 1 H, J = 8.8 and 3.5), 4.91 (bs, 1 H), 5.13 (bs, 1 H), 5.20 (m, 2 H), 5.31 (m, 1 H), 6.06 (d, 1 H, J = 11.1), 6.28 (d, 1 H, J = 11.1), 9.13 (d, 2 H, J = 2.1), 9.22 (t, 1 H, J = 2.1); ^{13}C NMR ($CDCl_3$) δ: 12.2 (q), 17.5 (q), 19.5 (q), 19.8 (q), 21.0 (q), 22.1 (t), 23.5 (t), 27.7 (t), 28.9 (t), 31.7 (t), 32.0 (t), 33.0 (d), 40.3 (d and t), 41.9 (t), 42.7 (d), 45.8 (s), 56.3 (d), 74.8 (d), 113.2 (t), 117.1 (d), 122.1 (d), 123.0 (d), 129.3 (d), 131.9 (d), 133.1 (s), 134.3 (s), 135.4 (d), 143.1 (s), 143.8 (s), 148.5 (s), 161.8 (s) (one doublet was not observed).

19. The submitter indicates that HPLC analysis of the product shows its purity to be 97.3% (with 0.3% of ergosteryl 3,5-dinitrobenzoate). HPLC conditions for this unchecked analysis are as follows: column: Chromegasphere SI-60 (3μ, 15-cm x 5-mm) (purchased from ES Industries); mobile phase: 5% EtOAc in heptane (1 mL/min); detection: 275 nm. The retention times of **3** and ergosteryl 3,5-dinitrobenzoate are 4.90 and 3.83 min, respectively. Those of the other impurities are 4.20, 4.55, 6.17, and

8.67 min. The checkers observed traces of pyridine in the ^{1}H NMR spectrum of the product.

20. The checkers cooled this solution in an ice-water bath.

21. The elemental analysis and the spectral properties of **4** are as follows: Anal. Calcd for $C_{28}H_{44}O$: C, 84.79; H, 11.18. Found: C, 84.89; H, 11.17. UV (EtOH) λ_{max} 264 nm (ε 18,450); ^{1}H NMR ($CDCl_3$) δ: 0.55 (s, 3 H), 0.82 (d, 3 H, J = 6.5), 0.84 (d, 3 H, J = 6.5), 0.91 (d, 3 H, J = 6.5), 1.01 (d, 3 H, J = 6.7), 1.2-2.5 (m, 19 H), 2.57 (dd, 1 H, J = 11.6 and 3.5), 2.81 (broad d, 1 H, J = 9.0), 3.95 (m, 1 H), 4.82 (bs, 1 H), 5.04 (bs, 1 H), 5.20 (m, 2 H), 6.05 (d, 1 H, J = 11.2), 6.23 (d, 1 H, J = 11.2); The OH hydrogen was not observed; ^{13}C NMR ($CDCl_3$) δ: 12.2 (q), 17.5 (q), 19.5 (q), 19.8 (q), 21.0 (q), 22.1 (t), 23.4 (t), 27.7 (t), 28.9 (t), 31.8 (t), 33.0 (d), 35.0 (t), 40.3 (d and t), 42.7 (d), 45.6 (s), 45.8 (t), 56.3 (d), 69.1 (d), 112.3 (t), 117.4 (d), 122.3 (d), 131.8 (d), 134.9 (s), 135.5 (d), 142.1 (s), 144.0 (s) (one doublet was not observed).

22. The submitter indicates that HPLC analysis of the product shows the purity to be 99.7% (with 0.1% of ergosterol). The HPLC conditions for this unchecked analysis are as follows: column: Chromegasphere SI-60 (3μ, 15-cm x 5-mm) (purchased from ES Industries); mobile phase: 5% EtOAc in heptane (2 mL/min); detection: 275 nm. The retention times of **1** and **4** are 19.54 and 16.27 min, respectively. The checkers observed signals due to a trace olefinic contaminant in the ^{1}H NMR of the product at δ 6.47 and 5.95.

Waste Disposal Information

All toxic materials were disposed of in accordance with "Prudent Practices in the Laboratory"; National Academy Press; Washington, DC, 1995.

3. Discussion

The overall yields of vitamin D derivatives[3] via the photolyses of pro-isomers (such as **1**) are, generally only 15-30% (typically on a few milligram scale). Difficulties in obtaining higher yields arise from the fact that the photolysis of the pro-isomer gives a photostationary state where the distribution of the products (pro-, pre-, lumi-, and tachy-isomers) depends on the photolyzing wavelength[4] (see Scheme). A shorter wavelength (<290 nm) leads to a predominance of the ring-opened products (pre- and tachy-isomers), and a longer wavelength (>300 nm) promotes the ring-closure reaction to the pro- and lumi-isomers. Furthermore, the pre-isomer and vitamin D are in thermal equilibrium in which the ratio is dependent on the temperature (Scheme).[5]

Scheme

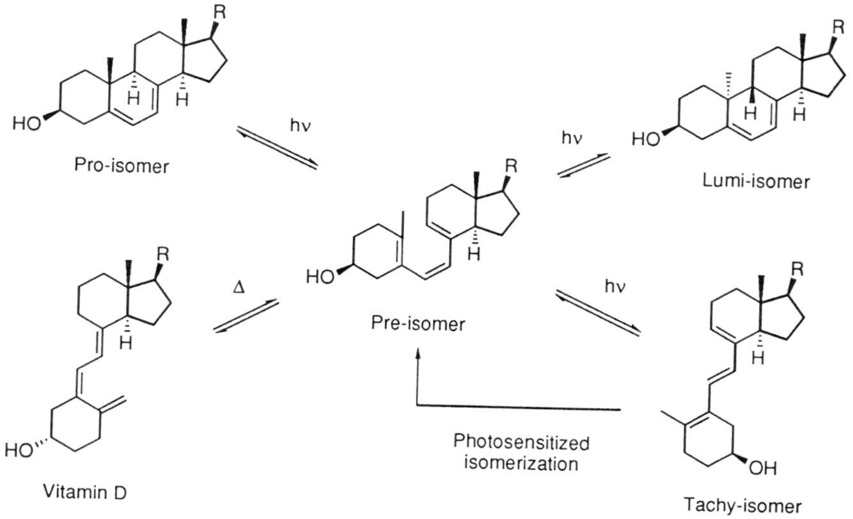

Pro-isomer

Lumi-isomer

Pre-isomer

Vitamin D

Tachy-isomer

Photosensitized isomerization

Higher yields of the pre-isomer have been achieved in the past by irradiating the pro-isomer with a narrow band of approximately 250 nm light (using a low pressure mercury lamp or a laser) and then selective isomerization of the tachy-isomer thus formed into pre-isomer, either by irradiation at approximately 350 nm[4b,c] or by photosensitized isomerization.[6] Vitamin D_3 was then isolated in 50% yield after thermal isomerization.[4c] However, the use of high-intensity light sources with narrow band spectra, for large scale applications is prohibitive because of their high cost; moreover, such sources require the use of specialized equipment.

The present procedure uses a medium pressure mercury lamp, which is an inexpensive, high-intensity light source commonly used in a synthetic laboratory. The 300-315 nm light that promotes the ring-closure reaction is effectively removed by adding a small amount of ethyl p-dimethylaminobenzoate which has a strong, relatively sharp absorption at 305 nm (ε 32,500). After the photosensitized isomerization and thermal equilibration, vitamin D_2 is isolated as the 3,5-dinitrobenzoate **3**.[2] Hydrolysis of this relatively stable derivative furnishes crystalline vitamin D_2 in high purity. This procedure, which uses readily available and inexpensive photolysis equipment, proceeds in good yields and does not require chromatographic purification of the product.

TABLE

THE CHEMICAL SHIFTS OF DIAGNOSTIC PEAKS IN ^{1}H NMR[a]

	18-C<u>H</u>$_3$	3-H	olefinic protons
Ergosterol (**1**)	0.62 (s)	3.62 (m)	5.39 (m), 5.57 (m)
Pre-isomer (**2**)	0.71 (s)	3.92 (m)	5.48 (m), 5.67 (d, 11 Hz), 5.93 (d, 11 Hz)
Tachy-isomer	0.69 (s)	3.92 (m)	5.67 (m), 6.00 (d, 16 Hz), 6.71 (d, 16 Hz)

[a] The chemical shifts are reported in ppm relative to C<u>H</u>Cl_3 (7.25) in $CDCl_3$ as an internal standard.

1. Roche Research Center, Hoffmann-La Roche Inc., Nutley, NJ 07110. The author would like to thank Mr. M. J. Petrin and Mr. R. C. West for the HPLC analyses.
2. For the original preparation of vitamin D_2, see: (a) Askew, F. A.; Bourdillon, R. B.; Bruce, H. M.; Callow, R. K.; Philpot, J. St. L.; Webster, T. A. *Proc. Roy. Soc. (London)* **1932**, *B109*, 488; (b) Windaus, A.; Linsert, O.; Lüttringhaus, A.; Weidlich, G. *Justus Liebigs Ann. Chem.* **1932**, *492*, 226.
3. For reviews, see: (a) Havinga, E. *Experientia* **1973**, *29*, 1181; (b) Norman, A. W. "Vitamin D: The Calcium Homeostatic Steroid Hormone"; Academic Press: New York, 1979; pp. 37-69.
4. (a) Havinga, E.; de Kock, R. J.; Rappoldt, M. P. *Tetrahedron* **1960**, *11*, 276; (b) Malatesta, V.; Willis, C.; Hackett, P. A. *J. Am. Chem. Soc.* **1981**, *103*, 6781; (c) Dauben, W. G.; Phillips, R. B. *J. Am. Chem. Soc.* **1982**, *104*, 355; (d) Dauben, W. G.; Phillips, R. B. *J. Am. Chem. Soc.* **1982**, *104*, 5780.
5. (a) Hanewald, K. H.; Rappoldt, M. P.; Roborgh, J. R. *Rec. Trav. Chim.* **1961**, *80*, 1003; (b) Okamura, W. H.; Elnagar, H. Y.; Ruther, M.; Dobreff, S. *J. Org. Chem.* **1993**, *58*, 600.
6. (a) Eyley, S. C.; Williams, D. H. *J. Chem. Soc. Chem. Commun.* **1975**, 858; (b) Stevens, R. D. S. U.S. Patent 4 686 023, 1987; *Chem. Abstr.* **1987**, *107*, 237124j.

Appendix

Chemical Abstracts Nomenclature (Collective Index Number); (Registry Number)

Vitamin D_2: Ergocalciferol (8); 9,10-Secoergosta-5,7,10 (19), 22-tetraen-3-ol, (3β)- (9); (50-14-6)

Ergosterol (8); Ergosta-5,7,22-trien-3-ol, (3β)- (9); (57-87-4)

Vitamin D_2 3,5-dinitrobenzoate: Ergocalciferol, 3,5-dinitrobenzoate (8,9); (4712-11-2)

Ethyl p-dimethylaminobenzoate: Benzoic acid, p-(dimethylamino)-, ethyl ester (8); Benzoic acid, 4-(dimethylamino)-, ethyl ester (9); (10287-53-3)

tert-Butyl methyl ether: Ether, tert-butyl methyl (8); Propane, 2-methoxy-2-methyl- (9); (1634-04-4)

9-Acetylanthracene: Ketone, 9-anthryl methyl (8); Ethanone, 1-(9-anthracenyl)- (9); (784-04-3)

3,5-Dinitrobenzoyl chloride: Benzoyl chloride, 3,5-dinitro- (8,9); (99-33-2)

5-PHENYLDIPYRROMETHANE AND 5,15-DIPHENYLPORPHYRIN

A. Pyrrole + benzaldehyde (CHO) $\xrightarrow{TFA}$ **1**

B. **1** + $(CH_3O)_3CH$ $\xrightarrow{Cl_3CCO_2H}$ **2**

Submitted by Ross W. Boyle,[1] Christian Bruckner,[2] Jeffrey Posakony,[3] Brian R. James,[3] and David Dolphin.[3]

Checked by Iwao Okamoto and Rick L. Danheiser.

1. Procedure

A. 5-Phenyldipyrromethane (**1**).[4] To a 250-mL, two-necked, round-bottomed flask equipped with a magnetic stir bar and fitted with a gas outlet connected to an oil bubbler are added freshly distilled pyrrole (150 mL, 2.16 mol) and benzaldehyde (6.0 mL, 59 mmol) (Note 1). The second neck of the flask is sealed with a rubber septum, and the mixture is deoxygenated by bubbling dry nitrogen through it for 15 min. Trifluoroacetic acid (0.45 mL, 5.8 mmol) is added in one portion by syringe, and the resulting mixture is stirred magnetically for 15 min at room temperature. Excess pyrrole (Note 2) is removed by rotary evaporation (5-20 mm) with warming (50-60°C) to yield a dark oil. The oil is taken up in a minimal amount (ca. 10 mL) of dichloromethane, and charged onto the top of a flash chromatography column (50 g of 230 - 400 mesh silica gel). The major fraction containing **1** and some higher oligomers is eluted with dichloromethane and collected (TLC is monitored with visualization using bromine vapor: Compound **1** turns bright red/orange, and the higher oligomers show beige to brown spots of lower R_f). Rotary evaporation of the solvent yields a tan oil that is transferred to a vacuum sublimation apparatus (50 mm x 200 mm) and subjected to high vacuum (0.01 mm). To allow residual pyrrole and solvent to escape, a slow heating rate (ca. 0.75°C/min) is maintained until visible sublimation sets in at approximately 120°C (Note 3). After sublimation ceases, the white crystalline sublimate consisting of **1** (8.57-8.83 g; 65-67%) is collected (Note 4).

B. 5,15-Diphenylporphyrin (**2**).[5] A 1-L, one-necked, round-bottomed flask equipped with a magnetic stir bar and a 250-mL, pressure-equalizing, dropping funnel fitted with an argon inlet adaper is charged with a solution of 5-phenyldipyrromethane (**1**) (0.5 g, 2.3 mmol) in 630 mL of dichloromethane (Note 5). Trimethyl orthoformate (18 mL, 0.165 mol) is added in one portion (Note 6). The flask is wrapped in aluminum foil, stirred magnetically, and a solution of trichloroacetic acid (8.83 g, 54 mmol) in 230 mL of dichloromethane is added dropwise over 15 min. The resulting solution is stirred in the dark for 4 hr, whereupon excess acid is quenched by adding 15.6 mL of pyridine in one portion. The solution is stirred for 17 hr, while still protected from light. Finally, air is bubbled into the solution for 10 min, and the reaction mixture is stirred for 4 hr open to air and light. The solvent is removed by rotary evaporation (25°C, 20 mm), and the residue is further concentrated under high vacuum (0.01 mm) for 17 hr. The solid residue is dissolved in 100 mL of dichloromethane, and silica gel (5 g, 230-400 mesh) is added to the solution. The solvent is evaporated and the crude product, adsorbed on silica gel, is loaded onto the top of a chromatography column packed with 50 g of 230-400 mesh silica gel. The product is eluted with 70:30 dichloromethane/hexane. The combined eluants are concentrated under reduced pressure, and the residue is triturated with 50 mL of methanol. The solid crystalline product is recovered by filtration (Note 7). The purple crystals are air-dried and recrystallized from a mixture of toluene (15 mL) and pyridine (0.1 mL) to give **2** (0.073-0.104 g; 14-20%) as lustrous purple plates (Note 8).

2. Notes

1. Pyrrole was obtained from Lancaster Synthesis Inc., and all other chemicals were obtained from Aldrich Chemical Company, Inc. and used as received.

2. Excess pyrrole is required to maximize formation of dipyrromethane. Lowering the pyrrole/benzaldehyde ratio favors formation of higher oligomers and polymers. Unreacted pyrrole is recovered by distillation after the reaction is complete and can be recycled.

3. A slow heating rate and application of vacuum are crucial to avoid bumping of the crude oil onto the cold finger.

4. The spectral and analytical properties for **1** are as follows: mp 102.0-102.5°C (lit.[6] 100.2-101.1°C), the submitters observed 104°C; ^1H NMR (300 MHz, $CDCl_3$) δ: 5.47 (s, 1 H), 5.92 (br s, 2 H), 6.15 (d, 2 H, J = 2.8), 6.69 (br s, 2 H), 7.20-7.34 (m, 5 H), 7.90 (br s, 2 H); ^{13}C NMR (75 MHz, $CDCl_3$) δ: 43.9, 107.1, 108.3, 117.1, 126.8, 128.3, 128.5, 132.3, 141.9; IR (CH_2Cl_2) cm^{-1}: 3440, 3020, 2970, 1590, 1555, 1485, 1440, 1420, 1390, 1110, 1080, 1020, 965, 880, 780; HRMS M^+ calcd for $C_{15}H_{14}N_2$: 222.1157, found: 222.1156; Anal. Calcd for $C_{15}H_{14}N_2$: C, 81.05; H, 6.35; N, 12.60. Found: C. 80.83; H, 6.38; N, 12.42.

5. High dilution conditions are necessary to maximize cyclotetramerization at the expense of linear polymerization.

6. Trimethyl orthoformate is used in excess relative to 5-phenyldipyrromethane to promote reaction between these two species, because dipyrromethanes alone, in common with pyrroles, undergo self-condensation under acidic conditions to yield polypyrroles.

7. The use of a fritted filter unit (Millipore) fitted with Millipore HVLP 04700 filter discs (0.45 μm) was found to be advantageous and allows essentially quantitative recovery of crystalline product.

8. Addition of pyridine to the recrystallization solvent is necessary to ensure that the porphyrin is present in the free base form during crystallization. Crystallization without pyridine results in recovery of product contaminated with porphyrin hydrochloride salts. The spectral and analytical properties for **2** are as follows: mp >300°C; ^1H NMR (300 MHz, $CDCl_3$) δ: -3.12 (br s, 2 H), 7.80-7.84 (m, 6 H), 8.28 (dd, 4 H), 9.09 (d, 4 H, J = 4.2), 9.40 (d, 4 H, J = 4.8), 10.32 (s, 2 H); ^{13}C NMR (75 MHz, $CDCl_3$) δ: 105.2, 119.0, 126.9, 127.6, 130.9, 131.5, 134.7, 141.2, 145.0, 147.0; IR (CH_2Cl_2) cm^{-1}: 3020, 2960, 1410, 1230, 950, 890, 850, 760; UV-vis ($CHCl_3$) λ_{max} (ε): 405nm (412000), 500nm (18100), 535nm (5150), 574nm (5720), 629nm (1450);

HRMS M^+ calcd for $C_{32}H_{22}N_4$: 462.1844, found: 462.1847. Anal. Calcd for $C_{32}H_{22}N_4$: C, 83.09; H, 4.79; N, 12.11. Found: C, 82.84; H, 4.53; N, 12.02.

Waste Disposal Information

All toxic materials were disposed of in accordance with "Prudent Practices in the Laboratory"; National Academy Press; Washington, DC, 1995.

3. Discussion

5,15-Diphenylporphyrin was, until recently, inaccessible in quantities suitable for use as a porphyrin model compound. The two classic synthetic porphyrins most commonly used in model studies, 2,3,7,8,12,13,17,18-octaethylporphyrin and 5,10,15,20-tetraphenylporphyrin, give access to stable, organic-solvent-soluble molecules with unsubstituted meso-positions and unsubstituted β-positions, respectively. 5,15-Diphenylporphyrin represents a unique model porphyrin, in that it is stable, freely soluble in a range of organic solvents (chloroform, dichloromethane, tetrahydrofuran), and presents two unsubstituted meso-positions and eight unsubstituted β-positions in the same molecule. The chemical properties peculiar to this molecule are only beginning to be explored, but they have already allowed the synthesis of novel ethynyl-linked porphyrin arrays as models of light-harvesting antenna systems,[7] and the development of a new bioconjugation method,[8] the key step of which was direct iodination of 5,15-diphenylporphyrin, a reaction that had either failed, or proved uncontrollable, on all other porphyrins.

To the submitter's knowledge the synthesis of 5,15-diphenylporphyrin has been reported on only two previous occasions. Both syntheses relied on the reaction of dipyrromethane, a relatively air-sensitive compound, with benzaldehyde. Treibs and

Haberle[9] reported a 3% yield for this reaction and noted it was "hard to access", while Manka and Lawrence[10] claimed a 92% yield. In the submitter's hands, however, the yield for the latter method could not be reproduced, and neither detailed experimental data, nor a measure of purity for the product, was available. The method presented here, therefore, represents the first reliable procedure for synthesizing analytically pure 5,15-diphenylporphyrin. The method requires only standard solvents and reagents. It should also be noted that this synthetic route contains a novel purification of 5-phenyl-dipyrromethane by sublimation, which obviates the need for large scale chromatography[4] to obtain gram quantities of this intermediate. Indeed, reactions giving 12 g of analytically pure product in 44% yield from pyrrole and benzaldehyde are practical. A yield of 54% has been reported for the synthesis of 5-phenyldipyrromethane from pyrrole and a triazolinedione ylide.[6] However, this method gave only milligram quantities of product and required purification by centrifugal chromatography. The condensation of 5-phenyldipyrromethane with trimethyl orthoformate to yield 5,15-diphenylporphyrin is also unusual as the only previous use of this method appears to have been in the synthesis of 5,15-diphenyl-2,3,7,8,12,13,17,18-octamethylporphyrin from 1,9-dicarboxy-2,3,7,8-tetramethyl-5-phenyldipyrromethane.[5] Note that in this latter reaction a yield of 13% was reported, while the submitter's adaptation of these conditions to the synthesis of 5,15-diphenylporphyrin from 5-phenyldipyrromethane resulted in a 21% yield.

1. Department of Biological and Chemical Sciences, University of Essex, Wivenhoe Park, Colchester CO4 3SQ, U.K.
2. Department of Chemistry, University of California, Berkeley CA 94720-1460.
3. Department of Chemistry, University of British Columbia, Vancouver, B.C. V6T 1Z1, Canada.

4. Procedure of Lee and Lindsey: Lee, C.-H.; Lindsey, J. S. *Tetrahedron*, **1994**, *50*, 11427-11440.

5. Adapted from the procedure of Baldwin, et al.: Baldwin, J. E.; Crossley, M. J.; Klose, T.; O'Rear, III, E. A.; Peters, M. K. *Tetrahedron* **1982**, *38*, 27-39.

6. Wilson, R. M.; Hengge, A. *J. Org. Chem.* **1987**, *52*, 2699-2707.

7. Lin, V. S. Y.; DiMagno, S. G.; Therien, M. J. *Science* **1994**, *264*, 1105-1111.

8. Boyle, R. W.; Johnson, C. K.; Dolphin, D. H. *J. Chem. Soc., Chem. Commun.* **1995**, 527-528.

9. Treibs, A.; Häberle, N. *Justus Liebigs Ann. Chem.* **1968**, *718*, 183-207.

10. Manka, J. S.; Lawrence, D. S. *Tetrahedron Lett.* **1989**, *30*, 6989-6992.

Appendix
Chemical Abstracts Nomenclature (Collective Index Number); (Registry Number)

5-Phenyldipyrromethane: 1H-Pyrrole, 2,2'-(phenylmethylene)bis- (12); (107798-98-1)

5,15-Diphenylporphyrin: Porphine, 5,15-diphenyl- (8,9); (22112-89-6)

Pyrrole: 1H-Pyrrole (9); (109-97-7)

Benzaldehyde (8,9); (100-52-7)

Trifluoroacetic acid: Acetic acid, trifluoro- (8,9); (76-05-1)

Trimethyl orthoformate: Orthoformic acid, trimethyl ester (8);

Methane, trimethoxy- (9); (149-73-5)

Trichloroacetic acid: Acetic acid, trichloro- (8,9) (76-03-9)

9,10-DIPHENYLPHENANTHRENE
(Phenanthrene, 9,10-diphenyl-)

CF_3SO_3H

Submitted by George A. Olah, Douglas A. Klumpp, Donald N. Baek, Gebhart Neyer, and Qi Wang.[1]
Checked by Vladimir Dragan and David J. Hart.

1. Procedure

Caution! Triflic acid is a highly corrosive liquid and should be handled carefully.

9,10-Diphenylphenanthrene. In a 250-mL, three-necked, round-bottomed flask equipped with a magnetic stirrer, a thermometer, and a septum is suspended 10.0 g (27.3 mmol) of benzopinacol (Note 1) in 40 mL of dry toluene (Note 2) under a nitrogen atmosphere. The suspension is cooled to 0°C using an ice-acetone bath, and 40 mL of freshly distilled trifluoromethanesulfonic acid (triflic acid) (Notes 3 and 4) is added in one portion via syringe (Note 5). The cold bath is removed, and the mixture is stirred for 24 hr at room temperature. The resulting heterogeneous mixture is poured over 100 g of ice, the organic layer is separated, and the aqueous layer is extracted with two 100-mL portions of toluene. The organic layers are combined, and washed with water (50 mL) and with brine (50 mL). The original solution is dried with anhydrous magnesium sulfate, filtered, and the filtrate is concentrated under reduced pressure to afford 9.2 g of crude product (mp 227-231°C) as a pale yellow solid. The crude product is recrystallized from 450 mL of toluene-ethanol (1:1) to give 6.7 g (74%) of pure 9,10-diphenylphenanthrene as a white solid (Note 6). Concentration of the mother liquor and recrystallization of the residue from 100 mL of toluene-ethanol (1:1) gives another 1.4 g (16%) of product (mp 237-238°C).

2. Notes

1. Benzopinacol was purchased from Aldrich Chemical Company, Inc., and used without further purification.

2. Toluene was purchased from Fisher Scientific Company and distilled from sodium prior to use. Benzene may be substituted for toluene with the following differences: (a) The initial solution is cooled to only 10°C, otherwise the benzene layer freezes. (b) The mixture is extracted with three 100-mL portions of benzene after pouring over ice. (c) Recrystallization from 430 mL of benzene-ethanol (1:1) gives a first crop of pure product in 80-90% yield. The submitters indicate that dichloromethane can be used as the solvent for the reaction and extractions.

3. Triflic acid was obtained from 3M (99% purity). The triflic acid must be distilled prior to its use, and the distillation should be performed under a dry, inert atmosphere. Triflic acid may be recycled using a published procedure.[2]

4. Experiments indicated that the acid strength of the reaction medium must be H_0 = -12 or stronger. Pure triflic acid has been estimated to be about H_0 = -14.[3] Contamination of the triflic acid by water results in a weakened acid system. When the acid strength drops below H_0 = -12, significant quantities of 2,2,2-triphenylacetophenone are produced from the pinacol rearrangement.

5. The temperature of the reaction mixture increases by approximately 15°C during the addition of triflic acid.

6. This material exhibited the following properties: mp 237-238°C (lit.[4] 238°C); ^{1}H NMR (300 MHz, $CDCl_3$) δ: 7.13-7.26 (m, 10 H), 7.5 (t with fine coupling, 2 H, J = 6.9), 7.6 (d with fine coupling, 2 H, J = 8.4), 7.7 (t with fine coupling, 2 H, J = 6.6), 8.82 (d, 2 H, J = 8.1); ^{13}C NMR (75 MHz, $CDCl_3$) δ: 122.9, 126.8, 126.9, 127.1, 128.0, 128.3, 130.4, 131.5, 132.3, 137.6, 139.9; HRMS calcd for $C_{26}H_{18}$ 330.1408, found 330.1408.

Waste Disposal Information

All toxic materials were disposed of in accordance with "Prudent Practices in the Laboratory"; National Academy Press; Washington, DC, 1995.

3. Discussion

When aromatic pinacols are reacted with an acid, products often arise from dehydration and rearrangement.[5] This general conversion is known as the pinacol rearrangement. The pinacol rearrangement may be promoted by both Brönsted and Lewis acids.[6] In the procedure described here, superacidic triflic acid is reacted with an aryl pinacol and a dehydrative cyclization occurs to give the substituted phenanthrene product. Related to this conversion, the chemistry of benzopinacol in sulfuric acid and triflic acid is contrasted in Scheme 1. We have proposed that the superacidic triflic acid causes the formation of diprotonated intermediates which promote the dehydrative cyclization.[4]

Scheme 1

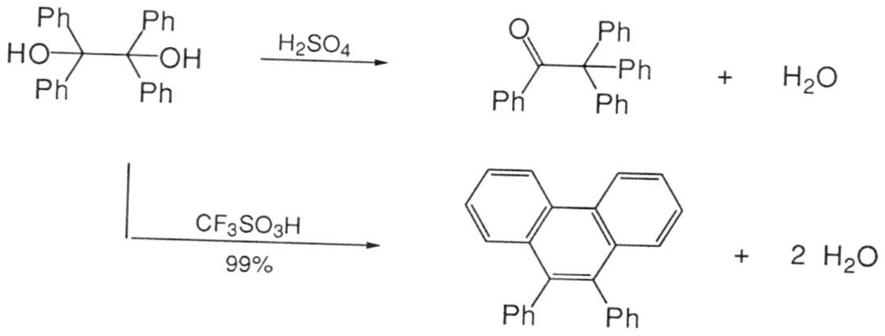

$$\text{Ph}_2\text{C(OH)–C(OH)Ph}_2 \xrightarrow{H_2SO_4} \text{PhCO–CPh}_3 + H_2O$$

$$\text{Ph}_2\text{C(OH)–C(OH)Ph}_2 \xrightarrow[99\%]{CF_3SO_3H} \text{9,10-diphenylphenanthrene} + 2\ H_2O$$

As described in an earlier report,[4] it has been demonstrated that the dehydrative cyclization of aryl pinacols in triflic acid is a general route to substituted phenanthrenes. Some of these conversions are described in the Table. Entries 5-7 show an interesting aspect of the conversion. When two of the aryl rings are substituted by strongly deactivating groups such as chlorine or bromine, the product is formed regioselectively. It may be that the chlorine or bromine deactivate the rings toward cyclization, and that the cyclization is occurring more rapidly through the phenyl or alkylphenyl rings. When two aryl rings are substituted by weakly deactivating groups such as fluorine, poor regiochemistry is observed, and cyclization occurs through both the phenyl and fluorophenyl rings, giving rise to complex product mixtures.

Aryl pinacols are readily obtained from the photolysis of substituted benzophenones in isopropyl alcohol. Photochemical coupling of benzophenones in isopropyl alcohol has even been reported to occur in direct sunlight.[7] It is known however, that some aryl ketones do not give the pinacol from photolysis in isopropyl alcohol.[8] Nevertheless, aryl pinacols have also been prepared from electrochemical reduction,[9] hydroxylation of olefins,[10] photoreduction of ketones by amines,[11] metal reduction,[12] and by other routes.[13] The phenanthrene ring system has often been prepared by the photochemical cyclization and oxidation of stilbenes,[14] and 9,10-diphenylphenanthrene has been previously synthesized by the reaction of 1,2-dichlorotetraphenylethane with excess aluminum chloride.[15] The present method provides a direct route to substituted phenanthrenes from readily available aryl ketones via the aryl pinacol.

1. Loker Hydrocarbon Research Institute, Department of Chemistry, University of Southern California, Los Angeles, CA 90089-1661.
2. Booth, B. L.; El-Fekky, T. A. *J. Chem. Soc., Perkin Trans I* **1979**, 2441.
3. Saito, S.; Saito, S.; Ohwada, T.; Shudo, K. *Chem. Pharm. Bull.* **1991**, *39*, 2718. For a reivew of triflic acid chemistry, see: Stang, P. J.; White, M. R. *Aldrichimica Acta* **1983**, *16*, 15.
4. Olah, G. A.; Klumpp, D. A.; Neyer, G.; Wang, Q. *Synthesis* **1996**, 321.
5. (a) Rickborn, B. In "Comprehensive Organic Synthesis"; Trost, B. M., Ed.; Pergamon: Oxford, 1991; Vol. 3, p. 721; (b) March, J. "Advanced Organic Chemistry Reactions, Mechanisms, and Structure"; 4th Ed.; Wiley Interscience: New York, 1992; p. 1072; (c) Collins, C. J. *Q. Rev., Chem. Soc.* **1960**, *14*, 357.
6. (a) Bartók, M.; Molnár, Á. In "The Chemistry of Functional Groups: The Chemistry of Ethers, Crown Ethers, Hydroxyl Groups and Their Sulphur Analogues"; Patai, S., Ed.; Wiley: New York, 1980; Suppl. E.; p. 721; (b) Bachmann, W. E. *Org. Synth., Coll. Vol. II* **1943**, 73.
7. Schönberg, A.; Mustafa, A. *Chem. Rev.* **1947**, *40*, 181.
8. Gilbert, A.; Baggott, J. E. "Essentials of Molecular Photochemistry"; Blackwell Scientific Publications: Boston, 1991; p. 302.
9. Stocker, J. H.; Jenevein, R. M. *J. Org. Chem.* **1969**, *34*, 2807.
10. Schröder, M. *Chem. Rev.* **1980**, *80*, 187.

11. (a) Cohen, S. G.; Cohen, J. I. *J. Phys. Chem.* **1968**, *72*, 3782; (b) Cohen, S. G.; Guttenplan, J. B. *Tetrahedron Lett.* **1968**, 5353; (c) Padwa, A.; Eisenhardt, W.; Gruber, R.; Pashayan, D. *J. Am. Chem. Soc.* **1969**, *91*, 1857.

12. (a) Rausch, M. D.; McEwen, W. E.; Kleinberg, J. *Chem. Rev.* **1957**, *57*, 417; (b) Clerici, A.; Porta, O. *Tetrahedron Lett.* **1982**, *23*, 3517; (c) Kegelman, M. R.; Brown, E V. *J. Am. Chem. Soc.* **1953**, *75*, 4649; (d) Namy, J. L.; Souppe, J.; Kagan, H. B. *Tetrahedron Lett.* **1983**, *24*, 765.

13. So, J.-H.; Park, M.-K.; Boudjouk, P. *J. Org. Chem.* **1988**, *53*, 5871.

14. Mallory, F. B.; Mallory, C. W. *Org. React.* **1984**, *30*, 1-456.

15. (a) Schmidlin, J.; Escher, R. *Chem. Ber.* **1910**, *43*, 1153; (b) Schoepfle, C. S.; Ryan, J. D. *J. Am. Chem. Soc.* **1932**, *54*, 3687.

Appendix
Chemical Abstracts Nomenclature (Collective Index Number); (Registry Number)

9,10-Diphenylphenanthrene: Phenanthrene, 9,10-diphenyl- (8,9); (602-15-3)

Benzopinacol: 1,2-Ethanediol, 1,1,2,2-tetraphenyl- (8,9); (464-72-2)

Trifluoromethanesulfonic acid: HIGHLY CORROSIVE: Methanesulfonic acid, trifluoro- (8,9); (1493-13-6)

TABLE. SUBSTITUTED PHENANTHRENES FROM THE DEHYDRATIVE CYCLICATION OF ARYL PINACOLS IN TRIFLIC ACID

Entry	Pinacol	Product	Yield[b]
1	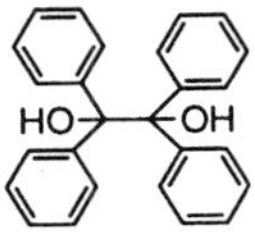		99%
2[a]	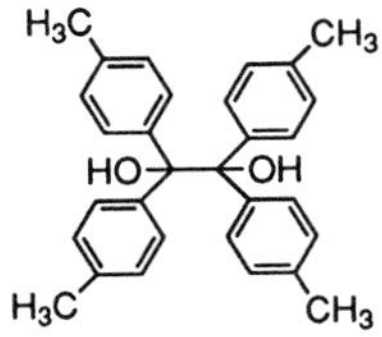	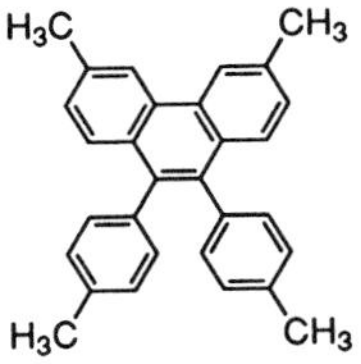	90%
3[a]	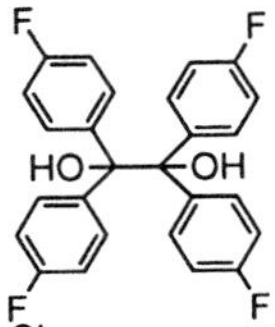	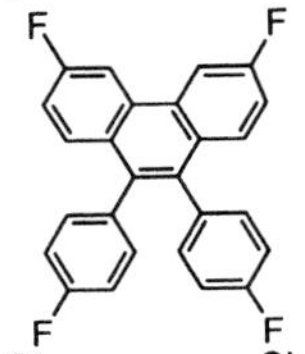	95%
4[a]	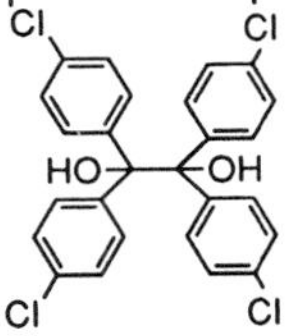	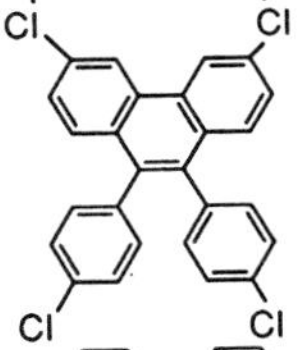	100%
5[a]	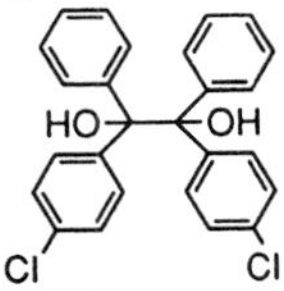	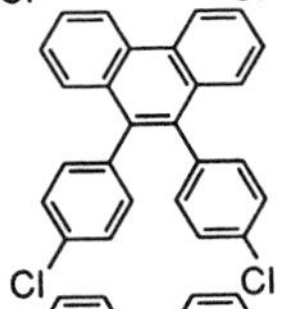	98%
6[a]	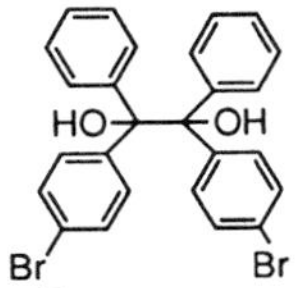	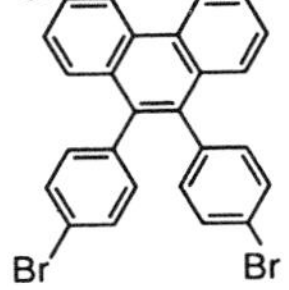	87%
7[a]	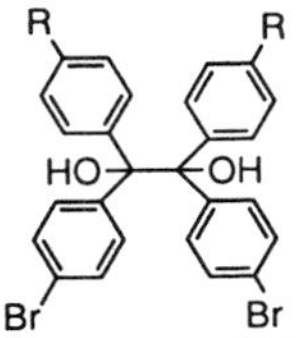 $R=(CH_2)_8CH_3$		94%

[a]Aryl pinacol prepared by the photolysis of the appropriately substituted benzophenone in isopropyl alcoh

[b]Yields reported for the crude product.

Unchecked Procedures

Accepted for checking during the period August 1, 1997 through September 1, 1998. An asterisk (*) indicates that the procedure has been subsequently checked.

Previously, *Organic Syntheses* has supplied these procedures upon request. However, because of the potential liability associated with procedures which have not been tested, we shall continue to list such procedures but requests for them should be directed to the submitters listed.

2805	Synthesis of 1,2:5,6-Dianhydro-3,4-O-isopropylidene-L-mannitol. D. A. Nugiel, K. Jacobs, A. C. Tabaka, and C. A. Teleha, The DuPont Merck Pharmaceutical Company, P.O. Box 80500, Wilmington, DE, 19880-0500.
2818	Catalytic Enantioselective Addition of Dialkylzincs to Aldehydes Using (2S)-(-)-3-exo-(Dimethylamino)isoborneol [(2S)-DAIB]. M. Kitamura, H. Oka, S. Suga, and R. Noyori, Department of Chemistry and Molecular Chirality Research Unit, Nagoya University, Chikusa, Nagoya 464-01, Japan.
2829R*	Allylindation in Aqueous Media: Methyl 3-(Hydroxymethyl)-4-methyl-2-methylenepentanoate. G. D. Bennett and L. A. Paquette, Department of Chemistry, The Ohio State University, Columbus, OH 43210.
2830R	The Synthesis of 2-Alkyl-4-Pyrones from Meldrum's Acid. M. T. Crimmins and D. G. Washburn, Venable and Kenan Laboratories of Chemistry, Department of Chemistry, University of North Carolina at Chapel Hill, Chapel Hill, NC 27599-3290.
2831	1,4,7,10-Tetraazacyclododecane. D. P. Reed and G. R. Weisman, Department of Chemistry, University of New Hampshire, Durham, NH 03842-3598.
2836	Stille Couplings Catalyzed by Palladium on Carbon with CuI as a Cocatalyst: Synthesis of 2-(4'-Acetylphenyl)thiophene. L. S. Liebeskind and E. Peña-Cabrera, Facultad de Química, Universidad de Guanajuato, Col. Noria Alta S/N, Guanajuato, Gto. 36000, Mexico.
2838	Dimethyl (Diazomethyl)phosphonate. J. J. Willemsen, D. G. Brown, E. J. Velthuisen, T. R. Hoye, and R. G. Brisbois, Department of Chemistry, Hamline University, St. Paul, MN 55104.
2840R	(3R,4S)-3-Hydroxy-4-phenylazetidin-2-one. I. Ojima, J. J. Walsh, I. Habus, Y. H. Park, and C. M. Sun, Department of Chemistry, State University of New York at Stony Brook, Stony Brook, NY 11794-3400.
2842	Bis(pinacolato)diboron. T. Ishiyama, M. Murata, T.-a. Ahiko, and N. Miyaura, Division of Molecular Chemistry, Graduate School of Engineering, Hokkaido University, Sapporo 060, Japan.

2843R	Diastereoselective Synthesis of cis-1,2-Disubstituted Cyclopropanols: trans-2-Methyl-1-(2-phenylethyl)-cyclopropanol. E. J. Corey and Rao Sidduri, Hoffmann-La Roche, Inc., 340 Kingsland St., Nutley, NJ 07110.
2846	Synthesis and [3+2] Cycloaddition of 2,2-Dialkoxy-1-methylene-cyclopropane: 6,6-Dimethyl-4,8-dioxa-1-methylenespiro[2.5]octane. M. Nakamura, X. Q. Wang, M. Isaka, S. Yamago, and E. Nakamura, Department of Chemistry, The University of Tokyo, Bunkyo-ku, Hongo, Tokyo 113, Japan.
2847	Synthesis of Penta-1,2-diene-4-one (Acetylallene). T. Constantieux and G. Buono, Laboratoire de Synthèse Asymétrique, (ENSSPICAM) U.M.R 6516, Faculté des Sciences et Techniques de St-Jérôme, Université d'Aix-Marseille, Avenue Escadrille Normandie-Niemen, 13397 Marseille Cedex 20, France.
2850R	Cyclopropene: A New Simple Synthesis and its Diels-Alder Reaction with Cyclopentadiene. P. Binger, P. Wedemann, and U. H. Brinker, Institut für Organische Chemie, Universität Wien, Währinger Str. 38, 1090 Wien, Austria.
2853	One-Pot Synthesis of Diethyl 3-Oxo-1-butenphosphonate. C. K. McClure, J. S. Link, R. J. Fischer, Department of Chemistry and Biochemistry, Montana State University, Bozeman MT 59717.
2854R	Preparation of [R-(R*,S*)]-β-Methyl-α-phenyl-1-pyrrolidineethanol. D. Zhao, C.-y. Chen, F. Xu, L. Tan, R. Tillyer, M. E. Pierce, and J. R. Moore, Process Research Department, Merck Research Laboratories, Division of Merck & Co., Inc., P.O. Box 2000, Rahway, NJ 07065.
2855*	(R)-(+)-2-Hydroxy-1,2,2-triphenylethyl Acetate. J. Macor, A. J. Sampognaro, P. R. Verhoest, and R. A. Mack, BMS PRI, Mail Stop H12-02, P.O. Box 4000, Princeton, NJ 08543-4000.
2856	(+)-cis-Dihydro-4-methyl-5-pentyl-2(3H)-furanone. S.-i. Fukuzawa and K. Seki, Department of Applied Chemistry, Chuo University, Bunkyo-ku, Tokyo 112, Japan.
2858	Synthesis of 2'-Deoxyribonucleosides: 3',5'-Di-O-Benzoylthymidine. D. R. Prudhomme, M. Park, Z. Wang, J. R. Buck, and C. J. Rizzo, Department of Chemistry, Box 1822, Station B, Vanderbilt University, Nashville, TN 37235.
2859	3,6-Dimethyl-9-ethylcarbazole (DMECZ). J. R. Buck, M. Park, Z. Wang, D. R. Prudhomme, and C. J. Rizzo, Department of Chemistry, Box 1822, Station B, Vanderbilt University, Nashville, TN 37235.

2860 (Phenyl)[2-(Trimethylsilyl)phenyl]iodonium Triflate. An Efficient and Mild Benzyne Precursor.
T. Kitamura, M. Todaka, and Y. Fujiwara, Department of Chemistry and Biochemistry, Graduate School of Engineering, Kyushu University, Hakozaki, Fukuoka 812-8581, Japan.

2861 α-Tosylbenzyl Isocyanide.
J. Sisko, M. Mellinger, P. W. Sheldrake, and N. H. Baine, SmithKline Beecham Pharmaceuticals, Synthetic Chemistry Department, P.O. Box 1539, King of Prussia, PA 19406.

2862R (3,4,5-Trifluorophenyl)boronic Acid-Catalyzed Amide Condensation of Carboxylic Acids and Amines: N-Benzyl-4-phenylbutyramide.
K. Ishihara, S. Ohara, and H. Yamamoto, School of Engineering, Nagoya University, Furo-cho, Chikusa, Nagoya 464-8603, Japan.

2864R Sulfinimines (Thiooximine S-Oxides): Asymmetric Synthesis of (R)-(+)-β-Phenylalanine from (S)-(+)-N-(Benzylidene)-p-Toluenesulfinamide.
G. V. Reddy, J. M. Szewczyk, D. L. Fanelli, D. M. Burns, and F. A. Davis, Department of Chemistry, Temple University, Philadelphia, PA 19122.

2869 (-)-(S)-2-(Benzyloxy)propanal.
D. Enders, S. von Berg, and B. Jandeleit, Institut für Organische Chemie, Technical University of Aachen, Professor Pirlet-Straße, 1 D-52074 Aachen, Germany.

2870 (-)-(E,S)-3-(Benzyloxy)-1-butenyl phenyl sulfone.
D. Enders, S. von Berg, and B. Jandeleit, Institut für Organische Chemie, Technical University of Aachen, Professor Pirlet-Straße, 1 D-52074 Aachen, Germany.

2871 (+)-(1R,2S,3R)-Tetracarbonyl[(1-3η)-1-(phenylsulfonyl)-but-2-en-1-yl]iron(1+)-tetrafluoroborate.
D. Enders, B. Jandeleit, and S. von Berg, Institut für Organische Chemie, Technical University of Aachen, Professor Pirlet-Straße, 1 D-52074 Aachen, Germany.

CUMULATIVE AUTHOR INDEX

FOR VOLUMES 75 AND 76

This index comprises the names of contributors to Volume **75** and **76**, only. For authors to previous volumes, see either indices in Collective Volumes I through IX or the single volume entitled *Organic Syntheses, Collective Volumes I-VIII, Cumulative Indices*, edited by J. P. Freeman.

CUMULATIVE SUBJECT INDEX
FOR VOLUMES 75 AND 76

This index comprises subject matter for Volume **75** and **76** only. For subjects in previous volumes, see either the indices in Collective Volumes I through IX or the single volume entitled *Organic Syntheses, Collective Volumes I-VIII, Cumulative Indices*, edited by J. P. Freeman.

The index lists the names of compounds in two forms. The first is the name used commonly in procedures. The second is the systematic name according to **Chemical Abstracts** nomenclature. Both are usually accompanied by registry numbers in parentheses. Also included are general terms for classes of compounds, types of reactions, special apparatus, and unfamiliar methods.

Most chemicals used in the procedure will appear in the index as written in the text. There generally will be entries for all starting materials, reagents, intermediates, important by-products, and final products. Entries in capital letters indicate compounds appearing in the title of the preparation.

ABNORMAL REACTIONS, SUPPRESSION OF, **76**, 228

Acetaldehyde; (75-07-0), **75**, 106

Acetaldehyde dimethyl acetal: Ethane, 1,1-dimethoxy-; (534-15-6), **75**, 46

Acetic acid, glacial; (64-19-7), **75**, 2, 225

Acetic anhydride: Acetic acid anhydride; (108-24-7), **76**, 70

Acetone: 2-Propanone; (67-64-1), **76**, 13

Acetonitrile: TOXIC: (75-05-8), **75**, 146; **76**, 24, 47, 67, 191

1-Acetoxy-3-(methoxymethoxy)butane: 1-Butanol, 3-(methoxymethoxy)-, acetate; (167563-42-0), **75**, 177

9-Acetylanthracene: Ethanone, 1-(9-anthracenyl)-; (784-04-3), **76**, 276

Acetyl chloride; (75-36-5), **75**, 177

1-Acetyl-1-cyclohexene: Ethanone, 1-(1-cyclohexen-1-yl)-; (932-66-1), **76**, 203

Acetylenes, terminal, cyanation of, **75**, 148

Agar plate preparation, **76**, 79

Aliquat 336: Ammonium, methyltrioctyl-, chloride; 1-Octanaminium, N-methyl-N,N-dioctyl-, chloride; (5137-55-3), **76**, 40

Alkyllithium solutions, to titrate, **76**, 68

Allene: 1,2-Propadiene; (463-49-0), **75**, 129

Allenes, stereodefined synthesis of, **76**, 185

ALLYLATION, CATALYTIC ASYMMETRIC, OF ALDEHYDES, **75**, 12
table, **75**, 17

Allyl bromide: 1-Propene, 3-bromo-; (106-95-6), **76**, 60, 221

L-ALLYLGLYCINE: 4-PENTENOIC ACID, 2-AMINO-, (R)-; (54594-06-8), **76**, 57

Allylic alcohols, enantioselective cyclopropanation, **76**, 97

Allylmagnesium bromide: Magnesium, bromo-2-propenyl-; (1730-25-2), **76**, 221

Allylmagnesium bromide (ethereal complex solution), **76**, 221

Allylmagnesium bromide (THF complex solution), **76**, 222

Allyltributylstannane: Stannane, tributyl-2-propenyl-; (24850-33-7), **75**, 12

Aminals, **76**, 30

α-Amino aldehydes:
Boc-protected, **76**, 117
N,N-Dibenzyl-protected, **76**, 115

α-Aminocarboxylic acids, **75**, 25

2-Amino-5-chlorobenzophenone: Methanone, (2-amino-5-chlorophenyl)phenyl-; (719-59-5), **76**, 142

(1S,2R)-(+)-2-Amino-1,2-diphenylethanol: Benzeneethanol, β-amino-α-phenyl-, [S-(R*,S*)]-; (23364-44-5), **75**, 45

(1S,2R)-1-AMINOINDAN-2-OL: 1H-INDEN-2-OL, 1-AMINO-2,3-DIHYDRO-, (1S-cis); (126456-43-7), **76**, 46

(R)-(-)-1-Amino-1-phenyl-2-methoxyethane: Benzenemethanamine, α-(methoxymethyl)-, (R)-; (64715-85-1), **75**, 19

(S)-2-Amino-3-phenylpropanol: Benzenepropanol, 2-amino-, (S)-; (3182-95-4), **76**, 113

α-Aminophosphonates, **75**, 25, 30
table, **75**, 30

α–Aminophosphonic acids and esters, **75**, 25

3-AMINOPROPYL CARBANION EQUIVALENT, **75**, 215

Ammonia, **76**, 66

Ammonium chloride; (12125-02-9), solid acid catalyst, **75**, 47

Ammonium hydroxide; (1336-21-6), **76**, 24, 63

Ammonium molybdate(VI) tetrahydrate: Molybdic acid, hexaammonium salt, tetrahydrate; (12027-67-7), **76**, 78

Ammonium sulfate, **76**, 79

Anti-inflammatory drugs, **76**, 173

L-Arginine hydrochloride: L-Arginine, monohydrochloride; (1119-34-2), **76**, 77

2-Aryl-2-cyclohexenones, table, **75**, 75

2-ARYLPROPIONIC ACIDS, PURE, SYNTHESIS OF, **76**, 169

Aseptic transfer, **76**, 80

Asymmetric reactions, catalytic, **76**, 10

ASYMMETRIC SYNTHESIS, **75**, 12, 19
OF α-AMINO ACIDS, **76**, 57

Azide, asymmetric introduction of, **75**, 34

Azeotropic drying, **76**, 67

Bacto-Agar, **76**, 79

Barton esters, **75**, 124

1,4-BENZADIYNE EQUIVALENT, **75**, 201

Benzaldehyde; (100-52-7), **76**, 24, 288

Benzeneboronic acid: Boronic acid, phenyl-; (98-80-6), **75**, 53

Benzenethiol: See Thiophenol, **75**, 100

Benzophenone: Methanone, diphenyl-; (119-61-9), **75**, 141, 224
as photosensitizer, **75**, 141

Benzopinacol: 1,2-Ethanediol, 1,1,2,2-tetraphenyl-; (464-72-2), **75**, 141; **76**, 294

Benzoquinolizines, **76**, 36

Benzoyl chloride; (98-88-4), **75**, 225

N-(2-Benzoyl-4-chlorophenyl)oxanilic acid ethyl ester: Acetic acid, [(2-benzoyl-4-chlorophenyl)amino]oxo-, ethyl ester; (19144-20-8), **76**, 142

4a(S),8a(R)-2-Benzoyl-1,3,4,4a,5,8a-hexahydro-6(2H)-isoquinolinone: 6(2H)-Isoquinolone, 2-benzoyl-1,3,4,4a,5,8a-hexahydro-, (4aS-cis)-; (52346-14-2), **75**, 225

N-Benzoyl meroquinene tert-butyl ester: 4-Piperidineacetic acid, 1-benzoyl-3-ethenyl-, 1,1-dimethylethyl ester, (3R-cis)-; (52346-13-1), **75**, 224

4a(S),8a(R)-2-BENZOYLOCTAHYDRO-6(2H)-ISOQUINOLINONE: 6(2H)-ISOQUINOLINONE, 2-BENZOYLOCTAHYDRO-, (4aS-cis)-; (52390-26-8), **75**, 223

3-Benzoyl-N-vinylpyrrolidin-2-one: 2-Pyrrolidinone, 3-benzoyl-1-ethenyl-, (±)-; (125330-80-5), **75**, 215

Benzyl alcohol: Benzenemethanol; (100-51-6), **76**, 111

Benzylamine: Benzenemethanamine ; (100-46-9), **75**, 107

N-BENZYL-2,3-AZETIDINEDIONE: 2,3-AZETIDINEDIONE, 1-(PHENYLMETHYL)-; (75986-07-1), **75**, 106

Benzyl bromide: Benzene, bromomethyl-; (100-39-0), **76**, 111, 113, 240

N-Benzylcinchonidinium chloride: Cinchonanium, 9-hydroxy-1-(phenylmethyl)-, chloride, (9S)-; (69221-14-3), **76**, 1

Benzyl (S)-2-(N,N-dibenzylamino)-3-phenylpropanoate: L-Phenylalanine, N,N-bis(phenylmethyl)-, phenylmethyl ester; (111138-83-1), **76**, 110

N-Benzyl-3-(Z/E)-ethylideneazetidin-2-one: 2-Azetidinone, 3-ethylidene-1-(phenylmethyl)-; (115870-02-5), **75**, 108

N-Benzyl-3-(1-hydroxyethyl)azetidin-2-one: 2-Azetidinone, 3-(1-hydroxyethyl)-1-(phenylmethyl)-; (R*,R*)- (89368-08-1); (R*,S*)-; (89368-09-2), **75**, 107

N-Benzylidenebenzylamine: Benzylamine, N-benzylidene-; Benzenemethanamine, N-(phenylmethylene)-; (780-25-6), **76**, 182

2-Benzyl-2-methylcyclohexanone: Cyclohexanone, 2-benzyl-2-methyl-; Cyclohexanone, 2-methyl-2-(phenylmethyl)-; 1206-21-9), **76**, 244

2-BENZYL-6-METHYLCYCLOHEXANONE: CYCLOHEXANONE, 2-METHYL-6-(PHENYLMETHYL)-; (24785-76-0), **76**, 239

Benzyloxyacetaldehyde: Acetaldehyde, (phenylmethoxy)-; (60656-87-3), **75**, 12

Benzyltriethylammonium chloride: Ammonium, benzyltriethyl-, chloride; Benzenemethanaminium, N,N,N-triethyl-, chloride; (56-37-1), **75**, 38

Benzyne precursors, **75**, 207

BIARYLS, UNSYMMETRICAL, SYNTHESIS OF, **75**, 53, 63
tables, **75**, 60, 67

Bicyclo[1.1.1]pentyl phenyl sulfide: Bicyclo[1.1.1]pentane, 1-(phenylthio)- (11); (98585-81-0), **75**, 99

(±)-1,1'-Bi-2-naphthol: (1,1'-Binaphthalene)-2.2'-diol; (602-09-5); **76**, 1, 6

(R)-1,1'-Bi-2-naphthol: [1,1'-Binaphthalene]-2,2'-diol, (R)-; (18531-94-7), **76**, 1, 6

(S)-(-) ([1,1'-BINAPHTHALENE]-2,2'-DIOL, (S)-: (S-BINOL); (18531-99-2), **75**, 12; **76**, 1, 6

(±)-1,1'-Bi-2-naphthyl ditriflate: Methanesulfonic acid, trifluoro-, (1,1'-binaphthalene)-2,2'-diyl ester, (±)-; (128575-34-8), **76**, 6

Biocatalytic transformations, **76**, xviii

4-BIPHENYLCARBOXALDEHYDE: [1,1'-BIPHENYL]-4-CARBOXALDEHYDE; (3218-36-8), **75**, 53

(R,R)-N,N'-Bis-(3,5-di-tert-butylsalicylidene)-1,2-cyclohexanediamine: Phenol, 2,2'-[1,2-cyclohexanediylbis(nitrilomethylidyne)]bis[4,6-bis(1,1-dimethylethyl)-[1R-(1α(E),2β(E))]]-; (135616-40-9), **75**, 4

(R,R)-N,N'-BIS-(3,5-DI-tert-BUTYLSALICYLIDENE)1,2-CYCLOHEXANEDIAMINO MANGANESE(III) CHLORIDE: MANGANESE, CHLORO[[2,2'-[1,2-CYCLOHEXANEDIYLBIS(NITRILOMETHYLIDYNE)] BIS[4,6-BIS(1,1-DIMETHYLETHYL)PHENOLATO]](2-)-N,N',O,O']-[SP-5-13-(1R-trans)];(138124-32-0), **75**, 1; **76**, 47

(R)-(+)- AND (S)-(-)-2,2'-BIS(DIPHENYLPHOSPHINO)-1,1'-BINAPHTHYL: PHOSPHINE, [1,1'-BINAPHTHALENE]-2,2'-DIYLBIS[DIPHENYL-, (R)-; (76189-55-4); (S)- (76189-56-5), **76**, **6**

[1,2-Bis(diphenylphosphino)ethane]nickel(II) chloride: Nickel, dichloro[ethanediylbis[diphenylphosphine]-P,P']-, (SP-4-2)-; (14647-23-5), **76**, 7

Bis(η-divinyltetramethyldisiloxane)tri-tert-butylphosphineplatinum(0): Platinum, [1,3-bis(η^2-ethenyl)-1,1,3,3-tetramethyldisiloxane][tris(1,1-dimethylethyl)phosphine]-; (104602-18-8), **75**, 79
preparation, **75**, 81

Bis(trichloromethyl) carbonate: See Triphosgene, **75**, 46

Boron tribromide: Borane, tribromo-; (10294-33-4), **75**, 129

Boron trifluoride etherate: Ethane, 1,1'-oxybis-, compd. with trifluoroborane (1:1); (109-63-7), **75**, 189; **76**, 13

Boronic acids, alkyl-, **76**, 96

Boroxines, (anhydrides), **76**, 92, 96

BROMIDES, EFFICIENT SYNTHESIS OF, **75**, 124

Bromine; (7726-95-6), **75**, 210

(2-Bromoallyl)diisopropoxyborane, **75**, 129, 133

4-Bromobenzaldehyde: Benzaldehyde, 4-bromo-; (1122-91-4), **75**, 53

1-Bromobutane: Butane, 1-bromo-; (109-65-9), **76**, 87

1-Bromo-5-chloropentane: Pentane, 1-bromo-5-chloro-; (54512-75-3), **76**, 222

1-Bromo-3-chloropropane: Propane, 1-bromo-3-chloro-; (109-70-6), **76**, 222

Bromofluorination of alkenes, **76**, 159

1-BROMO-2-FLUORO-2-PHENYLPROPANE: BENZENE, (2-BROMO-1-FLUORO-1-METHYLETHYL)-; (59974-27-5), **76**, 159

Bromoform: Methane, tribromo- (8,9); (75-25-2), **75**, 98

6-Bromo-1-hexene: 1-Hexene, 6-bromo-; (2695-47-8), **76**, 223

(S)-(-)-2-Bromo-3-hydroxypropanoic acid: Propanoic acid, 2-bromo-3-hydroxy-, (S)-; (70671-46-4), **75**, 37

Bromomalononitrile: Propanedinitrile, bromo-; (1885-22-9), **75**, 210
preparation, **75**, 211

2-Bromonaphthalene: Naphthalene, 2-bromo-; (580-13-2), **76**, 13

8-Bromo-1-octene, **76**, 224

2-Bromo-1-octen-3-ol: 1-Octen-3-ol, 1-bromo-, (E)-; (52418-90-3), **76**, 263

1-Bromo-3-phenylpropane: Benzene, (3-bromopropyl)-; (637-59-2), **75**, 155

(Z/E)-1-BROMO-1-PROPENE: 1-PROPENE, 1-BROMO-; (590-14-7), **76**, 214

4-(2-BROMO-2-PROPENYL)-4-METHYL-γ-BUTYROLACTONE: 2(3H)-FURANONE, 5-(2-BROMO-2-PROPENYL)DIHYDRO-5-METHYL-; (138416-14-5), **75**, 129

Bromotrichloromethane: Methane, bromotrichloro; (75-62-7), **75**, 125

1-Butanol, **76**, 48

2-Butanol; (78-92-2), **76**, 68

N-Boc-L-ALLYLGLYCINE: 4-PENTENOIC ACID, 2-[[(1,1-DIMETHYLETHOXYCARBONYL]AMINO]-, (R)-; (170899-08-8), **76**, 57

tert-Butyl alcohol: 2-Propanol, 2-methyl-; (75-65-0), **75**, 225

Butylboronic acid: Boronic acid, butyl-; (4426-47-5), **76**, 87

tert-Butyl chloride: Propane, 2-chloro-2-methyl-; (507-20-0), **75**, 107

(5S)-(5-O-tert-BUTYLDIMETHYLSILOXYMETHYL)FURAN-2(5H)-ONE: 2(5H)-FURANONE, 5-[[[(1,1-DIMETHYLETHYL)DIMETHYLSILYL]OXY] METHYL]-, (S)-; (105122-15-4), **75**, 140

(1R*,6S*,7S*)-4-tert-BUTYLDIMETHYLSILOXY)-6-(TRIMETHYLSILYL) BICYCLO[5.4.0]UNDEC-4-EN-2-ONE, **76**, 199

(tert-BUTYLDIMETHYLSILYL)ALLENE: SILANE, (1,1-DIMETHYLETHYL) DIMETHYL-1,2-PROPADIENYL-; (176545-76-9), **76**, 178

tert-Butyldimethylsilyl chloride: Silane, chloro(1,1-dimethylethyl)dimethyl-; (18162-48-6), **75**, 140; **76**, 179, 201

1-(tert-Butyldimethylsilyl)-1-(1-ethoxyethoxy)-1,2-propadiene, **76**, 201

1-(tert-Butyldimethylsilyl)-1-(1-ethoxyethoxy)-3-trimethylsilyl-1,2-propadiene (crude), **76**, 180

3-(tert-Butyldimethylsilyl)-2-propyn-1-ol: 2-Propyn-1-ol, 3-[(1,1-dimethylethyl)-dimethylsilyl]-; (120789-51-7), **76**, 179

(E)-1-(tert-Butyldimethylsilyl)-3-trimethylsilyl-2-propen-1-one: Silane, (1,1-dimethylethyl)dimethyl[1-oxo-3-(trimethylsilyl)-2-propenyl]-, (E)-; (83578-66-9), **76**, 202; (Z)-, **76**, 207

2-BUTYL-6-ETHENYL-5-METHOXY-1,4-BENZOQUINONE: 2,5-CYCLOHEXADIENE-1,4-DIONE, 5-BUTYL-3-ETHENYL-2-METHOXY-; (134863-12-0), **76**, 189

Butyllithium: Lithium, butyl-; (109-72-8), **75**, 22, 116, 201; **76**, 37, 59, 151, 179, 191, 201, 202, 203, 214, 230, 239

Butylmagnesium bromide: Magnesium, bromobutyl-; (693-03-8), **76**, 87

tert-Butyl methyl ether: Ether, tert-butyl methyl; Propane, 2-methoxy-2-methyl-; (1634-04-4), **76**, 276

tert-Butyl perbenzoate: Benzenecarboperoxoic acid, 1,1-dimethylethyl ester; (614-45-9), **75**, 161

1-BUTYL-1,2,3,4-TETRAHYDRO-1-NAPHTHOL, **76**, 228

(4R-trans)-2-Butyl-N,N,N',N'-tetramethyl[1,3,2]dioxaborolane-4,5-dicarboxamide: 1,3,2-Dioxaborolane-4,5-dicarboxamide, 2-butyl-N,N,N',N'-tetramethyl-, (4R-trans)-; (161344-85-0), **76**, 88

3-Butyn-2-ol: 3-Butyn-2-ol, (±)-; (65337-13-5), **75**, 79

Camphorsulfonic acid monohydrate (CSA): Bicyclo[2.2.1]heptane-1-methanesulfonic acid, 7,7-dimethyl-2-oxo-, (±)-; (5872-08-2), **75**, 46, 171; **76**, 178

Carbon tetrachloride, CANCER SUSPECT AGENT; Methane, tetrachloro-; (56-23-5), **76**, 124

CARBONYL ADDITIONS, CERIUM(III) CHLORIDE-PROMOTED, **76**, 237

Carbonyl compounds, reactions with organolithiums or Grignard reagents, **76**, 228

Cells, storage of, **76**, 80

Centrifugation, **76**, 78

Cerium(III) chloride, anhydrous: Cerium chloride; (7790-86-5), **76**, 229

Cerium(III) chloride heptahydrate: Cerium chloride ($CeCl_3$), heptahydrate; (18618-55-8), **76**, 228

Cerium(III) chloride monohydrate: Cerium chloride ($CeCl_3$), monohydrate; (64332-99-6), **76**, 228

Chiral :
- auxiliaries, **76**, 18, 73
- diamines, **76**, 30
- dioxaborolane ligands, **76**, 97
- Lewis acids, **76**, 18
- ligands, **76**, xvii, 53, 97
- NON-RACEMIC DIOLS, SYNTHESIS OF, **76**, 101
- reagents:
  - for enantioselective syntheses, **76**, xviii
  - from amino acids, **76**, 128
- sulfoxides, **76**, 42
- synthons, **76**, 82
- titanium catalysts, **76**, 19

Chlorobenzene: Benzene, chloro-; (108-90-7), **76**, 77

3-Chloro-2,2-bis(chloromethyl)propanoic acid: Propionic acid, 3-chloro-2,2-bis(chloromethyl)-; (17831-70-8), **75**, 90

3-Chloro-2,2-bis(chloromethyl)propan-1-ol: 1-Propanol, 3-chloro-2,2-bis(chloromethyl)- ;(813-99-0), **75**, 89

3-CHLORO-2-(CHLOROMETHYL)-1-PROPENE: 1-PROPENE, 3-CHLORO-2-(CHLOROMETHYL)-; (1871-57-4), **75**, 89

1-CHLORO-(2S,3S)-DIHYDROXYCYCLOHEXA-4,6-DIENE: 3,5-CYCLOHEXADIENE-1,2-DIOL, 3-CHLORO-, (1S-CIS)-; (65986-73-4), **76**, 77

6-CHLORO-1-HEXENE: 1-HEXENE, 6-CHLORO-; (928-89-2), **76**, 221

8-CHLORO-1-OCTENE: 1-OCTENE, 8-CHLORO-; (871-90-9), **76**, 221

m-Chloroperoxybenzoic acid: Benzocarboperoxoic acid, 3-chloro-; (937-14-4), **75**, 154

2-Chlorophenol: Phenol, 2-chloro; (95-57-8), **76**, 271

2-CHLOROPHENYL PHOSPHORODICHLORIDOTHIOATE: PHOSPHORODICHLORIDOTHIOIC ACID, O-(2-CHLOROPHENYL) ESTER; (68591-34-4), **76**, 271

Chloroplatinic acid hexahydrate: Platinate (2-), hexachloro-, dihydrogen, (OC-6-11)-; (16941-12-1), **75**, 81

N-Chlorosuccinimide: 2,5-Pyrrolidinedione, 1-chloro-; (128-09-6), **76**, 124

Chlorotrimethylsilane: Silane, chlorotrimethyl-; (75-77-4), **75**, 146; **76**, 24, 252

Chlorotriphenylmethane: Benzene, 1,1',1"-(chloromethylidyne)tris-; (76-83-5), **75**, 184

trans-Cinnamaldehyde: 2-Propenal, 3-phenyl-, (E)-; (14371-10-9), **76**, 214

Cinnamyl alcohol: 2-propen-1-ol, 3-phenyl-; (104-54-1), **76**, 89

Citric acid monohydrate: 1, 2, 3-Propanetricarboxylic acid, 2-hydroxy-, monohydrate;
(5949-29-1), **75**, 31

Clathrate, **76**, 14

Cobalt nitrate hexahydrate: Nitric acid, cobalt (2+ salt), hexahydrate;
(10026-22-9), **76**, 79

Copper; (7440-50-8), **76**, 94, 255

COPPER-CATALYZED CONJUGATE ADDITION OF ORGANOZINC
REAGENTS TO α,β-UNSATURATED KETONES, **76**, 252

Copper cyanide: See Cuprous cyanide, **75**, 146

Copper sulfate, **76**, 79

(E)-CROTYLSILANES, CHIRAL, **75**, 78

Crown ethers, as phase transfer catalysts, **75**, 103

Cuprous chloride: Copper iodide; (7758-89-6), **76**, 133

Cuprous cyanide: Copper cyanide; (544-92-3), **75**, 146; **76**, 253

Cuprous iodide: Copper iodide; (7681-65-4), **76**, 103, 264

CYANOALKYNES, SYNTHESIS OF, **75**, 146
table, **75**, 150

Cyclizations:
dehydrative, **76**, 296
reductive, with low-valent titanium reagents, **76**, 149

Cycloalkenones, α-substituted, **75**, 73

Cyclobutenediones, synthesis of, **76**, 196

Cycloheptatrienylium tetrafluoroborate: Aldrich: Tropylium tetrafluoroborate:
Cycloheptatrienylium, tetrafluoroborate (1-); (27081-10-3), **75**, 210

Cycloheptenones, all-cis, bicyclic, **76**, 209

Cycloheptenones, synthesis, using [3+4] annulation, **76**, 213

1,2-Cyclohexanedione; (765-87-7), **75**, 170

Cyclohexanone; (108-94-1), **75**, 116, 119

Cyclohexanone, 2-methyl-2-(phenylmethyl)-; (1206-21-9), **76**, 244

2-Cyclohexen-1-one, HIGHLY TOXIC; (930-68-7), **75**, 69; **76**, 253

1,3-Cyclopentanedione; (3859-41-4), **75**, 189

3-CYCLOPENTENE-1-CARBOXYLIC ACID; (7686-77-3), **75**, 195

3-Cyclopentene-1,1-dicarboxylic acid; (88326-51-6), **75**, 195

Cyclopropanations, enantioselective, **76**, 97

DABCO: See 1,4-Diazabicyclo[2.2.2]octane, **75**, 107; **76**, 7

DBU: See 1,8-Diazabicyclo[5.4.0]undec-7-ene, **75**, 31, 108

DEC-9-ENYL BROMIDE: 1-DECENE, 10-BROMO-; (62871-09-4), **75**, 124

(±)-trans-1,2-Diaminocyclohexane: 1,2-Cyclohexanediamine, trans-; (1121-22-8), **75**, 2
ee determination, **75**, 2

(R,R)-1,2-Diammoniumcyclohexane mono-(+)-tartrate: 1,2-Cyclohexanediamine, (1R-trans)-, [R-(R*,R*)-2,3-dihydroxybutanedioate (1:1); (39961-95-0), **75**, 2

Diaryne equivalents, **75**, 204

Diastereomeric purity, determination of, **76**, 70

1,4-Diazabicyclo[2.2.2]octane [DABCO]; (280-57-9), **75**, 106

1,8-Diazabicyclo[5.4.0]undec-7-ene [DBU]: Pyrimido[1,2-a]azepine, 2,3,4,6,7,8,9,10-octahydro-; (6674-22-2), **75**, 31, 108

Diazene, propargylic, **76**, 185

Dibenzo-18-crown-6: Dibenzo[b,k][1,4,7,10,13,16]hexaoxacyclooctadecin, 6,7,9,10,17,18,20,21-; (14187-32-7), **75**, 98

(S)-2-(N,N-DIBENZYLAMINO)-3-PHENYLPROPANAL: BENZENEPROPANAL, α-[BIS(PHENYLMETHYL)AMINO]-, (S)-; (111060-64-1), **76**, 110

(S)-2-(N,N-Dibenzylamino)-3-phenyl-1-propanol: Benzenepropanol, β-[bis(phenylmethyl)amino]-, (S)-; (111060-52-7), **76**, 111

1,1-Dibromo-2,2-bis(chloromethyl)cyclopropane: Cyclopropane, 1,1-dibromo-2,2-bis(chloromethyl)- (11); (98577-44-7), **75**, 98

1,2-Dibromoethane: Ethane, 1,2-dibromo-; (106-93-4), **76**, 24, 252

Di-tert-butyl dicarbonate: Dicarbonic acid, bis(1,1-dimethylethyl) ester; (24424-99-5), **76**, 65

2,4-Di-tert-butylphenol: Phenol, 2,4-bis(1,1-dimethylethyl)-; (96-76-4),**75**, 3

3,5-Di-tert-butylsalicylaldehyde: Benzaldehyde,3,5-bis(1,1-dimethylethyl)-2-hydroxy-; (37942-07-7),**75**, 3

Dichlorobis(triphenylphosphine)palladium(II): Palladium, dichlorobis(triphenylphosphine)-; (13965-03-2), **76**, 264

cis-1,4-Dichlorobut-2-ene: 2-Butene, 1,4-dichloro-, (Z)-; (1476-11-5), **75**, 195

8,8-DICYANOFULVENE: PROPANEDINITRILE, 2,4,6-CYCLOHEPTATRIEN-1-YLIDENE- ; (2860-54-0), **75**, 210

Dicyclohexylcarbodiimide: HIGHLY TOXIC. Cyclohexanamine, N,N'-methanetetraylbis-; (538-75-0), **75**, 124

(S,S)-1,2,3,4-DIEPOXYBUTANE: 2,2'-BIOXIRANE, [S-(R*,R*)]-; (30031-64-2), **76**, 101

Diethanolamine: Ethanol, 2,2'-iminobis-; (111-42-2), **76**, 88

Diethanol complex, **76**, 88

Diethylamine: Ethanamine, N-ethyl-; (109-89-7), **76**, 264

DIETHYL (R)-(-)-(1-AMINO-3-METHYLBUTYL)PHOSPHONATE: PHOSPHONIC ACID, (1-AMINO-3-METHYLBUTYL)-, DIETHYL ESTER, (R)-; (159171-46-7), **75**, 19

N,N-Diethylaniline: Benzenamine, N,N-diethyl-; (91-66-7), **76**, 29

Diethyl azodicarboxylate: Diazenedicarboxylic acid, diethyl ester; (1972-28-7), **76**, 180

Diethyl (R)-(-)-[1-((N-(R)-(1-phenyl-2-methoxyethyl)amino)-3-methylbutyl)]phosphonate: Phosphonic acid, [1-(2-methoxy-1-phenylethyl)amino]-3-methylbutyl]-, diethyl ester, [R-(R*,R*)]-; (159117-09-6), **75**, 20

Diethyl phosphite: Phosphonic acid, diethyl ester; (762-04-9), **75**, 22

Diethyl tartrate, **76**, 93

Diethylzinc: Zinc, diethyl-; (557-20-0), **76**, 89

6,7-DIHYDROCYCLOPENTA-1,3-DIOXIN-5(4H)-ONE: CYCLOPENTA-1,3-DIOXIN-5(4H)-ONE,6,7-DIHYDRO-; (102306-78-5), **75**, 189

3,4-Dihydro-2H-pyran: 2H-Pyran, 3,4-dihydro-; (110-87-2), **76**, 178

(2S,3S)-DIHYDROXY-1,4-DIPHENYLBUTANE: 2,3-BUTANEDIOL, 1,4-DIPHENYL-, [S-(R*,R*)]-; (133644-99-2), **76**, 101

1,3-Diiodobicyclo[1.1.1]pentane: Bicyclo[1.1.1]pentane,1,3-diiodo- (1); (105542-98-1), **75**, 100

Diiodomethane: Methane, diiodo; (75-11-6), **76**, 89

Diisopropylamine: 2-Propanamine, N-(1-methylethyl)- (9); (108-18-9), **75**, 116; **76**, 59, 203, 239

Diisopropyl ether: Propane, 2,2'-oxybis-; (108-20-3), **75**, 130

N,N-Diisopropylethylamine: 2-Propanamine, N-ethyl-N-(1-methylethyl)-; (7087-68-5), **75**, 177

Dilithium tetrachloromanganate: Manganate (2–), tetrachloro-, dilithium, (I-4)-; (57384-24-4), **76**, 240

(+)-[(7,7-Dimethoxycamphoryl)sulfonyl]imine: 3H-3a,6-Methano-2,1-benzisothiazole,4,5,6,7-tetrahydro-7,7-dimethoxy-8,8-dimethyl-, 2,2-dioxide, [3aS]-;(131863-80-4), **76**, 38, 39

(+)-[(8,8-Dimethoxycamphoryl)sulfonyl]imine, **76**, 40

(+)-[(8,8-Dimethoxycamphoryl)sulfonyl]oxaziridine: 4H-4a,7-Methanoxazirino[3,2-i][2,1]benzisothiazole, tetrahydro-8,8-dimethoxy-9,9-dimethyl-, 3,3-dioxide, [2R-(2α,4aα,7α,8aS*)]-; (131863-82-6), **76**, 38, 39

Dimethoxyethane: Ethane, 1,2-dimethoxy-; (110-71-4), **76**, 89

(3,4-Dimethoxyphenyl)acetonitrile: Benzeneacetonitrile, 3,4-dimethoxy-; (93-17-4), **76**, 133

2-(3,4-Dimethoxyphenyl)-N,N-dimethylacetamidine, **76**, 133

Dimethylamine: Methanamine, N-methyl-; (124-40-3), **76**, 93, 133

4-Dimethylaminopyridine: HIGHLY TOXIC: 4-Pyridinamine, N,N-dimethyl-; (1122-58-3), **75**, 184; **76**, 70

4-DIMETHYLAMINO-N-TRIPHENYLMETHYLPYRIDINIUM CHLORIDE: PYRIDINIUM, 4-(DIMETHYLAMINO)-1-(TRIPHENYLMETHYL)-, CHLORIDE; (78646-25-0), **75**, 184

(1R,2R)-(+)-N,N'-Dimethyl-1,2-bis(3-trifluoromethyl)phenyl-
1,2-ethanediamine, **76**, 127

Dimethyl carbonate: Carbonic acid, dimethyl ester; (616-38-6), **76**, 170

Dimethyl 3-cyclopentene-1,1-dicarboxylate, **75**, 197

N,N-Dimethylformamide: CANCER SUSPECT AGENT: Formamide, N,N-dimethyl- ;
(68-12-2), **75**, 162

N,N'-DIMETHYL-1,2-DIPHENYLETHYLENEDIAMINE: 1,2-ETHANEDIAMINE,
N,N'-DIMETHYL-1,2-DIPHENYL, (R*,S*)-; (60509-62-8); [R-(R*,R*)-;
(118628-68-5); [S-(R*,R*)]; (70749-06-3), **76**, 23

N,N-DIMETHYLHOMOVERATRYLAMINE: BENZENEETHANAMINE,
3,4-DIMETHOXY-N,N-DIMETHYL-; (3490-05-9), **76**, 133

(R,R)-Dimethyl O,O-isopropylidenetartrate: 1,3-Dioxolane-4,5-dicarboxylic
acid, 2,2-dimethyl-, dimethyl ester, (4R-trans)-; (37031-29-1), **76**, 13

Dimethyl malonate: Propanedioic acid, dimethyl ester; (108-59-8), **75**, 195

3,3-DIMETHYL-1-OXASPIRO[3.5]NONAN-2-ONE: 1-OXASPIRO[3.5]NONAN-2-ONE,
3,3-DIMETHYL-; (22741-15-7), **75**, 116

Dimethylphenylsilane: Silane, dimethylphenyl-; (766-77-8), **75**, 79

(±)-1-(Dimethylphenylsilyl)-1-buten-3-ol: 3-Buten-2-ol, 1-(dimethylphenylsilyl)-,
(E)-(±)-; (137120-08-2), **75**, 78

(3R)-1-(Dimethylphenylsilyl)-1-buten-3-ol: 3-Buten-2-ol, 1-(dimethylphenylsilyl)-,
[R-(E)]-;(133398-25-1), **75**, 79
ee determination, **75**, 82

(3S)-1-(Dimethylphenylsilyl)-1-buten-3-ol: 3-Buten-2-ol, 4-(dimethylphenylsilyl)-,
[S-(E)]-; (133398-24-0), **75**, 79
ee determination, **75**, 82

(3R)-1-(Dimethylphenylsilyl)-1-buten-3-ol acetate: 3-Buten-2-ol,
4-(dimethylphenylsilyl)-, acetate, [R-(E)]-; (129921-47-7), **75**, 79

2-(2,2-Dimethylpropanoyl)-1,3-dithiane: 1-Propanone, 1-(1,3-dithian-2-yl)-
2,2-dimethyl-; (73119-31-0), **76**, 37

(1S)-(2,2-Dimethylpropanoyl)-1,3-dithiane 1-oxide: 1-Propanone,
2,2-dimethyl-1-(1-oxido-1,3-dithian-2-yl)-, (1S-trans)-; (160496-17-3);
(1S-cis)-, **76**, 38

N,N'-Dimethylpropyleneurea [DMPU]: 2(1H)-Pyrimidinone, tetrahydro-1,3-dimethyl-;
(7226-23-5), **75**, 195

DIMETHYL SQUARATE: 3-CYCLOBUTENE-1,2-DIONE, 3,4-DIMETHOXY-; (5222-73-1), **76**, 189

Dimethyl sulfoxide: Methyl sulfoxide; Methane, sulfinylbis-; (67-68-5), **75**, 146; **76**, 112

(R,R)-Dimethyl tartrate: Butanedioic acid, 2,3-dihydroxy-, [R-(R*,R*)]-dimethyl ester; (608-68-4), **76**, 13

(4R,5R)-2,2-DIMETHYL-α,α,α',α'-TETRA(NAPHTH-2-YL)-1,3-DIOXOLANE-4,5-DIMETHANOL: [1,3-DIOXOLANE-4,5-DIMETHANOL, 2,2-DIMETHYL-α,α,α',α'-TETRA-2-NAPHTHALENYL-, (4R-TRANS)-]; (137365-09-4), **76**, 12

3,5-Dinitrobenzoyl chloride: Benzoyl chloride, 3,5-dinitro-; (99-33-2), **76**, 277

2,4-Dinitrofluorobenzene, **76**, 52

Diol metabolites, **76**, 82

Dioxaborolane ligand, **76**, 89

p-Dioxane, CANCER SUSPECT AGENT: 1,4-Dioxane; (123-91-1), **76**, 65

1,3-Dioxin vinylogous esters, **75**, 192

1,3-Diphenylacetone p-tosylhydrazone: p-Toluenesulfonic acid, (α-benzylphenethylidene)hydrazide; Benzenesulfonic acid, 4-methyl-, [2-phenyl-1-(phenylmethyl)ethylidene]hydrazide; (19816-88-7), **76**, 206

(4R,5S)-4-Diphenylhydroxymethyl-5-tert-butyldimethylsiloxymethylfuran-2(5H)-one, **75**, 45

(4R,5S)-4,5-Diphenyl-2-oxazolidinone: 2-Oxazolidinone, 4,5-diphenyl-, (4R-cis)-; (86286-50-2), **75**, 45

9,10-DIPHENYLPHENANTHRENE: PHENANTHRENE, 9,10-DIPHENYL-; (602-15-3), **76**, 294

Diphenylphosphine: Phosphine, diphenyl-; (829-85-6), **76**, 7

Diphenylphosphoryl azide: Phosphorazidic acid, diphenyl ester; (26386-88-9), **75**, 31

5,15-DIPHENYLPORPHYRIN: PORPHINE, 5,15-DIPHENYL-; (22112-89-6), **76**, 287

(4R,5S)-4,5-DIPHENYL-3-VINYL-2-OXAZOLIDINONE: 2-OXAZOLIDINONE, 3-ETHENYL-4,5-DIPHENYL-, (4R-cis)-; (143059-81-8), **75**, 45

(4S,5R)-4,5-Diphenyl-3-vinyl-2-oxazolidinone: 2-Oxazolidinone, 3-ethenyl-4,5-diphenyl-, (4S-cis)-; (128947-27-3), **75**, 45

Diphosgene, **75**, 48

2.2'-Dipyridyl: 2,2'-Bipyridine; (366-18-7), **76**, 68

Disodium hydrogen phosphate: Phosphoric acid, disodium salt; (7558-79-4), **76**, 78

1,3-Dithiane; (505-23-7), **76**, 37

(1S)-(–)-1,3-DITHIANE 1-OXIDE: 1,3-DITHIANE, 1-OXIDE, (S)-; (63865-78-1), **76**, 37

DiTOX (1,3-Dithiane 1-oxide), **76**, 42

1,3-Divinyltetramethyldisiloxane: Disiloxane, 1,3-diethenyl-1,1,3,3-tetramethyl-; (2627-95-4); **75**, 81

DMPU: See N,N'-Dimethylpropyleneurea, **75**, 195

Duff reaction, adaptation of, **75**, 6, 9

Enantiomeric composition, determination of, **76**, 29, 30, 51, 95

ENANTIOMERICALLY PURE α-N,N-DIBENZYLAMINO ALDEHYDES, SYNTHESIS OF, **76**, 110

Enantiomerically pure products, syntheses, **76**, xviii

Epoxidation, **75**, 153; **76**, 50, 53
    asymmetric, **75**, 9

EPOXIDATION CATALYST, ENANTIOSELECTIVE, **75**, 1
    Epoxides, optically acitive, **76**, 107
    Ergosterol: Ergosta-5,7,22-trien-3-ol, (3β)-; (57-87-4), **76**, 276

3-ETHENYL-4-METHOXYCYCLOBUTENE-1,2-DIONE: 3-CYCLOBUTENE-1,2-DIONE, 3-ETHENYL-4-METHOXY-; (124022-02-2), **76**, 189

1-(1-Ethoxyethoxy)-1,2-propadiene: 1,2-Propadiene, 1-(1-ethoxyethoxy)-; (20524-89-4), **76**, 200

1-(1-Ethoxyethoxy)-1-propyne: 1-Propyne, 3-(1-ethoxyethoxy)-; (18669-04-0), **76**, 200

ETHYL (R)-2-AZIDOPROPIONATE: PROPANOIC ACID, 2-AZIDO-, ETHYL ESTER, (R)-; (124988-44-9), **75**, 31
    assay of optical purity, **75**, 33

Ethyl benzoate: Benzoic acid, ethyl ester; (93-89-0), **75**, 215

Ethyl bromide: Bromoethane; (74-96-4), **75**, 38

Ethyl 5-bromovalerate: Pentanoic acid, 5-bromo-, ethyl ester; (14660-52-7), **76**, 255

ETHYL 5-CHLORO-3-PHENYLINDOLE-2-CARBOXYLATE: 1H-INDOLE-2-CARBOXYLIC ACID, 5-CHLORO-3-PHENYL-, ETHYL ESTER; (212139-32-2), **76**, 142

Ethyl p-dimethylaminobenzoate: Benzoic acid, 4-dimethylamino-, ethyl ester; (10287-53-3), **76**, 276

Ethyl 2,2-dimethylpropanoate: Propanoic acid, 2,2-dimethyl-, ethyl ester; (3938-95-2), **76**, 37

Ethylene: Ethene; (74-85-1), **76**, 24

Ethylenediaminetetraacetic acid, tetrasodium salt: Glycine, N,N'-1,2-ethanediylbis[N-carboxymethyl)-, tetrasodium salt, trihydrate; (67401-50-7), **76**, 78

Ethylene glycol dimethyl ether: Ethane, 1,2-dimethoxy-; (110-71-4), **76**, 143

ETHYL (R)-(+)-2,3-EPOXYPROPANOATE: See ETHYL GLYCIDATE, **75**, 37

ETHYL GLYCIDATE (ETHYL (R)-(+)-2,3-EPOXYPROPANOATE): OXIRANECARBOXYLIC ACID, ETHYL ESTER, (R)-; (111058-33-4), **75**, 37

Ethyl 5-iodovalerate: Pentanoic acid, 5-iodo-, ethyl ester; (41302-32-3), **76**, 252

Ethyl (S)-(-)-lactate: Propanoic acid, 2-hydroxy-, ethyl ester, (S)-; (687-47-8), **75**, 31

Ethyl levulinate: Pentanoic acid, 4-oxo-, ethyl ester; (539-88-8), **75**, 130

Ethyl oxalyl chloride: Acetic acid, chlorooxo-, ethyl ester; (4755-77-5), **76**, 142

(±)-3-Ethyl-1-oxaspiro[3.5]nonan-2-one, **75**, 116

ETHYL 5-(3-OXOCYCLOHEXYL)PENTANOATE, **76**, 252

Ethyl trifluoroacetate: Acetic acid, trifluoro-, ethyl ester; (383-63-1), **75**, 153

Ethyl trimethylacetate: Propanoic acid, 2,2-dimethyl-, ethyl ester; (3938-95-2), **76**, 37

Ethyl vinyl ether: Ethene, ethoxy-; (109-92-2), **76**, 200

Ferrous sulfate, **76**, 78

Fluorinated compounds, **76**, 154, 161

Furan; (110-00-9), **75**, 201

L-(S)-Glyceraldehyde acetonide (2,3-O-Isopropylidene-L-glyceraldehyde):
1,3-Dioxolane-4-carboxaldehyde, (S)-; (22323-80-4), **75**, 139

Glycine methyl ester; (616-34-20), **76**, 58

Glycine methyl ester hydrochloride: Glycine methyl ester, hydrochloride;
(5680-79-5), **76**, 66

Green Chemistry, **76**, 174

GRIGNARD REAGENTS, **76**, 87, 221, 222, 228

Haloalkenes, as synthons, **76**, 224

HALOBORATION, of 1-alkynes, **75**, 134
OF ALLENE, **75**, 129

Hanovia mercury lamp, **76**, 276

Hexafluoroisopropyl alcohol: 2-Propanol, 1,1,1,3,3,3-hexafluoro-; (920-66-1), **76**, 151

Hexamethyldisilane, **75**, 155

Hexamethylenetetramine: 1,3,5,7-Tetraazatricyclo[3.3.1.1^{3,7}]decane; (100-97-0),**75**, 3

Hexamethylphosphoric triamide, HIGHLY TOXIC: Phosphoric triamide,
hexamethyl-; (680-31-9), **76**, 201

1-Hexyne; (693-02-7), **76**, 191

Hunsdiecker reaction, **75**, 127

Hydratropic acids, syntheses, **76**, 173

Hydrazine monohydrate, HIGHLY TOXIC. CANCER SUSPECT AGENT: Hydrazine;
(302-01-2), **76**, 183

Hydrogen bromide: Hydrobromic acid; (10035-10-6), **75**, 37; **76**, 264

Hydrogenation, **75**, 21

Hydrogen peroxide; (7722-84-1), **76**, 40, 90

Hydrosilation, **75**, 85

(4R,5S)-4-HYDROXYMETHYL-(5-O-tert-BUTYLDIMETHYLSILOXYMETHYL)FURAN-
2(5H)-ONE: D-erythro-PENTONIC ACID, 2,3-DIDEOXY-5-O-
[(1,1-DIMETHYLETHYL)DIMETHYLSILYL]-3-HYDROXYMETHYL)-,
γ-LACTONE; (164848-06-0), **75**, 139

(S)-5-Hydroxymethylfuran-2(5H)-one: 2(5H)-Furanone, 5-(hydroxymethyl)-, (S)-; (78508-96-0), **75**, 140

4-HYDROXY-1,1,1,3,3-PENTAFLUORO-2-HEXANONE HYDRATE: 2,2,4-HEXANETRIOL,1,1,1,3,3-PENTAFLUORO-; (119333-90-3), **76**, 151

N-Hydroxythiopyridone: 2(1H)-Pyridinethione, 1-hydroxy-; (1121-30-8), **75**, 124

[(2-)-N,O,O'[2,2'-Iminobis[ethanolato]]]-2-butylboron, **76**, 88

Indene: 1H-Indene; (95-13-6), **76**, 47

(1S,2R)-Indene oxide: 6H-Indeno[1,2-b]oxirene, 1a,6a-dihydro-; (768-22-9), **76**, 47

Indigo test, **76**, 80

Indinavir (Crixivan®), **76**, 52

Indole: 1H-Indole; (120-72-9), **76**, 80

Indoloquinolizines, **76**, 36

Iodine; (7553-56-2), **75**, 69, 100; **76**, 13

Imidazole; (288-32-4), **75**, 140

4-Iodoanisole: Benzene, 1-iodo-4-methoxy-; (696-62-8), **75**, 61

2-IODO-2-CYCLOHEXEN-1-ONE: 2-CYCLOHEXEN-1-ONE, 2-IODO-; (33948-36-6), **75**, 69

Irradiation, **76**, 276

Irradiation apparatus, **75**, 141

Isobutyl chloride: See 2-Methylpropanoyl chloride, **75**, 118

ISOMERIZATION OF β-ALKYNYL ALLYLIC ALCOHOLS TO FURANS CATALYZED BY SILVER NITRATE ON SILICA GEL, **76**, 263

Isomerization, of a meso-diamine to the dl-isomer, **76**, 25

Isoprene: 1,3-Butadiene, 2-methyl-; (78-79-5), **76**, 25

2,3-O-Isopropylidene-L-glyceraldehyde: 1,3-Dioxolane-4-carboxaldehyde, 2,2-dimethyl-, L-; (22323-80-4), **75**, 139

2,3-O-Isopropylidene-L-threitol: 1,3-Dioxolane-4,5-dimethanol, 2,2-dimethyl-, (4S-trans)-; (50622-09-8), **76**, 102

2,3-O-Isopropylidene-L-threitol 1,4-bismethanesulfonate, **76**, 101

(3R*,4R*)- and (3R*,4S*)-4-Isopropyl-4-methyl-3-octyl-2-oxetanone, **75**, 119

Isovaleraldehyde: Butanal, 3-methyl-; (590-86-3), **75**, 20

Karl Fischer titration, **76**, 68

KETONES, ALDOLIZATION OF, **75**, 116

Kinetic resolution, **75**, 79

β-LACTONES, decarboxylation of, **75**, 120
regioselective fission of, **75**, 120
stereoselective reactions of, **75**, 120
SYNTHESIS OF, **75**, 116

LIGANDLESS PALLADIUM CATALYST, **75**, 61

Lipase Amano AK, **75**, 79

Lithium aluminum hydride: Aluminate (1-), tetrahydro-, lithium, (T-4); (16853-85-3), **75**, 80; **76**, 111

Lithium chloride; (7447-41-8), **76**, 58, 60, 241

Lithium diethyl phosphite, **75**, 20

Lithium diisopropylamide, **75**, 116; **76**, 59, 203, 239

Lithium dimethylcyanocuprate, **76**, 253

Lithium hydride; (7580-67-8), **75**, 195

Lithium hydroxide monohydrate; Lithium hydroxide, monohydrate; (1310-66-3), **75**, 195

Lithium methoxide: Methanol, lithium salt; (865-34-9), **76**, 58

LITHIUM PENTAFLUOROPROPEN-2-OLATE: 1-PROPEN-2-OL, 1,1,3,3,3-PENTAFLUORO-, LITHIUM SALT; (116019-90-0), **76**, 151

Lithium wire; (7439-93-2), **76**, 25

Magnesium; (7439-95-4), **75**, 107; **76**, 13, 87, 221

Malononitrile: HIGHLY TOXIC. Propanedinitrile; (109-77-3), **75**, 210

Manganese acetate tetrahydrate: Acetic acid, manganese (2+ salt), tetrahydrate; (6156-78-1), **75**, 4

Manganese(II) chloride tetrahydrate, **76**, 242

Manganese(II) chloride: Manganese chloride; (7773-01-5), **76**, 241, 242

Manganese(II) sulfate, **76**, 79

McMurry olefin synthesis, two extensions to, **76**, 145

l-Menthol, **76**, 29

Meroquinene tert-butyl ester: 4-Piperidineacetic acid, 3-ethenyl-, 1,1-dimethylethyl ester, (3R-cis)-; (52346-11-9), **75**, 225

Meroquinene esters, **75**, 231

Metal complexes, **76**, 18

Metallation conditions, **76**, 73

"Metals 44 solution", **76**, 78

Methanesulfonic acid; (75-75-2), **76**, 102

Methanesulfonyl chloride; (124-63-0), **75**, 108; **76**, 102

L-Methionine methyl ester hydrochloride: L-Methionine, methyl ester, hydrochloride; (2491-18-1), **76**, 123

(4R,5S)-3-(1-Methoxyethyl)-4,5-diphenyl-2-oxazolidinone: 2-Oxazolidinone, 3-(1-methoxyethyl)-4,5-diphenyl-, [4R-[3(R*),4α,5α]]-; 142977-52-4), **75**, 46

3-(METHOXYMETHOXY)-1-BUTANOL: 1-BUTANOL, 3-(METHOXYMETHOXY)-; (60405-27-8), **75**, 177

4-METHOXY-2'-METHYLBIPHENYL: 1,1'-BIPHENYL, 4'-METHOXY-2-METHYL-; (92495-54-0), **75**, 61

4-Methoxyphenylboronic acid: Boronic acid, (4-methoxyphenyl)-; (5720-07-0), **75**, 70

2-(4-METHOXYPHENYL)-2-CYCLOHEXEN-1-ONE: 2-CYCLOHEXEN-1-ONE, 2-(4-METHOXYPHENYL); (63828-70-6), **75**, 69

Methyl acrylate: 2-Propenoic acid, methyl ester; (96-33-3), **75**, 106

Methylamine: Methanamine; (74-89-5), **76**, 24

N-Methylbenzimine: Methanamine, N-(phenylmethylene)-; (622-29-7), **76**, 24

Methyl 2-(benzylamino)methyl-3-hydroxybutanoate: Butanoic acid, 3-hydroxy-2-[[(phenylmethyl)amino]methyl]-, methyl ester; (R*,R*)- (118559-03-8); (R*,S*)- (118558-99-9), **75**, 107

3-Methylbutan-2-one; (563-80-4), **75**, 119

Methyl tert-butyl ether: Propane, 2-methoxy-2-methyl- (9); (1634-04-4), **75**, 31

2-METHYLCYCLOHEXANONE: CYCLOHEXANONE, 2-METHYL-; (583-60-8), **76**, 240

Methyl 3-cyclopentene-1-carboxylate: 3-Cyclopentene-1-carboxylic acid, methyl ester: (58101-60-3), **75**, 197

(1'S,2'S)-METHYL-3O,4O-(1',2'-DIMETHOXYCYCLOHEXANE-1',2'-DIYL)-α-D-MANNOPYRANOSIDE: (α-D-MANNOPYRANOSIDE, METHYL 3,4-O-(1,2-DIMETHOXY-1,2-CYCLOHEXANEDIYL)-, [3[S(S)]]-); (163125-35-7), **75**, 170

(3R),(4E)-METHYL 3-(DIMETHYLPHENYLSILYL)-4-HEXENOATE: 4-HEXENOIC ACID, 3-(DIMETHYLPHENYLSILYL)-, METHYL ESTER, [R-(E)]-; (136174-52-2), **75**, 78
ee determination, **75**, 83

(3S),(4E)-METHYL 3-(DIMETHYLPHENYLSILYL)-4-HEXENOATE: 4-HEXENOIC ACID, 3-(DIMETHYLPHENYLSILYL)-, METHYL ESTER, [S-(E)]-; (136314-66-4), **75**, 78
ee determination, **75**, 82

4-Methyl-1,3-dioxane: 1,3-Dioxane, 4-methyl-; (1120-97-4), **75**, 177

7-Methylene-8-hexadecyn-6-ol: 8-Hexadecyn-6-ol, 7-methylene-; (170233-66-6), **76**, 264

Methyl formate: Formic acid, methyl; (107-31-3), **75**, 171

Methyl 3-hydroxy-2-methylenebutanoate: Butanoic acid, 3-hydroxy-2-methylene-, methyl ester; (18020-65-0), **75**, 106

Methyl iodide: Methane, iodo-; (74-88-4), **75**, 19

Methyl (4S)-4,5-O-isopropylidenepent-(2Z)-enoate: 2-Propenoic acid, 3-(2,2-dimethyl-1,3-dioxolan-4-yl)-, methyl ester, [S-(Z)]-; (81703-94-8), **75**, 140

Methyllithium: Lithium, methyl-; (917-54-4), **75**, 99; **76**, 193, 253

Methyl α-D-mannopyranoside; (617-04-9), **75**, 171

Methyl (S)-2-phthalimido-4-methylthiobutanoate: 2H-Isoindole-2-acetic acid, 1,3-dihydro-α-[2-(methylthio)ethyl]-1,3-dioxo-, methyl ester, (S)-; (39739-05-4), **76**, 123

METHYL (S)-2-PHTHALIMIDO-4-OXOBUTANOATE: 2H-ISOINDOLE-2-ACETIC ACID, 1,3-DIHYDRO-1,3-DIOXO-α-(2-OXOETHYL)-, METHYL ESTER, (S)-; (137278-36-5), **76**, 123

2-Methylpropanoyl chloride: Propanoyl chloride, 2-methyl-; (79-30-1), **75**, 118

1-Methyl-2-pyrrolidinone: 2-Pyrrolidinone, 1-methyl-; (872-50-4), **76**, 240

α-Methylstyrene: Benzene, (1-methylethenyl)-; (98-83-9), **76**, 159

Methyl (triphenylphosphoranylidene)acetate: Propanoic acid, 2-(triphenylphosphoranylidene)-, methyl ester; (2605-67-6), **75**, 139

Mineral salt bath (MSB), **76**, 77

Mitsunobu displacement, **76**, 185

4 Å Molecular sieves: Zeolites, 4 Å; (70955-01-0*), **75**, 12, 189

MONOALKYLATION, REGIOSELECTIVE, OF KETONES, VIA MANGANESE ENOLATES, **76**, 239

MONO-C-METHYLATION OF ARYLACETONITRILES AND METHYL ARYLACETATES BY DIMETHYL CARBONATE, **76**, 169

2-NAPHTHYLMAGNESIUM BROMIDE: MAGNESIUM, BROMO-2-NAPHTHALENYL-; (21473-01-8), **76**, 13

Nitric acid; (7697-37-2), **75**, 90

Nitriles, into tertiary amines, **76**, 133

Nitrilotriacetic acid, CANCER SUSPECT AGENT: Glycine, N-bis(carboxymethyl)-; (139-13-9), **76**, 78

o-Nitrobenzenesulfinic acid, **76**, 185

o-Nitrobenzenesulfonyl chloride: Benzenesulfonyl chloride, 2-nitro-; (1694-92-4), **76**, 183

o-Nitrobenzenesulfonyl hydrazide: Benzenesulfonic acid, 2-nitro-, hydrazide; (5906-99-0), **76**, 180, 183

Nitrogen oxides, **75**, 90

NMR Shift reagents, **76**, 18

1-Nonyne; (3452-09-3), **76**, 264

2,3,7,8,12,13,17,18-Octaethylporphyrin, **76**, 291

1-Octyn-3-ol; (818-72-4), **76**, 264

Oligonucleoside phosphorothioates, syntheses of, **76**, 272

Optically active diols, synthesis of, **76**, 106

Organocopper chemistry, **76**, 257

ORGANOLITHIUMS, **76**, 228

Organometallic chemistry, **76**, xx

Osmium oxide: See Osmium tetroxide, **75**, 109

Osmium tetroxide: Osmium oxide, (T-4)-; (20816-12-0), **75**, 108

Oxalyl chloride: Ethanedioyl dichloride; (79-37-8), **76**, 112

Oxalyl chloride-dimethyl sulfoxide (Swern reagent), **76**, 112

Oxidative degradation, of aromatic compounds, by *Pseudomonas*, **76**, 81

Palladium, 10% on carbon, **75**, 226

Palladium acetate: Acetic acid, palladium (2+) salt; (3375-31-3), **75**, 53, 61

Palladium(II) bis(benzonitrile)dichloride: Palladium, bis(benzonitrile)dichloro-; (14220-64-5), **75**, 70

PALLADIUM CATALYST, LIGANDLESS, **75**, 61

20% Palladium hydroxide on carbon, **75**, 21

Parr shaker, **75**, 24, 226

Pentaerythritol: 1,3-Propanediol, 2,2-bis(hydroxymethyl)-; (115-77-5),**75**, 89

Pentaerythrityl tetrachloride: Propane, 1,3-dichloro-2,2-bis(chloromethyl)-; (3228-99-7), **75**, 89

Pentaerythrityl trichlorohydrin: 1-Propanol, 3-chloro-2,2-bis(chloromethyl)-; (813-99-0); **75**, 89

(2-Pentanol; (6032-29-7), **76**, 255

2-PENTYL-3-METHYL-5-HEPTYLFURAN: FURAN, 5-HEPTYL-3-METHYL-2-PENTYL-; (170233-67-7), **76**, 263

cis-4a(S),8a(R)-PERHYDRO-6(2H)-ISOQUINOLINONES, **75**, 223

pH 7 Buffer, **76**, 253

Phase transfer catalysis, **76**, 273

Phenanthrenes, substituted, **76**, 296

1,10-Phenanthroline; (66-71-7), **76**, 91, 255

Phenol: HIGHLY TOXIC; (108-95-2), **75**, 118

Phenylacetonitrile: Benzeneacetonitrile; (140-29-4), **76**, 169

Phenylacetylene: Benzene, ethynyl-; (536-74-3), **75**, 146

(S)-Phenylalanine: L-Phenylalanine; (63-91-2), **76**, 110

Phenyl butanoate, **75**, 119

(2S,3S)-(+)-(3-PHENYLCYCLOPROPYL)METHANOL: CYCLOPROPANEMETHANOL, 2-PHENYL-, (1S-TRANS)-; (110659-58-0), **76**, 86

Phenyl decanoate, **75**, 119

5-PHENYLDIPYRROMETHANE: 1H-PYRROLE, 2,2'-(PHENYLMETHYL)BIS-; (107798-98-1), **76**, 287

PHENYL ESTER ENOLATES, IN SYNTHESIS OF β-LACTONES, **75**, 116

(R)-(-)-2-Phenylglycinol: Benzeneethanol, β-amino-, (R)-; (56613-80-0), **75**, 19

6-PHENYLHEX-2-YN-5-EN-4-OL: 1-HEXEN-4-YN-3-OL, 1-PHENYL-; (63124-68-5), **76**, 214

Phenylmagnesium bromide: Magnesium, bromophenyl-; (100-58-3), **76**, 103

N-[(1R)-Phenyl-(2R)-methoxyethyl]-isovaleraldehyde imine, **75**, 20

(S)-1-(PHENYLMETHOXY)-4-PENTEN-2-OL: 4-PENTEN-2-OL, 1-(PHENYLMETHOXY)-, (S)-; (88981-35-5), **75**, 12

Phenyl 2-methylpropanoate: Propanoic acid, 2-methyl-, phenyl ester; (20279-29-2), **75**, 116

2-PHENYLPROPIONIC ACID: BENZENEACETIC ACID, α-METHYL-; (492-37-5), **76**, 169

2-Phenylpropionitrile: Benzeneacetonitrile, α-methyl-; (1823-91-2), **76**, 171

3-Phenylpropyltriphenylphosphonium bromide: Phosphonium, triphenyl(3-phenylpropyl)-, bromide; (7484-37-9), **75**, 153
    preparation, **75**, 155

3-PHENYL-2-PROPYNENITRILE: 2-PROPYNENITRILE, 3-PHENYL-; (935-02-4), **75**, 146

4-Phenylpyridine N-oxide: Pyridine, 4-phenyl-, 1-oxide; (1131-61-9), **76**, 47

2-PHENYL-1-PYRROLINE: 2H-PYRROLE, 3,4-DIHYDRO-5-PHENYL-; (700-91-4), **75**, 215

Phosphorus pentoxide: Phosphorus oxide; (1314-56-3), **76**, 265

Phosphorus trichloride; (7719-12-2), **76**, 29

PHOTOINDUCED ADDITION, OF ALCOHOLS, **75**, 139

Photolysis, **75**, 125
  of ergosterol, **76**, 283

Photolysis products from ergosterol: pro-, pre-, lumi-, and tachy-, **76**, 283

Phthalic anhydride: 1,3-Isobenzofurandione; (85-44-9), **76**, 123

Pinacol: 2,3-Butanediol, 2,3-dimethyl-; (76-09-5) **75**, 98

Pinacol rearrangement, **76**, 296

N-Pivaloyl-o-toluidine: Propanamide, 2,2-dimethyl-N-(2-methylphenyl)-; (61495-04-3), **76**, 215

Polyphenylene, **76**, 81

Porphyrin model studies, **76**, 291

Potassium tert-butoxide: 2-Propanol, 2-methyl-, potassium salt; (865-47-4), **75**, 224; **76**, 200

Potassium dihydrogen phosphate: Phosphoric acid, monopotassium salt; (7778-77-0), **76**, 78

Potassium (R)-(+)-2.3-epoxypropanoate: Oxiranecarboxylic acid, potassium salt, (R)-; (82044-23-3), **75**, 37

Potassium glycidate: See Potassium (R)-(+)-2,3-epoxypropanoate, **75**, 38

Potassium hydride; (7693-26-7), **75**, 20

Preculture preparation, **76**, 77

4a,9a-Propano-4H-cyclopenta[5,6]pyrano[2,3-d]-1,3-dioxin-6,12(5H)-dione, **75**, 191

Propanoic acid: See Propionic acid, **75**, 80

1-Propanol; (71-23-8), **75**, 53

Propargyl alcohol: 2-Propyn-1-ol; (107-19-7), **76**, 178, 200

2-Propargyloxytetrahydropyran: 2H-Pyran, tetrahydro-2-(2-propynyloxy)-; (6089-04-9), **76**, 178

[1.1.1]PROPELLANE: Tricyclo[1.1.1.0^{1,3}]pentane; (35634-10-7), **75**, 98

Propionaldehyde: Propanal; (123-38-6), **76**, 152

Propionic acid: Propanoic acid ; (79-09-4), **75**, 80

Propyl alcohol: See 1-Propanol, **75**, 53

1-PROPYNYLLITHIUM: LITHIUM, 1-PROPYNYL-; (4529-04-8), **76**, 214

Protection, of trans-hydroxyl groups in sugars, **75**, 173
 table, **75**, 174
 of primary and secondary amines, **75**, 167, 232
 selective, of 1,3-diols, **75**, 177
 table, **75**, 182
 selective, of primary alcohols, **75**, 186

Protective groups:
 Boc-, **76**, 72
 1-ethoxyethyl, removal of, **76**, 207
 in phosphoramidite syntheses, **76**, 273
 phthaloyl, for amino groups, **76**, 129

(R,R)-(-)-Pseudoephedrine: Benzenemethanol, α-[1-(methylamino)ethyl]-, [R-(R*,R*)]-; (321-97-1), **76**, 58
 recovery, **76**, 63

(R,R)-(-)-Pseudoephedrine L-allylglycinamide: 4-Pentenamide, 2-amino-N-(2-hydroxy-1-methyl-2-phenylethyl)-N-methyl, [1S-[1R*(S*), 2R*]]-; (170642-23-6), **76**, 59

(R,R)-(-)-Pseudoephedrine D-allylglycinamide diacetate, **76**, 70

(R,R)-(-)-Pseudoephedrine L-allylglycinamide diacetate, **76**, 70

(R,R)-(-)-Pseudoephedrine glycinamide: Acetamide, 2-amino-N-(2-hydroxy-1-methyl-2-phenylethyl)-N-methyl-, [R-(R*,R*)]-; (170115-98-7), **76**, 57; [S-(R*,R*)]-; (170115-96-5), **76**, 71

(R,R)-(-)-Pseudoephedrine glycinamide diacetate, **76**, 70

(R,R)-(-)-Pseudoephedrine glycinamide monohydrate, **76**, 58

Pseudomonas putida 39/D, **76**, 77

Pyridine; (110-86-1),**75**, 69, 89, 210, 225; **76**, 142, 102, 277

Pyridinium chloride: Pyridinium hydrochloride: (628-13-7), **76**, 102

Pyridinium tetrafluoroborate: Pyridine, tetrafluoroborate (1-); (505-07-7), **75**, 211

Pyrrole: 1H-Pyrrole; (109-97-7), **76**, 288

1-PYRROLINES, 2-SUBSTITUTED, **75**, 215
table, **75**, 221
2,3-disubstituted, table, **75**, 222

Quinine: Cinchonan-9-ol, 6'-methoxy-, (8a, 9R)-; (130-95-0), **75**, 224

Quininone: Cinchon-9-one, 6'-methoxy-, (8a)-; (84-31-1), **75**, 224

Quinolizidines, **76**, 36

Racemic pinitol, **76**, 81

Reagents and compounds, useful, syntheses of, **76**, xxi

Rearrangements, **76**, xix

Resolution, OF 1,1'-BI-2-NAPHTHOL, **76**, 1
of a dl-diamine, **76**, 26
of trans-1,2-diaminocyclohexane, **75**, 9

Ritter reaction, **76**, 52

(Salen)metal complexes, **75**, 10

(Salen)Mn-catalyzed epoxidation reactions, **76**, 53

Salicyladehydes, 3,5-substituted, **75**, 10

Secondary amines, from nitriles, **76**, 137

L-Serine: See (S)-Serine, **75**, 37

(S)-Serine: L-Serine; (56-45-1), **75**, 37

Shaker, benchtop, orbital incubator, **76**, 77

Silver nitrate, ~ 10 wt.% on silica gel: Nitric acid, silver salt;
(7761-88-8), **76**, 265

Silver(I) oxide: Silver oxide (Ag_2O) (9); (20667-12-3), **75**, 70

Sodium; (7440-23-5), **75**, 79

Sodium bisulfite, **75**, 161; **76**, 255

Sodium borohydride: Borate (1–), tetrahydro-, sodium-; (16940-66-2), **76**, 134

Sodium hexamethyldisilazide: Silanamine, 1,1,1-trimethyl-
N-(trimethylsilyl)-, sodium salt; (1070-89-9), **76**, 37

Sodium hydride; (7646-69-7), **75**, 153, 215

Sodium hypochlorite: Hypochlorous acid, sodium salt; (7681-52-9), **76**, 47
for scavenging sulfur compounds, **76**, 124

Sodium iodide; (7681-82-5), **76**, 255

Sodium metaperiodate: Periodic acid, sodium salt; (7790-28-5), **75**, 108

Sodium nitrite: Nitrous acid, sodium salt; (7632-00-0), **75**, 37

Sodium omadine: See Sodium 2-pyridinethiol-1-oxide, **75**, 125

Sodium 2-pyridinethiol-1-oxide (sodium omadine), **75**, 125

Sodium sulfide, **76**, 95

Sodium sulfite: Sulfurous acid, disodium salt; (7757-83-7), **76**, 90

Sodium tetraborate decahydrate: Borax; (1303-96-4), **76**, 79

Sodium β-trimethylsilylethanesulfonate: Ethanesulfonic acid, 2-(trimethylsilyl)-,
sodium salt; (18143-40-3), **75**, 161

Sonication, **76**, 232

Squarates, alkyl, **76**, 193

Squaric acid: 3-Cyclobutene-1,2-dione, 3,4-dihydroxy-; (2892-51-5), **76**, 190

Stereochemical purity, determination of, **76**, 52

Stereodefined synthesis, of allenes, **76**, 185

Sterile loop, **76**, 77

"Streaking for isolation", **76**, 79

Sulfur dioxide, **75**, 90

Sulfuric acid, fuming: Sulfuric acid, mixt. with sulfur trioxide; (8014-95-7), **76**, 47

Superacidic triflic acid, **76**, 296

SUZUKI COUPLING, **75**, 69
ACCELERATED, **75**, 61
MODIFIED, **75**, 53

Synthetic transformations,**76**, xix

Synthons, haloalkenes, **76**, 224

TADDOLS, **76**, xvii, 14

DL-(±)-Tartaric acid: Butanedioic acid, 2,3-dihydroxy-, (R*,R*)-; (133-37-9), **76**, 25

L-Tartaric acid: Butanedioic acid, 2,3-dihydroxy-, [R-(R*,R*)]-; (87-69-4),**75**, 2; **76**, 25, 48

Tedlar bag, **75**, 131

Tertiary amines, from nitriles, **76**, 133

1,2,4,5-Tetrabromobenzene: Benzene, 1,2,4,5-tetrabromo-; (636-28-2), **75**, 201

Tetrabutylammonium bromide: 1-Butanaminium, N,N,N-tributyl-, bromide; (1643-19-2), **76**, 271

Tetrabutylammonium fluoride: 1-Butanaminium, N,N,N-tributyl-, fluoride; (429-41-4), **76**, 254

Tetraethylammonium bromide: Ethanaminium, N,N,N-triethyl-, bromide; (71-91-0), **76**, 264

anti-1,4,5,8-TETRAHYDROANTHRACENE-1,4:5,8-DIEPOXIDE:1,4:5,8-DIEPOXYANTHRACENE, 1,4,5,8-TETRAHYDRO-, (1α,4α,5β,8β)- ; (87207-46-3), **75**, 201

syn-1,4,5,8-TETRAHYDROANTHRACENE-1,4:5,8-DIEPOXIDE: 1,4:5,8-DIEPOXYANTHRACENE, 1,4,5,8-TETRAHYDRO-, (1α,4α,5α,8α)-; (87248-22-4), **75**, 201

Tetrakis(triphenylphosphine)palladium(0), alternatives to, **75**, 57

α-Tetralone: 1(2H)-Naphthalenone, 3,4-dihydro-; (529-34-0), **76**, 230

2,3,4,4-Tetramethoxy-2-cyclobuten-1-one, **76**, 192

1,1,2,2-Tetramethoxycyclohexane: Cyclohexane, 1,1,2,2-tetramethoxy-; (163125-34-6), **75**, 170

(R,R)-(+)-N,N,N',N'-Tetramethyltartaric acid diamide: Butanediamide, 2,3-dihydroxy-, N,N,N',N'-tetramethyl-, [R-(R*,R*)]-; (26549-65-5), **76**, 89
recovery, **76**, 90

5,10,15,20-Tetraphenylporphyrin, **76**, 291

Tetravinyltin; (1112-56-7), **76**, 193

4,4'-Thiobis[2-(2-tert-butyl-m-cresol): Phenol, 4,4'-thiobis[2-(1,1-dimethylethyl)-3-methyl-; (4120-97-2), **76**, 201, 202

Thionyl chloride; (7719-09-7), **75**, 89, 162

Thionyl chloride-dimethylformamide, **75**, 162

Thiophenol: Benzenethiol; (108-98-5), **75**, 100

Thiophosphoryl chloride, HIGHLY TOXIC; (3982-91-0), **76**, 271

L-Threitol 1,4-bismethanesulfonate: 1,2,3,4-Butanetetrol, 1,4-dimethanesulfonate, [S-(R*,R*)]-; (299-75-2), **76**, 101

Titanium(III) chloride; (7705-07-9), **76**, 143

Titanium(IV) isopropoxide: See Titanium tetraisopropoxide, **75**, 12

Titanium reagents, for reductive cyclizations, **76**, 149

Titanium tetraisopropoxide: 2-Propanol, titanium(4+) salt; (546-68-9), **75**, 12

o-Tolueneboronic acid: See o-Tolylboronic acid, **75**, 61

Toluene dioxygenase, **76**, 80

p-Toluenesulfonic acid monohydrate: Benzenesulfonic acid, 4-methyl-, monohydrate; (6192-52-5), **76**, 180, 200, 203

m-Toluoyl chloride: Benzoyl chloride, 3-methyl-; (1711-06-4), **75**, 2

o-Tolylboronic acid: o-Tolueneboronic acid ; (16419-60-6), **75**, 61

Transmetallation, **76**, 258

Tri-tert-butylphosphine: Phosphine, tris(1,1-dimethylethyl)-; (13716-12-6), **75**, 81

Trichloroacetic acid: Acetic acid, trichloro-; (76-03-9), **76**, 288

Trichloromethyl chloroformate: See Diphosgene, **75**, 48

Tricyclo[1.1.1.0^{1,3}]pentane: See [1.1.1]Propellane, **75**, 98

Triethylamine: Ethanamine, N,N-diethyl-; (121-44-8), **75**, 45, 108; **76**, 112, 123, 202, 264

Triethylamine trishydrofluoride: Ethanamine, N,N-diethyl-, trishydrofluoride; (73602-61-6), **76**, 159

Triflic anhydride: Methanesulfonic acid, trifluoro-, anhydride; (358-23-6), **76**, 81

Trifluoroacetic acid: Acetic acid, trifluoro-; (76-05-1), **75**, 130; **76**, 207, 288

Trifluoroacetic anhydride: Acetic acid, trifluoro-, anhydride; (407-25-0), **76**, 95, 190, 193

1,1,1-TRIFLUORO-2-ETHOXY-2,3-EPOXY-5-PHENYLPENTANE: OXIRANE, 2-ETHOXY-3-(2-PHENYLETHYL)-2-(TRIFLUOROMETHYL)-, cis-(±)-; (141937-91-9), **75**, 153

(Z)-1,1,1-TRIFLUORO-2-ETHOXY-5-PHENYL-2-PENTENE: BENZENE, (4-ETHOXY-5,5,5-TRIFLUORO-3-PENTENYL)-, (Z)-; (141708-71-6), **75**, 153

Trifluoromethanesulfonic acid: Methanesulfonic acid, trifluoro-; (1493-13-6), **76**, 294

Trimethyl borate: Boric acid, trimethyl ester; (121-43-7), **76**, 87

Trimethyl orthoacetate: Ethane, 1,1,1-trimethoxy-; (1445-45-0), **75**, 80

Trimethyl orthoformate: Methane, trimethoxy-; (149-73-5), **75**, 170, 171; **76**, 190, 288

[β-(TRIMETHYLSILYL)ACRYLOYL]SILANE, **76**, 199

Trimethylsilyl chloride: Silane, chlorotrimethyl-; (75-77-4), **76**, 24, 202, 252

2-TRIMETHYLSILYLETHANESULFONYL CHLORIDE: ETHANESULFONYL CHLORIDE, 2-(TRIMETHYLSILYL)-; (106018-85-3), **75**, 161

1,3,5-Trioxane; (110-88-3), **75**, 189

Triphenylarsine: Arsine, triphenyl-; (603-32-7), **75**, 70

Triphenylphosphine: Phosphine, triphenyl-; (603-35-0), **75**, 33, 53, 155; **76**, 180

Triphosgene: Carbonic acid, bis(trichloromethyl) ester; (32315-10-9), **75**, 45

Tris(chloromethyl)acetic acid: Propionic acid, 3-chloro-2,2-bis(chloromethyl)-; (17831-70-8); **75**, 90

Tropylium tetrafluoroborate: See Cycloheptatrienylium tetrafluoroborate, **75**, 210

Tryptones (Bacteriological), See: Peptones, Bacteriological; (73049-73-7*), **76**, 80

Undecenoic acid: 10-Undecenoic acid ; (112-38-9), **75**, 124

N-(10-Undecenoyloxy)pyridine-2-thione: 2(1H)-Pyridinethione, 1-[(1-oxo-10-undecenyl)oxy]-; (114050-28-1), **75**, 124

Vilsmeier-Haack reagent, **75**, 167

Vinyl acetate: Acetic acid ethenyl ester; (108-05-4), **75**, 79

N-Vinylation, **75**, 45

Vinyllithium: Lithium, ethenyl-; (917-57-7), **76**, 190, 193

N-Vinyl-2-pyrrolin-2-one: 2-Pyrrolidinone, 1-ethenyl-; (88-12-0), **75**, 215

Vinyltrimethylsilane: Silane, ethenyltrimethyl-; (754-05-2), **75**, 161

VITAMIN D_2: 9,10-SECOERGOSTA-5,7,10(19),22-TETRAEN-3-OL, (3β)-; (50-14-6), **76**, 275

Vitamin D_2 3,5-dinitrobenzoate: Ergocalciferol, 3,5-dinitrobenzoate; (4712-11-2), **76**, 276

Vortex mixer, **75**, 2

WITTIG OLEFINATION, OF PERFLUORO ALKYL CARBOXYLIC ESTERS, **75**, 153

Zeolites, 4 Å: See 4 Å Molecular sieves, **75**, 189

Zinc; (7440-66-6), **76**, 24, 143, 252

Zinc chloride; (7646-85-7) ,**75**, 177